LABORATORY ANIMALS: AN INTRODUCTION FOR NEW EXPERIMENTERS

LABORATORY ANIMALS: AN INTRODUCTION FOR NEW EXPERIMENTERS

Edited by

A. A. Tuffery
Department of Biological Sciences
N.E. Surrey College of Technology, Ewell, Surrey

A Wiley–Interscience Publication

Chichester · New York · Brisbane · Toronto · Singapore

Copyright © 1987 by John Wiley & Sons Ltd.
Reprinted with corrections November 1988.
Reprinted July 1990.

All rights reserved.

No part of this book may be reproduced by any means, or transmitted, or translated into a machine language without the written permission of the publisher.

Library of Congress Cataloging-in-Publication Data:

Laboratory animals

   'A Wiley–Interscience publication.'
   Bibliography: p.
   Includes index.
   1. Laboratory animals.   2. Animal experimentation.
3. Animals, Treatment of.   I. Tuffery, A.A.
QL55.L274   1987   619   86-24626

ISBN 0 471 91297 2

British Library Cataloguing in Publication Data:

Laboratory animals: an introduction for new experimenters.
   1. Laboratory animals
   I. Tuffery, A.A.
636.08'85   SF406

ISBN 0 471 91297 2

Printed in Great Britain by Antony Rowe Ltd

Contents

Contributors · vii

Acknowledgements · ix

Introduction: Use of Animals in Biomedical Research · 1
  A. A. Tuffery

Chapter 1   Ethical Aspects of Animal Experimentation · 5
    Jenny Remfry

Chapter 2   Law Relating to Animal Experimentation · 21
    Judith Hampson

Chapter 3   The Design of Experiments · 53
    M. R. Gamble

Chapter 4   The Supply of Laboratory Animals · 63
    H. Donnelly

Chapter 5   Quality in Laboratory Animals · 79
    C. Clough

Chapter 6   Principles of Animal Husbandry · 99
    Marie S. Wilson

Chapter 7   Hazards and Safety Aspects of Animal Work · 117
    M. W. Smith

Chapter 8	Animal Behaviour D. E. Blackman	129
Chapter 9	Animal Handling and Manipulations P. Scobie-Trumper	153
Chapter 10	Euthanasia C. J. Green	171
Chapter 11	General Aspects of the Administration of Drugs and other Substances H. B. Waynforth	179
Chapter 12	Feeding and Watering Marie E. Coates	203
Chapter 13	Non-Surgical Experimental Procedures P. A. Flecknell	225
Chapter 14	Anaesthesia and Analgesia C. J. Green	261
Chapter 15	Standards of Surgery for Experimental Animals H. B. Waynforth	303
Chapter 16	Laws Relevant to Animal Research in the United States B. E. Rollin	323
Bibliography		335
Index		337

Contributors

D. E. BLACKMAN Department of Psychology, University College, Cardiff CF1 1XL

G. CLOUGH Alanann Consultancy Services, PO Box 230, York YO1 1GG

MARIE E. COATES The Robens Institute, University of Surrey, Guildford, Surrey GU2 5XH

H. DONNELLY Laboratory Animal Science Unit, Department of Microbiology & Parasitology, Royal Veterinary College, Royal College St, London NW1 0TU

P. A. FLECKNELL Comparative Biology Centre, The Medical School, Framlington Place, Newcastle upon Tyne NE2 4HH

M. R. GAMBLE Research Department, The Boots Company PLC, Nottingham NG2 3AA

C. J. GREEN Division of Comparative Medicine, Clinical Research Centre, Watford Road, Harrow, Middx HA1 3UJ

JUDITH E. HAMPSON 17 Mackeson Road, London NW3 2LV

J. REMFRY Universities Federation for Animal Welfare, 8 Hamilton Close, South Mimms, Potters Bar, Herts EN6 3QD

B. E. ROLLIN Department of Philosophy, Colorado State University, Fort Collins, Colorado 80523, USA

P. Scobie-Trumper The Animal Unit, University of Surrey. Guildford, Surrey GU2 5XH

M.W. Smith Cambridge University Central Animal Services, Animal House, New Addinbrooks Hospital Site, Hills Road, Cambridge CB2 2QL

H. B. Waynforth Smith, Kline & French Research Ltd, The Frythe, Welwyn, Herts, AL6 9AR

Marie S. Wilson Smith Kline & French Research Ltd, The Frythe, Welwyn, Herts AL6 9AR

Acknowledgements

I would like first of all to acknowledge the work of the contributors to this volume and express my thanks to them for the time and energy they gave to this task. I am also indebted to many colleagues at my own college and in the laboratory animal world generally for their help and encouragement, and especially to those of the MRC Laboratory Animal Centre's Working Party on Courses for Animal Licensees who provided the inspiration and a great deal of the framework upon which this project was developed.

The skilled assistance of Gordon Holley (Graphics Section, NESCOT) who prepared the figures for Chapters 9 and 13 is very much appreciated.

Thanks are also due to Professor Stephen Clark, Liverpool University, for his advice afforded to Dr Remfry in the preparation of Chapter 1.

A grateful acknowledgement is due to Churchill Livingstone for permission to reproduce Table 12.1, from *UFAW Handbook* (4th edn) 1972.

Laboratory Animals: An Introduction for New Experimenters
Edited by A.A. Tuffery
© 1987 John Wiley & Sons Ltd

Introduction:
Use of Animals in Biomedical Research

A.A. TUFFERY

In Britain today there is considerable discussion (sometimes submerged by a less useful destructive activity on the part of some protagonists) on the merits and ethics of animal experimentation, and it is only right that the opening chapters of a volume specifically written for animal experimenters are concerned with ethical matters and related legislation. The deep-rooted national concern for these matters is reflected in the 1876 date of the legislation which has covered this area until the present time, and this country's inability to change it for over a century reflects our society's continued interest — it has been too much of a hot potato for any political party to pick up during all that time! At almost exactly the time this volume appears in print, a new Act will begin operating in the UK, and this will be discussed in some detail, together with USA and other overseas legislation.

Russell and Birch in their book *The Principles of Humane Experimental Technique*, published in 1954, outlined a policy for animal experimentation which has perhaps only in recent years come to be widely appreciated. These writers proposed a 'three Rs' policy, by which they meant to focus attention on the *replacement* of whole animals by alternative, non-sentient, material; on the *reduction* in the number of animals used where they cannot be replaced; and on *refinement* to lessen the incidence or severity of procedures which will have to be used.

Weihe (1985) has discussed the concept of 'alternatives' (Russell and Birch's first R) at some length, and he attempts to make a balanced assessment of the role of this particular concept in animal experimentation. But he also argues the case for the third R (refinement) and suggests that the possibilities offered by this idea need promotion. It is the third of the three Rs with which the most this book is primarily concerned.

Society in general must bear responsibility for what is allowed to be done to, or with, animals, and this includes eating them, hunting them, racing them, conserving them and experimenting upon them. If it decides to ban animal experiments it must accept the price of (1) a slower advance in medical, veterinary and biological understanding, (2) a reduced ability to investigate disease processes, (3) a greater chance of drug-induced disease and accidents, (4) a seriously reduced ability to investigate untoward toxic phenomena in man and animals and (5) an increased hazard as new vaccines and drugs are evaluated in host subjects. But if society accepts the need to conduct animal experiments, it has the right to expect that those who indulge in this activity should do it with every possible concern for their subjects' well-being, and that they should attempt to minimize the pain, stress and discomfort which is imposed for society's benefit. Dame Olga Uvarov (1985) has presented a valuable discussion paper covering many of these issues.

Society ought to be interested in what goes on in laboratories, and the animal welfare societies might well act as the conscience pricking the scientists who work with animals, but it is only the scientists themselves who can suggest the positive actions which will actually improve the humaneness of their experimentation. Improvements in cage design, management systems and the reduction of animal stress will only come from scientifically oriented observations. Society might provide the stimulus (and must provide the finance) for the undertaking of this work, but only the scientist can propose solutions likely to be of real benefit to the animals. Again, by introducing newcomers to this field to the techniques described, this book should make a contribution towards this end.

It is, of course, quite possible even now that some procedures currently performed on animals could be replaced by other experimental systems using non-sentient material and it is quite likely that even more replacement of sentient systems by alternative non-sentient systems will be developed in the future, but many thoughtful animal experimenters (as discussed by Weihe) feel that the prospects for such 'alternatives' are at present strictly limited. There will always be a need for some animal experimentation. 'Alternative' models for sentient systems must inevitably represent oversimplifications of complex situations; the animal body — human, canine or murine — is an incredible complex of interacting systems which no 'alternative' will ever be able to mimic completely and so an experimental investigation into one factor, for example the administration of a potential new drug, must at some stage involve a study of the effects of this procedure on the whole system, that is, the complete animal.

One might predict that, as our understanding of mechanisms of action develops, we may be able to refine the models being investigated and replace more and more whole animal systems. Nevertheless, at least at the early stages (and possibly at late or final stages) when making a safety evaluation of a new drug prior to release for widespread human use, it seems to me likely that animal experimentation will continue for a considerable time yet.

So if experiments on whole animals have to be conducted, then let us ensure that they are conducted in as humane a way as possible. Clearly, the more skilfully manipulations are made, the more expertly animals are inoculated, the more observantly animals are studied, then the more humane will be the experimental procedures. To hold an animal comfortably so as to make a smooth, painless intravenous injection, and to observe the animal's responses, requires instruction and training. This book is concerned with such training.

Britain's status in the world of fundamental and applied biological, medical and veterinary research would suggest that working under the limitations of the 1876 legislation has produced good research biologists, and our reputation as animal lovers and the home of the antivivisection movement would suggest that this has been achieved without losing at the same time a general concern by society for the animal's welfare. None the less, the animal welfare lobby (and with a more informed judgement, the animal licensees) have not regarded the past system of training as ideal. Various learned and professional societies have looked at this problem in the past few years and a direct approach to it was made at the writer's institution when short courses were set up for anyone beginning experimental work with animals. Since then other organizations have developed schemes with similar aims, and there are also many 'in-house' programmes run by commercial and research organizations of all types.

In 1977 the MRC Laboratory Animals Centre became concerned enough with the problems of training new workers to set up a working party to investigate this area and make some recommendations. The writer was invited to join this group, and the working party's final major recommendations, in the form of a complete syllabus for a short training course, was published in 1984 (Working Party on Courses for Animal Licensees, 1984). This syllabus was used as a basis for the preparation of this book — it could become a prescribed textbook for such courses and this, I hope, will be one of its major uses.

Hopefully, it will also become a useful, not too bulky, source of basic reference for research workers and animal technicians generally.

The scope of this work is deliberately limited to basics — dogs and cats are only briefly mentioned, and primates not at all.

The following chapters have been contributed by workers established in their fields and the reader will find some of the points discussed above reiterated and accompanied by practical advice for the achievement of humane, and scientifically sound, experimentation with laboratory animals.

Proper training and instruction is the major part of improving the humanity of animal experimentation; hopefully this book will contribute towards this end.

REFERENCES

Working Party on Courses for Animal Licensees (1984). Report of the Working Party on Courses for Animal Licensees, *Laboratory Animals*, **18**, 209–20.

Russell, W. M. S. and Birch, R. L. (1954). *The Principles of Humane Experimental Technique*, Methuen, London.

Uvarov, Dame Olga (1985). Research with animals: requirement, responsibility, welfare, *Laboratory Animals*, **19**, 51–75.

Weihe, W. H. (1985). Use and misuse of an imprecise concept: alternative methods in animal experiments, *Laboratory Animals*, **19**, 19–26.

Laboratory Animals: An Introduction for New Experimenters
Edited by A. A. Tuffery
© 1987 John Wiley & Sons Ltd

CHAPTER 1

Ethical Aspects of Animal Experimentation

JENNY REMFRY
UFAW, Potters Bar, Herts

INTRODUCTION

These days, scientists (and butchers, furriers, farmers and field sportsmen) are subject to abuse and attacks from groups of protesters who accuse them of cruel exploitation of animals. Are these accusations based on matters of fact or on questions of attitudes and ethical judgements? How seriously should scientists take them? How should they respond?

First, they should examine the facts of the case to see if the welfare of the animals has been ignored or neglected. If it has, they should take steps to put matters right. In many cases, the scientists have not committed the sins of which they are accused, and the animals are not suffering. But members of the general public may not know that. Public opinion has been affected by the animal rights campaign and many people are now questioning the basic assumptions of the scientists. This means that scientists should think carefully about their role in society and their relationship to the animals they use, even before they discuss their work with their nearest and dearest. If they wish to enter into arguments with animal rights campaigners, they must think even more deeply, because the concept of animal rights is a philosophical one and may take the thinkers into realms they have not previously explored.

PHILOSOPHY AND ETHICS

According to the *Encyclopaedia Britannica*, ethics is the systematic study of the nature of value concepts, 'good', 'bad', 'ought', 'right', 'wrong', etc., and of the general principles which justify us in applying them to actual situations. People who undertake this study are moral philosophers.

Just as physical science starts with the observation of physical objects, so moral philosophy starts with commonsense ethics — that is, the ethical judgements which at our best we feel constrained to make in the everyday task of making moral decisions. The aim of the moral philosopher is to bring these basic data into a system which is consistent with rational thought.

Philosophy is rarely taught in school, at any rate in Britain, so discussions about man's right to use animals in biomedical research can catch the scientist off balance.

Rational thought does not come naturally. It is not necessary for survival and most animals manage perfectly well without it. So even people who have been taught to think rationally about science may find it difficult to think rationally about their relation to animals.

Socrates helped his students to think by involving them in dialogues about matters which were of importance to them. This is still the most effective method. Just as a scientist should hold a dialogue with a statistician before finalizing his experimental protocol, so ideally should a biologist planning a potentially painful experiment on animals hold a dialogue with a philosopher or at least with someone who is prepared to discuss it in ethical terms.

Such a dialogue should help scientists to consider their work in terms of their rights and duties *vis-à-vis* mankind, science and animals. But first, they need to be clear about their own assumptions about and attitudes towards animals.

ASSUMPTIONS ABOUT ANIMALS

Children are given furry toys and led to believe that all animals are cuddly. Most children soon learn that in fact some animals bite or scratch, others kick, and the big ones are best avoided. These experiences, and the assumptions of their parents, lead to a variety of attitudes towards animals in society. For example, most people assume that it is necessary to keep animals on farms in order to satisfy our need for food, but a growing minority reject that view and become vegetarian. Most people assume that cats and dogs are ideal household pets, and yet thousands are abandoned and have to be killed by humane societies each month. Many wild animals are regarded as pests because they share man's food preferences, but naturalists may find the same animals of intrinsic interest.

Attitudes towards laboratory animals are just as varied. At one extreme there are scientists who in the name of freedom of science assume they have the right to use animals in the pursuit of knowledge without regard to the suffering caused. Most British scientists accept that animals need to be protected against the abuse that could result from this view and work willingly within the constraints of legislation (Animals (Scientific Procedures) 1986) which limits the use of painful experiments to genuine scientific and medical purposes, makes anaesthesia mandatory for surgical experiments and prohibits the public from watching experiments for amusement.

Those animal protection societies which have a policy on animal experimentation are against painful experiments. They may, however, be eager to use the medicines and vaccines resulting from animal work if the animals in their clinics or refuges need treatment.

Antivivisectionists believe that man has no right and usually not even a necessity to use animals for experiments. They argue that since the products and medicines now available are effective and safe, there is no need to produce new ones; that animals are in any case poor indicators of adverse reactions in man (e.g. the case of thalidomide); that experiments are often painful, trivial, wasteful and repetitive; and that anyway non-animal alternatives are probably available. The antivivisectionist societies have until recently campaigned peacefully for the abolition of vivisection, but without success.

The Animal Liberation Front was the first of the activist groups to gain public attention in the 1970s. Sometimes these groups are, or were, attached to antivivisection societies, but since the activist methods have become more violent and the risk to human life has increased, the respectable societies have had to dissociate themselves from them. There are, in addition, animal rights groups who share the abolitionist aims of the activists, but who do not condone violence.

The attitudes of people who are not members of any of these societies are variable and ambivalent. There are so many laws and regulations in the UK requiring the use of laboratory animals that it could be assumed that society thinks it right and proper to keep animals for laboratory use. According to public opinion polls, the majority of the population approves the use of animals for medical research, but fewer give support to their use in other fields (FRAME, 1984). This is probably because there is little understanding of the importance of fundamental research in the development of new medical or surgical techniques.

THINKING ABOUT ANIMALS

The ways people think about animals in their relation to man have been most strongly influenced in the West by Christianity and Greek philosophy. These both assumed that there was a basic difference between man and other species of animals. More recently, this assumption has been questioned, and philosophers have reconsidered the moral status of animals, sometimes with rather startling results.

The Bible and after

The best starting point is the Bible, because that is the book that has had the greatest influence. The creation stories in *Genesis* were intended to show man's relation to the animals: one of dominion over the rest of creation.

Then God said, 'Let us make men in our image and likeness to rule the fish of the sea, the birds of heaven, the cattle, all wild animals on earth, and all reptiles that crawl upon the earth' (*Genesis* 1).

Throughout the Old Testament, a good working relationship is assumed between man and his domestic animals. It is also assumed that a man living according to the will of God will be considerate to his animals:

A righteous man cares for his beast
But a wicked man is cruel at heart

(*Proverbs* 12)

In the New Testament, the teaching about animals, insofar as it exists, reinforces that in the Old Testament. The message is that man has rights over animals, but also moral duties towards them as stewards of God's creation.

Later Christian literature is influenced by Greek philosophy and the concept of stewardship becomes less important. Instead, man is exhorted to be kind to animals because of the destructive effect of cruelty on the human soul.

In Greek philosophy, the presence of a reasoning soul was seen to be man's highest attribute. All men, even those who were not Greek, were seen as having the potential of a rational soul. Animals did not have this potential and so were of little moral concern.

The idea of human rights arose from the philosophy of humanism. Under Greek and Roman law, only certain people had rights — those who had been granted citizenship. Humanist philosophers thought that all people must be recognized as having intrinsic worth and as being ends-in-themselves. They should all therefore be considered equal under the law. In most of Europe, common people were indeed given basic legal rights, although other rights such as religious freedom and the right to vote for elected representation in government came much later. Animals were different: although it was recognized that they were capable of feeling and emotion, they were considered to have no moral consciousness and so therefore could have no rights.

With René Descartes (1596–1650), the humanist attitudes towards animals became exaggerated into a mechanical philosophy. It was considered that animals had no souls and were therefore incapable of rational thought or of consciousness. Only humans can say '*Cogito ergo sum*' — 'I think therefore I am'. The reactions of animals to pain are mere reflexes, the actions of automata. This reasoning was used to justify the physiological experiments carried out on the continent in the eighteenth and nineteenth centuries, such as those performed by Claude Bernard on conscious dogs.

In Britain, Cartesian philosophy did not take a strong hold. Humanism itself, with its emphasis on the superiority of man and the importance of Christian virtues, was being undermined by the philosophical movement of utilitarianism. This taught that the only good is pleasure and the only evil is pain, and men should act so as to produce the greatest balance of pleasure over pain. It was recognized that animals feel pleasure and pain in a way similar to man, so it followed that they should be given similar consideration.

Jeremy Bentham (1748–1832) is the utilitarian most commonly quoted in relation to animals. As an atheist he was able to break free from assumptions about the superiority of man. Also he was writing at a time when black slaves had been freed by the French but were still in bondage in the British colonies.

> The day may come when the rest of the animal creation may acquire those rights which never could have been witholden from them but by the hand of tyranny. The French have already discovered that the blackness of the skin is no reason why a human being should be abandoned without redress to the caprice of the tormentor. It may one day come to be recognised that the number of the legs, the villosity of the skin, or the termination of the os sacrum are reasons equally insufficient for abandoning a sensitive being to the same fate. What else is it that should trace the insuperable line? Is it the faculty of reason, or perhaps the faculty of discourse? But a full-grown horse or dog is beyond comparison a more rational, as well as a more conversable animal, than an infant of a day or a week or even a month, old. But suppose they were otherwise, what would it avail? The question is not Can they reason? nor Can they talk? but, Can they suffer?
> (*Introduction to the Principles of Morals and Legislation*)

What were those rights that Bentham wished to extend to animals? Not the right to a long life, because death was not seen as a 'pain' by utilitarians, and anyway the loss of a pig's life would be far outweighed by the pleasure of the people eating it. The right envisaged would be that to seek and enjoy pleasure; but man could take the right away if it would lead to a greater good or the good of a greater number.

So the greatest benefit to animals of this philosophy was a recognition that animals can suffer and so should be given some consideration. Such an insight was greatly needed at the time: during the seventeenth and eighteenth centuries cruelty to animals was commonplace in England, and Englishmen amused themselves with the sports of bull-baiting, bear-baiting, cock-fighting and cock-throwing (throwing sticks at cocks until they were killed). C. D. Niven in his *History of the Humane Movement* (1967) writes that by the end of the eighteenth century 'the treatment of horses must have been pitiless, because England was regarded by Europeans as the hell for horses'.

The nineteenth century

Real improvements in attitudes to animals came as part of the Christian evangelical revival in the nineteenth century which led the way to far-reaching social reform. The abolition of slavery, the ending of child labour and the founding of the RSPCA were all parts of this reform. Charles Darwin, although he shocked so many religious people by asserting man's kinship to the apes, helped the process by showing that man is an animal and that therefore man's moral concern should extend to the other animals. He was also one of several distinguished scientists to support the passage of the Cruelty to Animals Act of 1876.

By the end of the nineteenth century, attitudes towards animals had changed so much that sentimentality was in fashion, and pictures by Landseer depicting heartbroken dogs guarding the corpses of their masters, or stags at bay, were in great demand.

The antivivisection movement may be seen as part of this process. In 1875, Frances Power Cobbe founded the Victoria Street Society in London to campaign for the legal restriction of 'vivisection'. Their main demand was that surgical experiments should be carried out only under anaesthesia and that the animal should not be allowed to recover consciousness. This was technically possible because the anaesthetic properties of chloroform had been discovered, although some scientists, particularly on the continent of Europe, preferred not to use it.

The Cruelty to Animals Bill, published in 1875, satisfied their demands. But as a result of pressure from scientists, clauses were added to permit the recovery of animals under some circumstances, or the omission of anaesthesia in others. In anger, Miss Cobbe founded the first truly antivivisection society — the Society for the Protection of Animals from Vivisection. Others soon followed: the German League Against Scientific Animal Torture in 1879, the Société Contre la Vivisection in 1882, the American Anti-Vivisection Society in 1883.

Many others have been formed since, and their fortunes rise and fall, but none has so far achieved its objective of total abolition.

The twentieth century

In 1926, Charles Hume founded the University of London Animal Welfare Society (later the Universities Federation for Animal Welfare), in an attempt to make people think rationally about their attitude to animals. The activities of the antivivisectionists had put biologists onto the defensive so that dialogue had become impossible and 'animal welfare" had become the object of ridicule (Hume, 1960). Hume said that what animal welfare needed was educated people with cool heads and warm hearts, prepared to look at the suffering of all animals (including 'pests') and to find practical ways of alleviating it.

One of the problems in establishing rational attitudes towards animals is that we cannot be sure that they feel pain, fear, etc., in the same way that we do. Even if the nervous pathways are the same as in man, the degree of self-awareness necessary to suffer pain may be absent. Hume maintained that where there is doubt, an animal should be given the benefit of the doubt. So, we should assume that mammals feel pain like us, that birds feel pain of some sort, but that the degree of suffering is likely to be less in reptiles, amphibians, fish, crustacea, molluscs, insects and worms. The science of animal behaviour has helped and useful reviews of contributions to the study of pain are given in Dawkins (1980) and Wood-Gush *et al.* (1981).

By collaborating with concerned scientists, Hume was able to publish the first edition of the *UFAW Handbook on the Care and Management of Laboratory Animals* in 1947. This is now in its sixth edition. Later, he commissioned Russell and Birch to write *The Principles of Humane Experimental Technique* (1959), and this popularized the concept of the three Rs: replacement, reduction and refinement.

In the USA, useful collaboration with scientists has been achieved by the Animal Welfare Institute and the Scientists Center for Animal Welfare. For example, see Dodds and Orlans (1982).

ANIMAL LIBERATION

The liberation movements of the 1960s in the USA were important in the fight to enable black people in the south to exercise the civil rights which in theory they already enjoyed. The movement spread to fight against the prejudice and discrimination suffered by women and certain minority groups. Then in the 1970s a few people started claiming that animals were also a minority group in need of 'liberation'.

Probably the most influential work was Peter Singer's *Animal Liberation* (1976) — the work of a philosopher explaining in popular terms why we need to give greater consideration to animals. The arguments and language he used had an intoxicating effect on younger members of the antivivisection societies: they used them to give intellectual respectability to their gut feelings and made animal rights a rallying cry. The book also stimulated other philosophers to turn their attention to man's relation to the other animals, and to discuss seriously what sort of rights animals might have.

Singer is a utilitarian and his starting point is Bentham's 'the question is . . . can they suffer?' For him, animal liberation means that equal consideration should be given to human and non-human animals.

Singer develops the argument along these lines:

1. Animals feel pain like humans do, although like babies they may not be able to communicate very effectively about it.
2. There is no moral justification for regarding the pain or pleasure felt by animals as being less important than that felt by humans.
3. Humans or animals with higher intelligence do not necessarily suffer more than those with lower intelligence.

4. If it is wrong to kill a human, it must be wrong to kill an animal. Why should a brain-damaged child or hopelessly senile adult be kept alive and a normal healthy pig be killed? In the absence of religious beliefs on the special value of human life, such discrimination against animals is speciesist, just as discrimination against women in sexist.
5. Similar beings should be given similar rights to life.

Two separate conclusions can be logically reached from these arguments: that all animals should be kept alive, even if they would be better off dead; or that humans should be killed when they cease to be healthy or useful. Singer agrees that both positions are unsatisfactory and he suggests a middle road which neither cheapens human life nor overvalues animal life: 'What we must do is bring non-human animals within the sphere of moral concern and cease to treat their lives as expendable for whatever trivial purposes we may have' (Chapter 1).

Singer concedes that not all lives are of equal worth. Animals such as humans, with self-awareness, intelligence, the capacity of meaningful relation with others and ability to plan for the future, have a greater value than animals without. These more valuable animals should not be spared from equal suffering, but their lives should be spared in situations where a choice must be made.

The practical conclusions for anyone sympathetic to Singer's arguments are:

1. Stop eating meat, because it is unnecessary.
2. Use animals in experiments only if it is necessary — that is, if the purpose is important, there are no non-animal alternatives and if the good to be gained is greater than the evil to the experimental animals.
3. Consider carefully your use of animals in sport, entertainment and for companionship.
4. Develop a greater respect for wildlife and, if it is necessary to reduce their numbers, find a method which causes no suffering.

ANIMAL RIGHTS

Singer did not specify the rights that animals have, but the subject is explored in *Animal Rights and Human Obligations* (Regan and Singer, 1976) and in Clark (1977) and Rollin (1981).

The animal rights position is expressed in the most extreme form by Regan, who considers that almost all of man's associations with animals are not in the animals' best interests, and are thus exploitative. He is not a utilitarian, and so cannot justify any exploitation on the grounds of usefulness. For example, keeping an animal as a pet is probably not in the animal's best interest, and neutering it would certainly violate at least one of its basic rights.

Very few philosophers in the UK accept this position. One reason for this is the difference between the US and English legal systems and in particular the different emphasis placed on 'rights'.

In the USA, the purpose of the law is seen as the protection of rights. Since laws exist to protect animals, even against their owners, it follows that animals have rights. At present, these laws protect animals against suffering and cruelty, but they could in theory be extended to protect the life of an animal, on the grounds that animals have a right to realize their natural goals.

In Great Britain, the animal protection laws are not seen as conferring rights on animals but as placing duties on humans. Human adults who are able to plead for themselves have legal rights, because it is assumed that they take moral responsibility for their actions. The judiciary can grant legal rights to infants and other people unable to plead for themselves, but it rarely does so to animals.

There was a period in medieval Europe when in some countries wild or domestic animals were given legal status and provided with counsel so that they could be tried in court (Hill, 1955). If found guilty of crime they were banished, imprisoned or sentenced to death. This idea of animals being legally answerable for their own actions now seems incongruous, and if they are not legally answerable it seems logical to admit they have no legal rights. However, most people do agree that animals have certain natural rights. How are these to be safeguarded in the absence of legal rights?

In his book *The Status of Animals in the Christian Religion*, C. W. Hume (1957) stated that the way forward was for a greater neighbourliness towards animals, and he urged the church to encourage this. Recently, some theologians have taken up the challenge (e.g. Linzey, 1976) and services for animals and animal welfare are now becoming more common.

In 1980 the then Dean of Westminster, Edward Carpenter, published a report of his working party on *Animals and Ethics*. They expressed concern that man's stewardship of animals had become so exploitative, particularly in intensive farming systems, and urged that Christians should treat animals with greater respect and dignity as part of God's creation. They discussed whether animals had rights and concluded that while it was prudent to treat animals as if they had rights, no animal could possibly have absolute rights, because of the conflict of interest between different species. Does a cat have greater rights than a mouse? What are the rights of the tsetse-fly in relation to the rights of cattle and humans?

In the first of the Hume Memorial Lectures, the moral philosopher Canon G. R. Dunstan also addressed the question, and expressed his view that arguments about animal rights are fruitless. We have duties towards animals which can be specified and upheld by law, and which should perhaps be extended, but this does not imply that animals have rights.

ETHICS AND THE BIOLOGIST

Dialogue with the Animal Activists?

We have seen that there are two opposing philosophical approaches to man's relation with animals: the humanist, which says that there is an essential difference between man and other animals and that man's interests are more important, and the utilitarian, which says that animals which can suffer deserve equal consideration with man. The concept of animal rights has developed in the USA from the movement to give civil rights to minorities; it has no legal basis in the UK. Scientists are likely to take a humanist position, but can usefully reflect on the question of animal suffering in order to clarify their own moral stance in relation to the animals they use.

The activist rallying round the flag of animal rights do not see these questions as ones of private morality. They are aiming to alter public opinion so that certain things which are at present accepted by society, such as eating meat, fur-farming, hunting, shooting and fishing, and the use of animals in laboratories, will be made illegal. Not only are they prepared to forego the advantages of modern medicine, vaccines and surgery themselves, but they intend to deny them to the whole of society.

Dialogues between biologists who follow the humanist tradition and animal rights campaigners are thus difficult and usually unrewarding. Biologists should not use this difficulty as an excuse for avoiding ethical questions. Rather it should be used as a stimulus to examine accepted practices and to ask how improvements could be made or alternatives found.

Is science value-free?

The purpose of science is the acquisition of knowledge. Knowledge used wisely leads to an expansion of the human spirit, improvements in standards of living, increased life expectancy and the stimulation of commerce. Used unwisely it can be an instrument of evil; it can be the cause of overexploitation of natural resources and damage to the environment, and it can tempt the unscrupulous to profiteering.

Science is supposed to be value-free, but moral considerations must apply if the knowledge can be obtained only at the cost of suffering to humans or animals. So even if knowledge itself is ethically neutral, its acquisition and application is rarely completely value-free.

For example, in biomedical research animals are often unsatisfactory models or surrogates for man and it would frequently be easier to use humans as the experimental subjects. In the USA prisoners have been allowed to volunteer for this purpose until quite recently. Cosmetic manufacturers commonly use their employees to test their products. Terminally ill patients are often willing to try out new potentially life-saving medicines.

The unacceptable face of human experimentation was shown by the Nazis, who used Russian, Jews and gypsies in dangerous and painful experiments. The use of hospital patients is now subject to agreements made in the Declaration of Helsinki of 1964. This recommended that hospitals wanting to use patients as subjects should set up ethical committees to examine the proposals, protect the interest of the patients and ensure their safety. A precaution usually taken is that no new substance may be tested on a human until it has been shown to be safe in animals. The committees are prepared to take greater risks only if the new substance is thought to be particularly valuable and there is reason to think that it would be safe in man, or if the patient has only a short time to live.

Similarly, work involving animals cannot be value-free if it has clear implications for human health or welfare or if it is likely to cause pain or suffering to animals. Even fundamental research, by providing a firmer basis for medical research, or by making major breakthroughs in knowledge, may have great future practical value. Examples are given in Paton (1984) and Sechzer (1983).

THE MORAL DUTIES OF THE BIOLOGIST

A well-educated biologist will have a respect for life and a sympathetic attitude towards animals. Towards their own experimental animals they have specific duties:

1. *Responsibility for general welfare*
   Animals being kept for scientific purposes should be comfortably housed, properly fed and cared for by qualified people. The experimenter will probably not do this personally, but nevertheless in the UK a licensee is personally responsible for the welfare of the animals used under the licence. There is an extensive literature, for example UFAW (1986).

2. *Calculation of ends and means*
   In planning an experiment, the first question should always be: Is the experiment really necessary? A search of the literature may reveal that it has been done already or that the question being asked is irrelevant. In the case of new medicines or consumer products, another company may already be testing an identical substance, and sharing information could save a lot of time and money as well as animals.

   For experiments involving living animals, Martin and Bateson (1986) have produced a scheme for weighing the importance of the project against the anticipated suffering to the animals, so that proposals can be categorized as justifiable, unjustifiable or borderline. It was designed for animal behaviourists, but could be used for other disciplines if the costs to the animals could be calculated. A scheme for assessing levels of pain and suffering in the commonly used laboratory species has been proposed by Morton and Griffiths (1985).

3. *The three Rs*

As part of the ends and means calculation, the concept of the three Rs should be applied. They were introduced by Russell and Birch in 1959 as the principles of humane experimental technique, the authors suggesting that the principles could be summarized as replacement, reduction and refinement.

Replacement. When possible, living animals should be replaced by non-sentient material such as tissue cultures and computer models; mammals should be replaced by animals with less well-developed nervous systems; whole animals by decerebrate ones, or by isolated organ systems.

The Fund for the Replacement of Animals in Medical Experiments (FRAME) was set up in 1969; one of its functions is to evaluate and make known the various techniques which are being developed which do not involve living animals and which may be replacements for them. There have been some exciting developments, such as the use of human cell cultures for the production of polio and rabies vaccines, and the use of *in vitro* tests as prescreens in safety testing of products. For a general discussion see Smyth (1978) and Rowan (1984). For the role of replacement techniques in toxicity testing see Balls, Riddell and Worden (1983).

The number of animals used under licence in the UK annually is falling and this is probably partly due to the use of replacements, but there remain large areas such as animal behaviour, experimental surgery and pain research where it is difficult to imagine effective replacements ever being found (Home Office, 1985).

Reduction. If live animals must be used in potentially painful experiments, then the number should be the minimum necessary to achieve the desired level of dependability. In the past 30 years, great reductions have been achieved, first by the use of healthy animals — this can halve the number required for a long-term experiment — and then by using animals of known genetic constitution. Thirdly, proper statistical design and analysis may reveal that the number required is less (or greater) than at first thought. The body of knowledge known as laboratory animal science has contributed greatly to these achievements. For an overview see Remfry (1985).

Refinement. In the UK it is rarely permitted to use living animals for the acquisition of manual skill. This should not be a problem for the budding experimental surgeon, since formal training courses and systems of apprenticeship will be followed; but for scientists coming to animal work from non-surgical disciplines, the feel of living tissue and the realities of bleeding vessels may be unfamiliar. Ways must be found through the use of cadavers and by attachment to experienced surgeons to gain experience, so that when the actual experiment is performed it causes as little trauma as possible. In recovery experiments this is particularly important, because recovery and healing will be more rapid if anaesthesia, analgesia, pre- and postoperative care, the choice of instruments and the performance of the surgical techniques are all 'refined'.

In other fields, similar considerations apply. Routine injections can cause pain if inexpertly done. Injections of vaccine adjuvant into the footpad of rodents is painful and other sites should be considered. In behavioural work, rewards rather than punishments should be used to train animals. The animal should always be able to switch off a stimulus if it becomes unbearable. Further examples are discussed by Hampson and Silcock (1985).

Recently it has become more common amongst biologists to use the term 'alternative' to describe not only those techniques which replace the use of animals, but also those that make it possible to reduce numbers or, by refinement, to reduce pain (see Balls, 1985). This probably causes confusion in the minds of non-scientists.

4. *Accountability*

Scientists are accountable for their work to their employers or supervisors, and to any funding body supporting the work. If the work is published, the scientists become accountable to their professional colleagues. The public, or even the scientist two doors down the passage, may have little idea of what the work is about even though they will probably be financing it through their taxes. If animals are involved the lack of information may lead to suspicion and gossip.

Any scientist using animals must therefore be quite sure that the animals are being acquired and cared for and the work carried out in accordance with animal protection and other relevant laws. Departmental codes of practice and institutional policies must be followed.

Several countries produce their own guides to the care and use of animals (Canadian Council on Animal Care 1980, 1984; Institute of Laboratory Animal Resources, 1985). In the UK, guidelines are being prepared by the Royal Society and UFAW. These are, or will be, published documents that are available to interested members of the public, so that they can see the standards that biologists are expected to work to.

The use of ethical committees to monitor all animal work in an institute should also help to promote the welfare of the experimental animals. Hampson discusses the role and effectiveness of such committees in Chapter 2. The Biological Council *Guidelines* (1984) and a review by Britt (1985) are also worth consulting.

In order to avoid accusation of secrecy and to explain the purposes behind the work being done, influential members of the community (Members of Parliament, school headmasters, clergy) could be invited to visit the laboratories. If they approve of what they see, they will spread the word round.

CONCLUSION

Laws, codes and guidelines are all useful in setting up a framework within which the scientist can work. But in the final analysis, the welfare of the animals will depend on the conscience and integrity of the scientists. If those scientists are well informed, conscientious and have considered carefully their duties and responsibilities, then the animals should have little to fear.

REFERENCES

Balls, M. (1985). Animal experimentation: the search for valid and acceptable alternatives. In *Animal Experimentation: Improvements and Alternatives*, Supplement to *ATLA*, pp. 53–62, FRAME, Nottingham.

Balls, M., Riddell, Rosemary and Warden, A. N. (1983). *Animals and Alternatives in Toxicity Testing* (Proceedings of Meeting held to discuss Reports of FRAME Toxicity Committee), Academic Press, London and New York.

Biological Council (1984). *Guidelines on the Use of Living Animals in Scientific Investigations*. Available from Institute of Biology, London.

Britt, D. P. (1985). Research review (ethical) committees for animal experimentation, Supplement to *UFAW News-Sheet*, April 1985, UFAW, Potters Bar.

Canadian Council on Animal Care (1980–1984). *Guide to the Care and Use of Experimental Animals*, 2 vols, CCAC, Ottawa.

Carpenter, Edward (1980). *Animals and Ethics — Report of a Working Party*, Watkins, London and Dulverton.

Clarke, S. R. L. (1977). *The Moral Status of Animals*, Clarendon Press, Oxford.

Dawkins, Marian Stamp (1980). *Animal Suffering — the Science of Animal Welfare*, Chapman and Hall, London and New York.

Dodds, Jean W. and Orlans, F. Barbara (eds) (1982) *Scientific Perspectives on Animal Welfare*, Academic Press, New York and London.

Dunstan, G. R. (1983). *Science and Sensibility. The First Hume Memorial Lecture*, UFAW, Potters Bar.

FRAME News (1984). Gallup Poll in Surrey, Nov/Dec 1984, FRAME, Nottingham.

Hampson, Judith E. and Silcock, Sheila R. (1985). Pain in laboratory animals — the case for refinement. In *Animal Experimentation: Improvements and Alternatives*, Supplement to *ATLA*, pp. 19–24, FRAME, Nottingham.

Hill, Rosalind (1955). *Both Small and Great Beasts*, UFAW, Potters Bar.

Home Office (1985). *Statistics of Experiments on Living Animals, Great Britain 1984*, HMSO, London.

Hume, C. W. (1957). *The Status of Animals in the Christian Religion*, UFAW, Potters Bar, reprinted 1980.

Hume, C. W. (1960). The vivisection controversy in Britain, *UFAW Courier* No. 17. Reprinted in *Man and Beast*, 1962 and 1982, and as Supplement to *UFAW News-Sheet*, April 1984.

Institute of Laboratory Animal Resources (1985). *Guide for the Care and Use of Laboratory Animals*, US Dept of Health, Education & Welfare, NIH, Bethesda.

Linzey, Andrew (1976). *Animal Rights — a Christian Assessment of Man's Treatment of Animals*, SCM Press, London.

Martin, P. and Bateson, P. (1986). Appendix to *Measuring Behaviour*, Cambridge University Press, Cambridge (in press).

Morton, D. B. and Griffiths, P. H. M. (1985). Guidelines on the recognition of pain, distress and discomfort in experimental animals and an hypothesis for assessment, *Vet Rec. H.*, **116**, 431–6.

Niven, C. D. (1967). *History of the Humane Movement*, Johnson, London.

Paton, William (1984). *Man and Mouse — Animals in Medical Research*, Oxford University Press, Oxford and New York.

Regan, T. and Singer, P. (eds) (1976). *Animal Rights and Human Obligations*, Prentice-Hall, Englewood Cliffs.

Remfry, Jenny (1985). Recent developments in laboratory animal science. In *Animal Experimentation: Improvements and Alternatives*, Supplement to *ATLA*, pp. 25–30, FRAME, Nottingham.

Rollin, Bernard E. (1981). *Animal Rights and Human Morality*, Prometheus Books, Buffalo.

Rowan, A. N. (1984). *Of Mice, Models and Men — A Critical Evaluation of Animal Research*, State University of New York Press, Albany.

Russell, W. M. S. and Birch, R. L. (1959). *The Principles of Humane Experimental Technique*, Methuen, London (out of print).

Sechzer, Jeri A. (1983). *The Role of Animals in Biomedical Research*, Vol. 406 of *Annals of the New York Academy of Sciences*, New York.

Singer, Peter (1976). *Animal Liberation*, Jonathan Cape, London.

Smyth, D. H. (1978). *Alternatives to Animal Experiments*, Scolar Press in association with the Research Defence Society, London.

UFAW (1986). *The UFAW Handbook on the Care and Management of Laboratory Animals*, Longmans, London and New York (in press).

Wood-Gush, D. M. G., Dawkins, Marion and Ewbank, R. (1981). *Self-Awareness in Domesticated Animals* (Symposium Proceedings), UFAW, Potters Bar.

Laboratory Animals: An Introduction for New Experimenters
Edited by A. A. Tuffery
© 1987 John Wiley & Sons Ltd

CHAPTER 2

Law Relating to Animal Experimentation

JUDITH HAMPSON
London

As public concern about animal experimentation continues to grow, national governments and international bodies are under increasing pressure to improve laboratory animal protection controls.

Nowhere is the problem of controlling animal experimentation greater than in the USA. Flagrant abuses of inadequate existing controls continue to be exposed by a growing and coherent animal rights movement which has, to date, largely stayed clear of violence and destructive activities and has thus gained public credibility. Concerted efforts are now being made to deal with the immense problems in a country which uses many millions of experimental animals per year. The future system of control will depend heavily upon institutional animal care and use committees made up of scientists and animal care personnel, but with lay membership.

Thus the USA is following to some extent the lead set up by Canada, which operates a voluntary system of control in which such committees play a central part. Sweden adopted regional ethical committees into its legislation in 1979 and the other Scandinavian countries seem likely to follow its lead. Many countries in western Europe are currently reviewing their legislation while a Convention has been adopted by the Council of Europe, laying down minimum controls to be implemented by the 21 member countries. An EEC Directive, based on the Convention but with stricter provisions, was passed by the European Parliament in September 1986 and has now been approved by the Council of Ministers.

The illegal activities of animal liberationists in some European countries, notably the United Kingdom and West Germany, do little to facilitate the reaching of a reasonable consensus on this complex and difficult issue.

Notable progress is being made in the Netherlands, where good dialogue exists between pragmatic antivivisectionists (many of whom are students in the biological sciences), the scientific community and government ministers. In

Holland modern legislation is currently being implemented which is very much dependent upon administrative machinery and the goodwill and cooperation of all those involved to make it work.

By contrast, West Germany has a strong animal welfare act, in force since 1972 and recently updated. However, there is no proper machinery to implement it and many of its provisions are largely ignored. In West Germany there is a poor liaison between the antivivisection movement and the scientific community; deep polarization and extremism is the result.

The United Kingdom seems to stand at a crossroads. It is steering a course somewhere between the flexible approach of the Netherlands and the strict statute approach of West Germany. The UK has recently passed new legislation which is dependent for its workability upon administration within a flexible framework. This framework has found the support of pragmatic animal welfarists, while it has been roundly condemned by many abolitionists and animal rights groups.

It is against this background of continued debate and legislative reform that the European Commission has drawn up its Directive. When this comes into force it will become necessary for all twelve member countries of the EEC to bring their national legislation into line with its provisions.

LEGISLATION IN THE UNITED KINGDOM

History of the 1876 Cruelty to Animals Act

The oldest piece of legislation controlling animal experiments, the UK Cruelty to Animals Act 1876, has just been repealed and replaced by the Animals (Scientific Procedures) Act 1986. This law was passed largely as a result of pressure instigated by British antivivisectionists, who were appalled by the 'horror stories' of the mid 1800s reaching England from the continent, where French and German physiologists and doctors were avidly pursuing the 'new scientific method'. The objective of the Act was to prevent similar abuses occurring in the UK, and in this aim it obtained not only the support of the British government but of the scientific community itself.

The Act was passed after a twelve-month investigation conducted by a Royal Commission (1875) which examined 53 witnessess, including 47 medical men. The evidence presented in favour of the necessity to perform experiments on living animals in the pursuit and teaching of physiology and medicine had been overwhelming. However, the need to control the practice so as to prevent abuse was also made clear, and the Commission, accepting the indispensability of experimentation to science and medicine, concluded that the needs of science must be reconciled with the just claims of humanity and that through legislation, the practice of 'vivisection' must be made fully accountable and acceptable to public opinion.

The Government subsequently drafted legislation which was substantially amended in the House of Lords as a result of pressure applied by almost the entire medical profession. The amendments had the effect of lifting most of the bill's restrictions through a series of exemption certificates.

The antivivisectionists saw the resulting Act as a 'vivisector's charter', permitting most of that which they had sought to prohibit. The scientists saw it as a necessary reassurance to an agitated public which undid the damage done by antivivisectionists and exercised some control while leaving the essential progress of science unimpeded.

Throughout the last 109 years that the Act was in operation this polarization deepened, but despite its shortcomings, the 1876 Act functioned well, exercising a degree of control in the UK which has been lacking in other countries without restrictive legislation, such as the USA. Its effectiveness was largely due to its administration by the Home Office, which found it necessary, over the years, to stretch the application of the Act far beyond the literal meaning of the nineteenth-century wording.

Main provisions and administration of the 1876 Act

The new British legislation provides a very comprehensive system of controls which depend heavily upon administration by the Home Office for their effective implementation. In order fully to understand the new provisions it is useful to look at the system of control which preceded them. The 1876 licensing system is still in operation in the UK and will not be fully replaced by the new project-licensing system until about 1990.

The 1876 Act applied to all experiments on vertebrate animals likely to cause pain and the only reason for which experiments could be performed was that they must be conducted 'with a view to the advancement by new discovery of physiological knowledge or of knowledge which will be useful for saving or prolonging life or alleviating suffering' (in both humans or animals). This clause was open to wide interpretation and permits experiments for the acquisition of 'pure' knowledge.

A number of restrictions were laid upon animal experimenters. Experiments were prohibited except on completely anaesthetized animals which were killed before recovery from the anaesthetic. They were also not permitted to be performed in the illustration of lectures or other teaching, or on equines. Experiments not exceeding these conditions could be performed under simple licence. All these restrictions could, however, be dispensed with by means of certificates attached to the licences. These are summarized in Table 2.1. The Home Secretary could append conditions to licences and the nine standard conditions in Table 2.2 were invariably attached.

Table 2.1 Certificates dispensing with main restrictions of the 1876 Act

Certificate	Provisions	Conditions applying
Licence only—no certificate	Animal to be fully anaesthetized throughout procedure and to be killed before recovery from the effects of the anaesthetic.	1–3, 7–9
A	Allows the dispensation of anaesthesia where this would frustrate the object of the experiment. *Never* allowed for *any* surgical procedure more severe than superficial venesection.	1–3, 7–9
B	Animal to be anaesthetized throughout the procedure but where considered necessary, allowed to recover from the anaesthetic's effects and to live until the object of the experiment has been attained.	1–3, 7–9 plus 4 and 5
C	Allows experiments to be performed under full anaesthesia and without any recovery from the effects of the anaesthetic for the purpose of illustrating lectures and demonstrations to *bona fide* students, for the aquisition of knowledge for purposes permitted under the Act.	1–3, 7–9 plus 6
D	Provision originally made in order to permit the verification of former scientific discoveries. This has not proved necessary and no such certificate has ever been issued.	
E	Only issued in conjunction with Certificate A and, subject to those conditions, allows the use of dogs and cats in experiments where the animals are not anaesthetized.	As appropriate to other certificates issued in conjunction
EE	Only issued in conjunction with Certificate B and, subject to those conditions, allows the use of dogs and cats in experiments where recovery from anaesthesia is required.	
F	Allows the use of horses, mules and asses.	

Proposals for reform of the 1876 Act

Ever since its passage, attempts were continually made to reform the 1876 Act, both by antivivisectionists, whose aim is the total abolition of animal experiments, and by pragmatic reformers, who aim in the short term to see experimentation more tightly controlled.

Since 1876 there have been two major official inquiries into the administration of the Act: a Royal Commission which sat from 1906 to 1912 and took very extensive evidence, and a Departmental Committee, under the chairmanship

Table 2.2 Conditions attached to all licences

The Home Secretary has power (Section B) to append any conditions to licences he requires. These must be consistent with the Act's main provisions. Ten standard conditions are invariably attached to all licenses, and apply to the licence and appropriate certificate.

Condition	Provisions
1	Specifies the place at which the experiments are to be performed.
2	Requires that no experiments under certificate are to be performed until notification is received that the certificate has not been disallowed.
3—the 'pain condition'	This requires that: (a) If an animal at any time during any experiment is found to be suffering pain which is *either* severe *or* likely to endure, and if the main result of the experiment has been attained, the animal shall forthwith be painlessly killed; (b) If an animal at any time during any such experiment is found to be suffering severe pain which is likely to endure, such animal shall forthwith be painlessly killed; (c) If any animal appears to an Inspector to be suffering considerable pain, and if such Inspector directs such animal to be destroyed, it shall forthwith be painlessly killed.
4—'limitation condition'	Restricts Certificate B to procedures of no greater surgical severity than simple innoculation or superficial venesection.
5—'anaesthesia condition'	Requires experiments performed under Certificate B to be: (a) carried out under anaesthetics of sufficient power to prevent the animal from feeling pain; (b) the animals upon which such experiments are performed shall be treated with strict aseptic precautions, and if these fail and pain results, the animal shall be immediately killed under anaesthesia.
6	Requires humane killing of the animal before recovery from anaesthesia under Certificate C.
7	Prohibits the use (except on decerebrate animals) of curare or other similar muscle relaxants without special permission from the Secretary of State. Forty-eight hours' prior notice must be given to the Inspector so that he may attend such experiments if he wishes. Such substances are not deemed to be anaesthetics under the Act and their use is restricted to use in conjunction with full anaesthesia throughout the experiment.
8	Requires each licensee to keep a record of his/her experiments and send a report to the Secretary of State at the end of each year.[a]
9	Requires each licensee to send to the Secretary of State information relating to any experiments performed by him/her which have been described in a printed publication. The part of this condition which imposed restrictions on the filming of experiments has now been rescinded.

[a] On the basis of these records the Home Office produces its *Statistics of Experiments on Living Animals* at the end of each year. These are obtainable from HMSO.

of Sir Sydney Littlewood, which reported to Government in 1965. Administrative changes resulted from both inquiries, but the law itself was not amended.

A spate of unsuccessful bills introduced into both Houses of Parliament throughout the 1960s and 1970s finally prompted a former President of the Research Defence Society, Lord Halsbury, to introduce a detailed private members' bill into the House of Lords in 1979. This resulted in the setting up of a Select Committee of the Lords, including a number of eminent scientists. The Committee produced a detailed report, forming the basis for a prospective new law. The bill made no further progress, but many of the report's findings were noted by the Government, which had made, in 1979, an election manifesto pledge to update the 1876 Act. In that same year the Home Secretary invited his newly reconstituted Animal Experimentation Advisory Committee to draw up the framework of a new law. Building upon the work already done by the Halsbury Select Committee, this body found it unnecessary to take evidence, and it produced a detailed framework by 1981. This formed the basis of the proposals contained in the Government's White Paper of May 1983 (Cmnd 8883).

Considerable note was also taken of the proposals put forward jointly by the Committee for the Reform of Animal Experimentation (CRAE), the British Veterinary Association (BVA) and the Fund for the Replacement of Animals in Medical Experiments (FRAME).

The White Paper was intended to be a discussion document and the government proposals were later amended substantially.

The Animals (Scientific Procedures) Act 1986

In May 1985 the Government published its revised proposals in the form of a Supplementary White Paper (Cmnd 9521). These formed the basis of the bill which was introduced into the House of Lords in November 1985. Progress through both Houses of Parliament was rapid and the bill completed its passage in May 1986, having received a number of amendments urged by the CRAE/BVA/FRAME liaison group and the RSPCA. It received Royal Assent in June of that year. The new law will repeal the 1876 Act when the administration has fully turned over to the new licensing system. This process will probably take about three years to complete.

The Act is an enabling piece of legislation. It prohibits very little, but it aims to apply much stricter controls over what is done by establishing a detailed system of licensing and setting up a chain of accountability and responsibility within the framework of the Act's administration. How that administration will work is explained in detail in a document issued by the Home Office — *Guidance on the Operation of the New Legislation to Replace the Cruelty to Animals Act 1876.*

The new system of control introduces a dual licensing system in place of the old system of personal licences and certificates. The personal licence, like the licence under the 1876 Act, is a licence of competence, granted to persons judged by the Inspectorate to have appropriate qualifications, training and experience to carry out listed procedures using specified techniques on named species.

The licensee will also be required either to possess a project licence, or will be working on a project employing several licensees under the direction of a project licence holder. After the first six months of holding a personal licence, a licensee is free to work on any project for which (s)he has the appropriate authority.

This new system of personal and project license gives the Home Office much greater control over pain at the licensing level. It enables a reasonable assessment of the amount of pain or distress likely to be involved in a project and the restriction of that pain or distress by imposition of severity-limiting conditions on the project licence. The system also provides the possibility of stricter control over the purposes for which research is carried out. This is achieved by linkage of the amount of pain/distress allowed to an assessment of the importance of the research, a kind of cost/benefit analysis. The severity is graded according to licensee's and inspector's assessment as to whether it will result in pain or distress, which is classed as mild, moderate or substantial.

Permissible purposes are widely drawn so as to include all possible procedures within the scope of the new controls. However, in determining whether and on what terms to grant a project licence, the Secretary of State will weigh the prospective adverse effects on the animals concerned against the benefit likely to accrue as a result of the programme specified in the licence (Clause 5(4)). Thus the Home Secretary is made responsible to the public, through Parliament, for justifying what is licensed, a responsibility which successive Home Secretaries have been loath to accept but a provision for which the CRAE/BVA/FRAME group and the RSPCA have pushed very hard over the past three years. This entirely new departure from the old system of control (previously a licence would be granted for any procedure likely to result in the acquisition of new knowledge) will enable the Secretary of State, if so advised by his official advisers, to tighten up on the purposes for which research is carried out either by imposing stringent restrictions over certain categories of research, or even refusing altogether to license them. The system goes some way towards meeting the demands of more extreme animal rights groups who seek the banning of certain procedures such as LD50 and Draize tests, defence experiments and psychological and cosmetics research, though as yet the significance of the new controls seems not to have been grasped either by these groups or by the scientific community at large.

In effect, the difficult judgements required to implement this new system will be made by a series of official advisers to the Secretary of State. The range of advice now available to the Home Secretary has been increased through the sponsorship of personal licences, the strengthening of the Inspectorate (a measure also necessitated by the increased bureaucratic workload), by increased specialization within that Inspectorate, by the establishment of a panel of specialist independent advisers (assessors) available for consultation by inspectors for the granting of novel or difficult project licences, and by the setting up of an Animal Procedures Committee with statutory authority which can act on its own initiative and will, among other things, review retrospectively many project licences, especially those in the category of substantial suffering, and advise on improvement of the licensing system. The Committee will also have referred to it all applications for licences to perform cosmetic tests, to test the effect of tobacco products on animals and to perform experiments for the acquisition of manual skill in microsurgery.

This committee will make administration of the new law much more publicly accountable. It will play an essential role in monitoring the operation of the new controls and, should the Home Secretary choose not to accept its advice, (s)he will be required to make the reasons public. The Home Secretary, in appointing this new committee, shall have regard to providing adequate representation of both user and animal welfare interests. It will consist of a chairman (currently a professor of law) and twelve members, at least two-thirds of whom are doctors, veterinarians or biological scientists. Not more than half shall be licensees. Three professional animal welfarists and two lay members serve on the present committee.

If it works well, the system will both foster the refinement of research protocols and restrict research activity in those areas already identified as causing major public concern.

In granting project licences the Home Secretary will wish to satisfy himself both that the work cannot be carried out using less severe techniques or endpoints than those proposed, and that non-sentient 'alternatives' which could replace the use of animals wholly or in part of the project are not available.

Once a licence is granted the next crucial level of control must be implemented in the laboratory itself. It has been noted elsewhere in this discussion (see p. 22) that there is little merit in a strong statute law if the machinery does not exist to implement its provisions. Under the new UK system three persons responsible for implementing and monitoring controls in the laboratory will be named and will have statutory authority and responsibility. The first will be the person responsible for ensuring that premises meet the requirements of their certificate of registration, such as general conditions of husbandry and care, cleanliness, ventilation systems and so forth. For the first time in the UK, non-experimental establishments such as those producing sera or breeding animals for supply to laboratories are required to be registered and are open to Home Office inspection.

The second named person will be a veterinarian, either employed by the laboratory or available locally to be called in as necessary. This person will be responsible for ensuring general maintenance of health and welfare and for advising on measures to be taken to relieve animal suffering or to kill animals where necessary during experimental procedures. The third named person, a kind of 'animals' advocate', has similar responsibilities to the veterinarian but is available to monitor the animals on a day-to-day basis. This person will usually be a senior animal technician or curator, and will possess a high degree of experience and training.

The latter two named persons are given a statutory responsibility to inform the personal licensee if animals are found to be suffering unexpectedly or in excess of the permitted severity conditions, or are suffering severely enough to merit implementation of the Termination Condition. The latter is more stringent than the condition imposed under the 1876 Act and requires that any animal suffering severely must immediately be killed. There is no longer the proviso that an assessment be made as to whether the suffering is likely to endure.

The personal licensee has a statutory duty to take appropriate action to alleviate the suffering or kill the animals and to inform the inspector in cases where the permitted level of severity has been exceeded unexpectedly. If the personal licensee is not available, either of the two named persons has authority to take whatever action is considered necessary.

It is recognized that estimating levels of suffering prior to commencing a project and identifying and interpreting levels of distress and pain in the laboratory are two quite different things. Licensees are expected to familiarize themselves with signs of pain, discomfort and distress in the species they are using, and to consult where necessary with colleagues in order to obtain expert advice. To assist licensees in making these difficult subjective judgements general principles for the recognition of pain and suffering have been issued (Association of Veterinary Teachers and Research Workers, 1986; Morton and Griffiths, 1985) and the Home Office recommends that licensees consult these guidelines.

The Home Office will also shortly be issuing a *Code of Practice*, outlining prescribed methods of husbandry as well as guidance on the recognition of suffering and the means of its alleviation.

It has also issued detailed guidance on the operation of the new legislation which explains in detail how the Act will be administered and the responsibilities of those operating under it as licensees, advisers and persons with other statutory responsibilities.

Perhaps the greatest weakness of the system is that it provides no formal machinery for ensuring that licensees receive adequate training. Voluntary training schemes do exist but are undersubscribed, especially by the academic sector. Government incentives to expand the use of such schemes are urgently needed.

There are three other novel features contained in the new system of controls. The source of supply of animals has been tightened up, special attention being paid to preventing the use of stray or stolen cats and dogs. All such animals will now have to be purpose-bred unless special (and very limited) exemptions are granted. Companion animals may also be privately homed after an experiment provided that a veterinarian certifies that they are sufficiently restored to health.

As under the 1876 Act, some animals may be reused in a subsequent procedure, but a new and controversial departure is that this now applies also to animals which have undergone general anaesthesia. The objective is to reduce overall the numbers of animals used and the provision does not result in additional suffering to individual animals since the second procedure must be carried out entirely under terminal anaesthesia. Guidance notes on reuse are being issued and special permission of the Secretary of State is required for reuse.

Guidelines on husbandry and care have been drawn up by the Royal Society in collaboration with the Universities Federation for Animal Welfare. These are currently under consideration by the Home Office Advisory Committee and are likely to be incorporated into the Code of Practice.

Finally, the old prohibition on the attainment of manual skill has been lifted but the Home Office intends this to be strictly limited and will, for the present, allow it only on terminally anaesthetized rodents for the purposes of training in microsurgery.

There are stiffer penalties for breaches of the new law.

There are also numerous acts and codes specifying animal tests in assuring the safety in use of a wide variety of products and to ensure safety of workers manufacturing such products. In addition to national laws, most products are subjected also to transport, import and export regulations as well as to EEC directives, all of which may require specific animal tests. Some of these are described briefly in Chapter 4.

LEGISLATION IN THE USA — THE ANIMAL WELFARE ACT 1966

The USA has very little control over the actual use of animals in research and, until recently, enforcement of controls over the care of animals in institutions has been almost non-existent. The result has been great variability of conditions between institutions, some maintaining high standards while others could only be described as deplorable.

Over the last few years, meticulous underground investigation and illegal break-ins carried out by a coherent and articulate animal rights movement have made the worst offenders highly visible and it has primarily been this activity which has resulted in the changes now taking place in the USA.

Background to American legislation

In the latter half of the nineteenth century numerous attempts to pass federal legislation based on the British model failed. The humane movement then directed the main thrust of its campaign at state legislation and most states passed anticruelty laws as a result, but the practice of vivisection was usually exempted.

After the Second World War an already strong biomedical community became an even more formidable opposition to control of animal experiments with an explosion of federally funded research effort. There was little public support for any moves which smacked of opposition to medical research.

That federal legislation happened at all was due less to government desire to interfere with the research community than to the difficulties arising from the phenomenal increase in the demand for laboratory animals. This demand led, due to the efforts of the National Society for Medical Research, to the passage, in several states, of pound seizure laws, allowing or even mandating the release of unclaimed animals from municipal pounds and private shelters to laboratories. The first such law was passed in Minnesota in 1949, and others soon followed. These laws struck right at the heart of a humane movement whose brief it was to provide a last sanctuary for abandoned and unwanted animals or, as a last resort, a quick and humane death.

It took many years for the humane movement to strike back effectively. The first pound seizure law to be repealed was in Rhode Island, 1972, but the turning point came with the repeal of the Metcalf-Hatch Act in New York State in 1979; after that several states were soon to follow suit.

While this battle raged between the biomedical research and humane communities new attempts were made to introduce federal legislation to control research, based on the British 1876 Act. Researchers were able to resist these attempts, claiming that British research had been much restricted by controlling legislation. However, two events occurred in the mid sixties which shifted public concern in favour of legislation.

First an allegedly stolen family dog was traced to a laboratory. The congressman who had aided the pet's owners in their search then put forward a bill to regulate cat and dog dealers (Stevens, 1978). Secondly, the squalid conditions revealed by a raid on a Maryland dealer's facility were publicized in *Life Magazine* in February 1966.

Bills in both Houses followed these two events, resulting, after much debate, in the passing of the Laboratory Animal Welfare Act on 24 August 1966.

The Laboratory Animal Welfare Act 1966 (amended 1970; 1976; 1985)

The history of the American legislation meant that it was formulated primarily to control the supply of cats and dogs and promote better conditions of husbandry and care in facilities. Birds, rats and mice, which make up more than 80 per cent of laboratory animals used, are excluded from its provisions because of inadequate resources to enforce the Act. The question of the purposes for which animals are used and the treatment they receive while actually under experiment was never intended to be addressed by the legislation, which specifically states:

> nothing in the Act shall be construed as authorising the Secretary to promulgate rules, regulations or orders with regard to design, outlines, guidelines or performance of actual research or experimentation by a research facility.

However, the 1970 amendment does require the 'appropriate use of anaesthetic, analgesic or tranquilising drugs.' These provisions are neither properly defined nor enforced.

The Act is administered by the US Department of Agriculture (USDA), which registers premises and requires reports from institutions detailing their use of laboratory animals. However, no reliable statistics are yet available for the USA. The total number of animals used has been variously estimated at between 25 and 70 million (Rowan, 1984; Orlans, 1984; OTA, 1986). Official sources now estimate that around 35–50 million laboratory animals are bred in the United States each year with some 25–30 million actually used. Surveys are currently underway to verify this. Animal rights advocates dispute the lower figure, claiming that annual usage is nearer 60 million.

Enforcement of the provisions of the Act relies upon inspection by the USDA's Animal and Plant Health Inspection Service (APHIS), which has neither the staff nor the budget to secure compliance even with the limited provisions of the Act relating to such a massive research activity. The role of APHIS is primarily livestock disease control. The USDA was never keen to have jurisdiction over legislation relating to laboratory animals and the motivation to discharge its responsibilities has not been high. Most inspectors are insufficiently trained for the job, though this is now beginning to change.

Moreover, the reporting system introduced in 1970 to provide Congress with a means of exerting control over pain and distress is ineffective. It does not provide APHIS with sufficient information to ensure that pain relief is in accordance with professionally acceptable standards of care because neither pain nor distress are adequately defined, nor are the explanations given for withholding pain relief required to be meaningful and comprehensive. For the same reasons, the system fails to meet the other objective of Congress in introducing it, which was to increase public accountability of research (Solomon and Loveheim, 1982).

In December 1985 the President signed the bill which became public law 99–198. Stemming from proposals originally put forward in the Dole and

Brown private bills, this Act contains 'Improved Standards for the Laboratory Animals Act', which amend the Act in the following ways:

1. Each registered facility is to establish an Institutional Animal Committee with at least *one* non-institutional (lay) representative.
2. This committee is to carry out semi-annual inspections of the facility.
3. Provision must be made for the exercising of dogs.
4. Primates must be provided with 'a physical environment adequate to promote (their) psychological well being.'
5. Painful procedures are to be avoided wherever possible and painkillers must be used where appropriate, unless they interfere with the objectives of the experiment.
6. An information centre is to be established at the National Agricultural Library, in cooperation with the National Library of Medicine, for data exchange and dissemination of information on non-animal alternatives. This is intended to aid the reduction of duplicative testing.

The inspectorate role of the USDA is expanded under the new provisions. Among other things, the Agricultural Secretary 'shall inspect each research facility at least once each year and, in the case of deficiencies . . . shall conduct such follow up inspections as may be necessary . . .' Under these provisions it is expected that the inspection budget will be increased and the level of training of inspectors improved.

These improvements still relate largely to the care and husbandry of laboratory animals rather than to the actual conduct of research. The Secretary of Agriculture is still precluded from promulgating rules about the design of actual research except where such rules relate to new annual reporting requirements or to minimization of suffering through humane care and treatment, including provisions for adequate veterinary care and the appropriate use of anaesthetics, analgesics, tranquillizing drugs and humane euthanasia. Inspectors are expressly forbidden from interfering with progress of research during their visits.

The new amendments, which apply to all registered facilities, put into the law provisions which had previously been applied only to institutions receiving federal funding. These provisions are applied through Public Health Service (PHS) policies which require institutions with PHS funding to conform not only with the Animal Welfare Act but also with guidelines issued by the National Institutes of Health (NIH) to all institutions receiving NIH grants. As NIH policies are updated, other granting agencies have tended to follow suit. The translation of these policies into the legislation itself should mean that all institutions begin to conform to the same standards.

The original NIH *Guide for the Care and Use of Laboratory Animals* (1963) was revised for the sixth time in 1985 and included detailed standards for buildings, equipment and housing, qualifications for personnel and detailed recommendations on the care and treatment of animals including provision for veterinary care, anaesthesia, analgesia, postoperative nursing, humane euthanasia and factors affecting environmental health. Multiple survival surgery is discouraged but not disallowed.

The *Guide* is purely advisory and most of its suggestions are expressed in terms of what is reasonable or appropriate. It was always intended to be flexible and indeed its provisions would be difficult to enforce. What is considered to be 'adequate' veterinary care, for example, is not defined. NIH carries out periodic site visits but these again have had more of an advisory than an enforcement role.

The status of the *Guide* was enhanced, however, and its provisions substantially strengthened, by the passage, in December 1985, of the NIH Reauthorization Bill. This requires the Department of Health and Human Services, through NIH, to establish guidelines for proper care and treatment of laboratory animals in *all* institutions, not just those which are federally funded.

These changes are the result of an extensive two-year survey by NIH on the implementation of its policies, involving site visits of fifteen institutions and study of 340 written and oral comments. The four major areas of change are:

> Institutions must designate clear lines of authority and responsibility for those involved in animal care and use. Two named officials must be designated, one with ultimate responsibility for the institution's animal programme, the other a veterinarian.

> Institutions must have an animal care and use committee which is to be involved with all aspects of the research programme. It must include an individual unaffiliated with the institution, a non-scientist and a veterinarian experienced in care and use of laboratory animals.

> Detailed information on the animal use programme and animal care is to be filed as a written Animal Welfare Assurance with the NIH Office for Protection from Research Risks (OPRR).

> The committees are to review and approve these sections of applications for funding which relate to the care and use of animals. Funding will be dependent upon this documentation.

The new legislation and revisions of PHS policy result in large part from revelations of appalling conditions and breaches of existing regulations at several major American institutions. The most dramatic and colourful of these have been brought to light as a result of activities by animal rights activists.

The first was the trial of Dr Edward Taub, of the Institute of Behavioural Research, who was convicted in 1981 by a Maryland court of failure to provide adequate veterinary care to non-human primates. Though the conviction was later quashed on a technicality, it did result in the withdrawal of NIH funding and it had to be admitted that both NIH and USDA monitoring systems had failed.

The dreadful conditions at IBR had been uncovered as a result of infiltration into the laboratory by an animal activist. While some maintained that it was an isolated case, animal protectionists suspected that it might be the tip of the iceberg. Subsequent events seem to have proved them right.

The last four years have witnessed numerous exposures of gross failures to comply with the law and existing regulations, often at prestigious institutions. In 1984 excerpts from videotapes stolen by the animal liberation front from the head injury trauma laboratory at the University of Pennsylvania were networked across America on national news. They depicted researchers failing to maintain adequate anaesthesia during surgery, smoking while operating on the brains of baboons, making fun of the experimental subjects and removing a helmet, after the supposed infliction of a precise injury meant to mimic injury in road accidents, with a hammer and chisel.

The lack of sensitivity of the researchers seen in these tapes has shocked the bulk of the research community not only in the USA but throughout the world. The work, which has continued for 13 years at a cost of 12 million tax dollars, is now suspended and unlikely to continue. NIH funding has been withdrawn indefinitely and the university fined $4000 for failure to comply with the Animal Welfare Act.

Animal activists raided the City of Hope Medical Center in Duarte, Southern California in December 1984, removing animals and documentation of serious violations of the law. These included dogs and cats found unanaesthetized and dying lingering deaths after subjection to extreme procedures. A subsequent NIH investigation resulted in the suspension of 31 research projects and an $11,000 fine was levied by USDA.

The first action taken against an institution by NIH since the new guidelines were issued in January 1986 was against Columbia University. This, the third suspension of funding in six months, again followed the production of photographic and other documentary evidence of violations following an ALF raid. The University was fined and ordered to correct filthy conditions before its funding was restored in mid 1986.

Despite its source, neither NIH nor federal investigating agencies have been able to ignore such evidence, but it does seem that these isolated actions may simply have highlighted deficiencies which are widespread. A recent publication by the Animal Welfare Institute (AWI, 1985) documents massive noncompliance with the Animal Welfare Act from evidence available, through the Freedom of Information Act, in the reports of USDA inspectors. Facts uncovered revealed abuse or neglect in 82.7 per cent of the facilities investigated during a four-year period, in which time funding to most of the institutions increased. Less than one-fifth of them complied even with the basic provisions of the Animal Welfare Act.

An example of the type of case cited is University College of California,

Berkeley where an extensive report by animal rights groups has revealed, among other things, broken primate cages with protruding wires, no ventilation, grossly inadequate veterinary care (refusal of access to campus veterinarians), food being delivered in waste trays due to absence of feeders, blocked drains, insect infestation and generally filthy conditions. After four years of refusal by the campus veterinarian to sign the declaration of compliance with the Animal Welfare Act and seven site visits by NIH which had not resulted in any withdrawal of massive federal funding, the facility was finally ordered by USDA in 1984 to 'cease to desist' from violating the Act and fined £12,000 by USDA. Complaints continue to be made up to the present time.

UCB had failed to gain accreditation by the American Association for Accreditation of Laboratory Animal Care (AAALAC), a voluntary scheme to which about 400 institutions subscribe, but many institutions cited in the AWI publication did have AAALAC accreditation. It has in the past been widely claimed within the scientific community that AAALAC accreditation is a guarantee of the highest standards of laboratory animal husbandry and care.

The future in the USA

The United States must find ways of dealing with its problems, if only because a determined and well-organized animal protection movement will continue to expose abuses. Humane reform bills seem likely to continue to be pressed in almost every session of Congress.

Despite NIH's reluctance to take on a policing role and its continued insistence that a voluntary system of control is the most realistic approach to promoting the proper care and use of laboratory animals, it is clear from recent events that NIH will be expected to do rather more than rely on goodwill in ensuring compliance with regulations in the future.

NIH and OPPR are currently sponsoring a series of workshops on the implementation of new PHS policies. It is now recognized that NIH will take stern action in future cases of abuse and a general 'clean up' operation is underway in many institutions. But many argue that the machinery of enforcement is still lacking; for example, though the law itself was the direct result of a stolen animal being taken across state lines for sale to a laboratory, USDA reports reveal frequent failure to identify cats and dogs purchased from random source dealers, though proper record-keeping is a specific requirement of the law.

While some changes are being made to reduce painful research the main focus of the control system is still on care rather than use.

Under the new NIH guidelines the use of anaesthesia and pain-relieving techniques still remains at the discretion of the investigator. There is still no restriction on experimental techniques and procedures. There is no species-specific detail about behavioural and environmental needs.

It has been claimed with some justification that the *Guide* represents little more than an introduction to the problem (McCardle, 1985). It does not compare with the comprehensive 300-page, two-volume guidelines produced by the Canadian Council on Animal Care (see p. 39). While the NIH *Guide* acknowledges the need for proper care, it does not provide adequate information nor incentives to ensure that it will be maintained.

Animal Care and Use Committees are mandated which do not require animal technicians, those most qualified to advise on care, to be members. The lay person and the non-scientist may be persons of any persuasion, chosen by the institution itself. They could be one and the same person, or the non-institutional member could be a researcher from somewhere else. There is no requirement for an animal welfare representative.

The *Guide* lays down no rigid rules, multiple surgery is discouraged rather than prohibited and its use in teaching is not discussed. Even prolonged restraint is still allowed if considered an essential part of the research. Exercise for dogs is prescribed only for long-term projects and their general care is little improved. There is no proper guidance on the monitoring of anaesthesia during surgery with use of paralytics. Guidelines on aseptic surgery are substantially relaxed in the case of rodents. Physical methods of euthanasia are allowed despite the fact that they may be inexpertly carried out by untrained personnel.

It is difficult to see, however, even with all the goodwill in the world, how effective control might be exerted in this vast country. The USA has 1200 research institutions scattered over 3.6 million square miles. A system such as that operated in the UK could not possibly be made to work in the USA. Yet somehow USDA must be capable of effectively monitoring what is going on across America. The quality of information returned to the registering authority and the amount of attention paid to it by USDA is in obvious need of drastic improvement.

The frequency and quality of USDA inspection is likely to improve under the new system. It is still unlikely to be adequate for the task but it will be reinforced by the improved monitoring by NIH in federally funded laboratories and by the establishment of animal care and use committees in all facilities. How well these committees work will depend largely upon the calibre and range of expertise of those called to sit upon them.

Tougher penalties, such as immediate withdrawal of funding, must be implemented for non-compliance in the future. It has also been suggested (Rowan, 1984) that a national committee should be set up under the legislation to provide a forum for discussion between researchers and other personnel responsible for animal care in the laboratories. Some of these personnel, such as senior laboratory animal technicians, should act as 'animals' advocates' and they should play a prominent role in the institutional committees. Both research practices and ethical questions would be considered by this committee. Such a dialogue would lead to improvements both in animal welfare and in American biomedical science.

THE ROLE OF ETHICAL COMMITTEES IN SYSTEMS OF CONTROL

Two countries, Sweden and Canada, have local ethical committees as a central feature of their regulatory systems.

In Canada the system is essentially one of self-regulation, designed by the scientific community to obviate the need for strict legislative control.

The general belief among Canadian scientists has been that national legislation can enforce only minimal standards and that optimal conditions are best effected through voluntary control. In addition, the sparse scattering of research institutions over thousands of miles in this vast country would make monitoring by a local inspectorate virtually impossible.

Thus in 1968 the Canadian Council on Animal Care (CCAC) was established under the aegis of the Association of Universities and Colleges of Canada with terms of reference to make improvements to the procurement and production of laboratory animals, to facilities, to levels of care and to the experimental procedures themselves.

Between 1968 and 1970 the Council was responsible for the setting up of animal care committees in each of Canada's major research institutions.

They are given very specific terms of reference relating to all matters of care, husbandry and humane procedure, in detailed guidelines laid down by the CCAC (1980–84). Compliance with these guidelines is a prerequisite for research grants from all the major granting agencies and for research contracts from federal government departments, though there has been no action taken against institutions which have persistently failed to comply.

In Sweden, local ethical committees are regionally rather than institutionally based. Again, the system was established by the scientific community itself, on a joint initiative by researchers from Uppsala University, the Swedish Medical Research Council and the Scandinavian Laboratory Animal Science Association (Scand-LAS).

AIMS AND PRACTICE OF ETHICAL COMMITTEES

Questions asked by ethical committees will include: whether non-sentient alternatives might be used in whole or in part to obtain the information required; if not, whether less painful or less stressful procedures might be substituted for those proposed (e.g. substitution of non-lethal tests for death endpoints); if all the possibilities for reduction of pain and/or distress through the use of anaesthesia, analgesia and tranquillization have been fully explored; if the investigator is competent to perform the proposed procedures and to supply the necessary care; if the facilities are suitable for maintenance, surgical aftercare and any other necessary provision; and, if the animals have to be killed, whether or not the most humane method of euthanasia has been selected.

The CCAC guidelines clearly state that certain procedures known to inflict excessive pain should *never* be carried out. These include the use of muscle relaxants without anaesthesia and traumatizing procedures, including crushing, striking or beating unanaesthetized animals or allowing animals so treated under anaesthesia to recover from it.

Certain other procedures known to cause substantial suffering may be conducted only when there is a clear benefit anticipated from the research.

The Canadian committees, being institutionally based, are able to take on duties exceeding project assessment and are an integral part of the animal care and project follow-up system.

The CCAC has a national assessment programme, which aims to carry out periodic surveillance checks (usually one every three years), to give support and advice to the institutional care committees and to ensure uniformity between institutions. Critics have complained that these checks have been insufficiently frequent to ensure compliance with CCAC guidelines.

The large Swedish committees meet only once or twice per year. Projects are assessed by panels of three members. The application is sent to a scientific member who calls together a technician and a lay member. If agreement can be reached by these three and the applicant, the research may commence immediately. The full committee is notified by post of the details, and decisions are later reviewed by the full committee.

Applications are graded by project assessment, invasive and non-invasive. Only the former category is automatically selected for review, though the committee has access to all applications and may disagree with the grading.

The question of whether a sufficient number of projects is selected for review is a criticism of ethical committee systems. Until recently in North America only those projects requiring grants have been reviewed. Once projects are approved the work may continue for several years, so an effective follow-up system is vital to the implementation of committee decisions.

EFFECTIVENESS OR ETHICAL COMMITTEES — ADVANTAGES AND DISADVANTAGES

Membership

Deciding on composition of ethical committees is one of the most difficult aspects. It is generally felt by those involved in both the Canadian and Swedish systems that inclusion of non-animal users and lay members is crucial, both to sharpening the thinking of researchers at project-planning stage and to public accountability of the committees.

It is also difficult to decide what kind of lay members are suitable. In order to be effective they must be articulate and persistent and not likely to be intimidated by scientific expertise and jargon.

The CCAC guidelines suggest that the lay member should be a non-animal user, for example an academic in the field of humanities, within the institution. The CFHS holds a rather stronger view, maintaining that lay members should not be affiliated with the institution and should be known to have animal welfare interests.

In 1985 the CFHS issued very detailed (50 pages) *Guidelines for Lay Members of Animal Care Committees*, offering excellent advice on how to review protocols, with particular regard to the issue of pain.

In Sweden, where lay members make up one-third of committee composition, there have been considerable problems, due in large part to the fact that some of the lay members proposed by animal welfare bodies have turned out to be antivivisectionists.

Effectiveness at cost/benefit analysis

A central question which must be asked is how effectively do ethical committees perform the cost/benefit analysis? In neither Sweden nor Canada do the committees judge the scientific merit of the work itself; this is the province of government departments and granting agencies, thus the fundamental question of whether the work should be allowed at all does not arise, since its ultimate value is not weighed into the balance.

In other words, these committees are not really 'ethical' at all; their judgements are about how best to conduct work whose validity is already tacitly accepted. They are more properly called animal care and use committees than ethical committees.

This limitation is welcomed by the scientific community, while animal advocates will argue that, at best, the committees have limited effectiveness and, at worst, they are merely a whitewash.

Canadian committees have functioned with varying degrees of effectiveness. Some are conscientious and meet regularly while others exist only on paper. Even in Ontario and Alberta, where legislation sets down regulations on composition and rules of operation, this legislation has been ignored by many institutions (CFHS, 1985b).

A recent case in Canada involved the prosecution of researchers at the University of Western Ontario for cruelty to a baboon. The animal had been confined to a restraining chair for six months. This case has sparked off a heated debate in Canada and elsewhere about the effectiveness of animal care and use committees. The research had been approved by such a committee and was defended by the campus veterinarian.

Effectiveness as a voluntary system of control

In Canada the animal care committees play a central part in a voluntary scheme of control which was deliberately intended to obviate the need for legislation. In Sweden the original intention was that the committees should be voluntary. Their aim was to encourage researchers to consult with their peers and subject their projects to in-depth analysis in order to secure humane and rational design. The feeling among the scheme's instigators is that since the committees were incorporated into the legislative system and have become the responsibility of the Ministry of Agriculture, these aims have been imperfectly fulfilled.

In Canada extremists from either side were deliberately excluded from the outset. The CCAC guidelines state: 'the polarised views of those at the extreme ends of the spectrum of the question of animal usage have, of necessity, been precluded; i.e. those wishing to conduct experiments with little or no constraints regarding infliction of pain and those opposed, on humanitarian principles, to any suffering whatsoever.

The Canadians, believing that legislation cannot enforce sensitivity, have opted for a system in which the active participation of researchers is intended to foster it. The Canadian system is profoundly mistrusted by those who claim that research cannot be monitored by those conducting it and only guardedly commended by moderate animal welfare groups, who believe that legislation is needed to strengthen the system and prohibit objectionable practices such as the long-term chairing of primates. The point they make is not that such abuses are widespread, but that so long as they occur at all, there must be official sanction against them.

Lessons for Other Countries

In the UK some universities and commercial institutions have project assessment committees, but they do not consider ethics and so duck the central issue.

In the USA, currently struggling to formulate a workable system of control, the effectiveness of animal care committees has been under intensive discussion. Several workshops on how to run them have been facilitated by the Scientists Center for Animal Welfare (SCAW, Washington, DC). The role of animal care and use committees has been greatly strengthened by controls recently implemented in the USA (see pp. 30–38). If these are to be effective in reducing pain and suffering in experiments, the emphasis must be on animal use just as much as on animal care. Failure in this respect has been a frequent criticism of Canadian as well as American committees.

The crucial question now is whether or not other countries can learn from the gains and pitfalls experienced by Canada and Sweden and whether they could usefully establish committees to suit their own particular needs, either as a central feature or a useful adjunct to their own systems of control.

The USA, lacking any coherent system of research monitoring, seems to have little choice but to go down this road.

In the UK there is general resistance to the idea of local committees. Though subjected to 109 years of legal restraint, scientists working under the British legislation have hitherto had a relatively free hand to decide on the scientific merits and ethical costs of their research. The 1876 Act made little provision for any cost/benefit analysis, though the 1986 law which replaces it makes this provision explicit. Researchers are naturally loath to relinquish this freedom, and point to the effectiveness of the British Home Office Inspectorate in securing control. The Inspectorate itself prefers control at the national level and does not favour the idea of what it sees as duplication of central effort by local committees.

However, under the new UK licensing system research protocols, especially those involving risk of substantial suffering, will be subjected to more rigorous ethical review at the licensing level. It is well worth considering whether local ethical committees could play a useful role in this process.

Full discussion of projects at the local level would provide a useful data base from which the Home Office would make its cost/benefit analyses in difficult cases.

If such committees were to make an effective contribution to the ethical debate they would have to include lay members and perhaps even such persons as ethical philosophers and professional animal welfarists. The main difficulties would lie in avoiding the obstacles to smooth running which have been experienced in Sweden due to the introduction of extremists and in protection of confidentiality of projects.

The benefits of such a system include not only improved laboratory animal welfare but increased sensitivity among scientists, reduction of the gap in understanding between scientists and the public and a reduction of distrust and suspicion among all those concerned in the animal welfare debate.

At a time when militant activities continue to widen that gap of understanding and deepen the distrust, it may well be that the UK cannot afford to pass up this opportunity for a new approach.

CURRENT GENERAL INITIATIVES IN EUROPE

The Council of Europe Convention

A European Convention for the Protection of Vertebrate Animals Used for Experimental and Other Scientific Purposes was adopted by the Committee of Ministers of the Council of Europe on 31 May 1985. This document is now open for ratification by the 21 member countries of the Council, each of which will then have to amend its national legislation to bring it into line with the provisions of the Convention. It is a weak document in terms of animal protection, though for some countries with little or no legislation adoption of these provisions would be an improvement. For countries with detailed legislation, the Convention will make little difference to current practice and will not require modification of laws already in force.

The Convention does not prohibit the infliction of severe and enduring pain, though special authorization is required for procedures of this severity. Anaesthesia and analgesia need not be used if their use is incompatible with the aims of the experiment.

However, in the area of general animal care, implementation of the Convention would make considerable improvements in many countries. Each institution must name the person responsible for animal care and a veterinarian must also be named. In addition to user establishments, breeding and supplying establishments must also be registered and provisions for animal marking and the keeping of records should go some way towards eliminating the use of stolen pet animals, which is reported to be a sizeable problem in some European countries. The use of stray cats and dogs is prohibited.

Detailed annexes set down standards for general animal husbandry.

Provision is made for the collection of annual statistics. This will be extremely useful, since at the present time only a few countries (the UK, the Netherlands, the Scandinavian countries and Switzerland) produce such information. Statistical information is to be published centrally by the Secretary General of the Council of Europe.

There was considerable debate during the drafting of the Convention as to whether a standing committee should be set up to monitor its implementation and to suggest amendments from time to time. This view was incorporated in a proposal put forward by the Rapporteur following Public Hearings on animal experimentation held in Strasbourg in December 1982. The report (Council of Europe Parliamentary Assembly, 1983), which became known as the Bassinet Report, strongly expressed the view that the Convention should be seen as an *accord evolutif* — its provisions to be continuously updated as experience was gained.

Several countries, including the UK, opposed the measure, but very recently a new clause has been added (Article 30) which makes provision for multilateral consultations to examine the Convention's application and the advisability of revising or extending it. These meetings are to take place every five years, or more frequently if a majority of contracting parties so request.

The EEC Directive

Perhaps of greater significance is the recent initiative taken by the European Economic Community to draw up an EC Directive. This document, produced by the Commission, has now been the subject of reports both of the Economic and Social Committee and the Committee of the Environment, Consumer Protection and Public Health. The latter has considered over a hundred suggested amendments. The Directive has been adopted by the Council of Ministers after consideration by the Parliament in September 1986.

The Directive is a stronger document, in terms of animal protection, than the European Convention. It does, for example, include some provisions to control the degree of pain and suffering allowable in scientific procedures.

When passed, the Directive will be binding on EEC member countries, whose national laws must be brought into line with it.

Organizations of Medical Sciences — guidelines

In addition to these European proposals, international guiding principles for biomedical research involving animals have been drawn up by the Council for International Organisations of Medical Sciences (CIOMS, 1985), a body established under the auspices of the World Health Organisation (WHO) and the United Nations Educational, Scientific and Cultural Organisation (UNESCO). The proposals were reviewed at a round table discussion between interested bodies (including health administrations, academic and industrial scientists and a very small number of animal welfare bodies) from many countries, which met in Geneva in December 1983 (Bankowski and Howard-Jones, 1984).

These basic principles are designed to set out a framework upon which more specific legislative or regulatory systems could be built. In drawing them up, the CIOMS took regard of the vastly different legal systems and cultural backgrounds existing in various countries throughout the world with regard, *inter alia*, to attitudes towards animals, safety testing and training of personnel. During discussion at this meeting the acute problems of laboratory animal care in the third world became particularly apparent.

INTERNATIONAL DUPLICATION OF DATA

A major problem relating to animal experimentation is that research and testing is such a vast enterprise that data are continuously duplicated in different countries. In the area of toxicity testing considerable attention has been paid to this problem, with particular regard to duplication of data in fulfilment of national and international safety testing regulations.

Currently, attempts are being made to achieve greater harmonization of testing practices. The *Principles of Good Laboratory Practice* issued in 1981 by the Organisation for Economic Co-operation and Development (OECD) made some moves in this direction. Current moves within the European Parliament are likely to be in the direction of mutual acceptability of safety data and greater use of data banks.

CONCLUSIONS

Pressure to tighten controls continues to increase throughout Europe and North America.

No system currently in operation works perfectly and some fail to be implemented at all or are designed only to exert very minimal controls.

Brief experience has now been gained in various countries and this should be sufficient to devise workable legislative systems which are practicable and which meet the particular needs of each country.

This experience has shown that voluntary systems such as that operated in Canada are insufficient to enforce strict controls, while strong legislation such as that of West Germany is of little value if the administrative machinery does not exist to implement it.

Any successful system will require the involvement of animal care personnel to implement its provisions on a day-to-day basis. To be really effective, a control system must be so designed as to foster open discussion and cooperation between scientists, animal welfare campaigners, the public and the controlling authority.

Benefits to be gained from devising and implementing effective modern controls should be improved laboratory animal welfare, sharper science, increased sensitivity among scientists and a reduction in the gap of understanding between all parties involved in the rational laboratory animal welfare debate.

REFERENCES

Association of Veterinary Teachers and Research Workers, (1986). Guidelines for the recognition and assessment of pain in animals. Vet. Record **118**, 334–338.

AWI (1985). *Beyond the Laboratory Door*, Animal Welfare Institute, Washington DC.

Bankowski, Z. and Howard-Jones, N. (eds) (1984). *Biomedical Research Involving Animals; Proposed International Guiding Principles*, Proceedings of the XVIIth CIOMS Round Table Conference.

CCAC (1980–84). *Guide to the Care and Use of Experimental Animals*, 2 vols, Canadian Council on Animal Care, Ottawa.

CFHS (1985a). *Guidelines for Lay Members of Animal Care Committees*, Canadian Federation of Humane Societies, Ottawa.

CFHS (1985b). *A Brief to the Committee on Ethics in Experimentation, Medical Research Council of Canada*, Canadian Federation of Humane Societies, Ottawa.

CIOMS (1985) International Guiding Principles for Biomedical Research Involving Animals. Geneva, 1985.

Cmnd 8883 (1983). *Scientific Procedures on Living Animals (1983)*, Cmnd 8883, HMSO, London.

Cmnd 9521 (1985). *Scientific Procedures on Living Animals (1985)*, Cmnd 9521, HMSO, London.

Home Office (1986). *Guidance on the Operation of the New Legislation to Replace the Cruelty to. Animals Act 1876*, HMSO, London.

Littlewood, Sir S. (1965). *Report of the Departmental Committee on Experiments on Animals*, Cmnd 2641, HMSO, London.

Loew, F. M. (1981). Alleviation of pain: the researcher's obligation, *Lab. Animal.*, **10**, 36–8.

McCardle, (1985). The emperor's new clothes — new government care guidelines and laboratory animals, a failure in disguise, *The Animals Agenda*, **V**, 12.

Morton, D. B. and Griffiths, P. H. M. (1985). Guidelines on the recognition of pain, distress and discomfort in experimental animals and on hypothesis for assessment, Vet. Record, **116**, 431–6.

NIH (1985). *Guide for the Care and Use of Laboratory Animals*, National Institutes of Health Publn No. 85–23, NIH, Bethesda.

NIH (1984). *Guide for Grants and Contracts, Laboratory Animal Welfare Special Edition*, vol. 12, No. 5, National Institutes of Health, Bethesda.

Orlans, F. Barbara (1984). What institutional animal research committees can do to improve humane care, *Lab. Animal.*, **13**, 24–9.

OTA (1986). *Report on Alternatives to Animal Use in Research, Testing and Education*, Office of Technology assessment report.

Council of Europe Parliamentary Assembly (1983). Report, Doc. 504999, 26 April 1983.

Rowan, A. (1984). *Of Mice, Models and Men — a Critical Evaluation of Animal Research*, State University of New York Press.

Solomon, M. and Lovenheim, P. C. (1982). Reporting requirements under the Animal Welfare Act: their inadequacies and the public's right to know, *Internat. J. for Study of Animal Problems*, **3**, 210–18.

Stevens, C. (1978). Laboratory animal welfare. In E. S. Leavitt (ed.) *Animals and their Legal Rights*, Animal Welfare Institute, Washington DC, pp. 46–58.

APPENDIX—LEGISLATION CONTROLLING ANIMAL EXPERIMENTATION IN WESTERN EUROPE

Within the 21 member countries of the Council of Europe, legislative constraints over animal experimentation are widely variable. Some countries have very little legislation, and for most there are no detailed statistics available for experiments conducted annually.

In several countries new legislation is pending. The following table is intended only as a guide to the legislation in force at the time of writing.

Country	Controlling legislation	Administrative authority	Main controls
Austria	Federal Law on Animal Experimentation (1974)	*Universities* Federal Minister for Science & Research *Trade & industry* Ministry of Trade, Commerce & Industry *Public health, food & veterinary science* Ministry of Health & Environmental Protection	Permits granted to heads of institutions specifying type of experiment allowed; inspection by persons from relevant authority; adequate anaesthesia required unless it frustrates object of experiment; surgery on vertebrates restricted to suitably qualified scientists; records detailing purpose, number and origin of animals must be kept for two years.
Cyprus	No legislation—experiments carried out in line with UK 1876 Act	Formal permission of Director of Department of Veterinary Sciences required	Permission issued only to those with technical qualifications; covers mainly vaccine testing and control; other work is rare.
Belgium	Animal Protection Act (1975)	Ministry of Justice and Ministry of Agriculture	Licence from Office of Veterinary Medical Inspection to directors of laboratories; inspection by State Veterinary Inspectors; anaesthesia required unless it frustrates object of experiment; 1981 decree requires biological, medical and veterinary students to be trained in 'alternative' methods; new legislation under discussion.

Denmark	Animal Experiments (No. 220/1977)	Ministry of Justice. A board appointed by the Ministry is responsible for administration; represents all interest groups including animal welfare	Individual licence to qualified persons for experiments likely to cause pain and suffering; anaesthesia required for all experiments likely to cause pain, but can be dispensed with; licenses not required for procedures causing no more than minor and momentary suffering; 'lower' animals to be used if possible; only vertebrates covered; animals must not be used where 'alternatives' have equal relevance; records of numbers, species and purposes to be kept and presented annually to controlling board; method of euthanasia for dogs, cats and non-human primates must be declared. Order amendment to bring into line with Swedish system. New law (1986) will include strict provisions on animal supply and local ethical committee.
Eire	As UK legislation	Ministry of Health	Licences, conditions and inspection by Ministry of Health; basis of control as for UK
Federal Republic of Germany	Animal Protection Law (1972) parts 5 and 6	Ministry of Food, Agriculture and Forestry	Licences issued by local authorities to heads of institutions only; orders issued by central Ministry are effected by local authorities; all painful/injurious procedures subject to licensing; Ministry team of veterinary inspectors; anaesthesia required for all surgical procedures, but can be dispensed with if it frustrates object of experiment; surgery restricted to suitably qualified persons; 'alternatives' to be used where feasible. Legislation under review.
France	1968 Order under Decree for Animal Experiments (1963), No. 68-139. Constitutes articles R24-31 of the article 454 of the Penal Code	Ministry of Agriculture and other ministries (has advisory inter-ministerial commission)	Individual authorization from relevant government department; inspection by Ministry of Agriculture veterinarians or Ministry of Public Health pharmacists; anaesthesia or equivalent analgesia required unless it frustrates object of experiment; experiments without anaesthesia restricted to one only. Surveys underway to find a consensus for possible legislation.

Greece	Law 1197, concerning the Protection of Animals (1981) article 4	Ministry of Agriculture (aided by Consultative Committee with the Veterinary Service of the Ministry)	Licence required for experiments causing pain or suffering; anaesthesia required for surgical experiments, administered by veterinary surgeon; surgical experiments restricted to graduates in medical, veterinary or biological sciences; no inspection.
Iceland	Stj.tíð 8, No. 77/1973	Chief Veterinary Officer, Ministry of Culture and Education	Experiments only allowed by special permit, granted only to persons with medical or veterinary training, or persons working in institutes with personnel with such training; untrained persons must delegate to trained persons; specific instructions attached to permits, including proper provision for anaesthesia and humane euthanasia at end of experiment; all procedures must be conducted at approved institutes; records must be kept. Legislation currently under revision.
Italy	Animal Protection Law (1931), amended 1941	Ministries for Health and Culture	Experiments performed only by named, suitably qualified individuals in authorized institutes; director holds responsibility; inspection by medical and veterinary officers of provincial health authorities; adequate anaesthesia required unless it frustrates object of experiment; only warm-blooded vertebrates covered; annual report required by pertinant authority. Services undermanned, no prospected change.
Grand Duchy of Luxembourg	Law for the Protection and Welfare of Animals (1983)	Ministry of Agriculture	Licenses issued by Ministry of Health; inspection by Ministry of Agriculture veterinarians. Same provisions as for German and Swiss laws.
Liechtenstein	Animal Welfare Act (1936), article 3 prohibits vivisection		New law in draft will allow governmental use of animals in exceptional circumstances.
Malta	No information available		

The Netherlands	Law for Animal Experiments (1977)	Ministry of Public Health: Veterinary Public Health's Department of Animal Experimentation	Retrospective licenses to institutions (no project assessment); compulsory for all painful experiments; animals in pain must be euthanized once the experiment is satisfied; anaesthesia can be dispensed with if it frustrates object of experiment, but required for all surgery; 'alternatives' must be used where available and 'lower' vertebrates in place of 'higher' ones where possible; no cats, dogs, equines or primates used if other species will suffice; source of dogs and cats recorded; strict rules on supply but exemptions allowed; inspection and supervision by two State Veterinary Inspectors and team of 35 regional inspectors; Inspectorate requires annual returns including numbers, purposes, species and estimate of degree of discomfort; detailed statistics produced; central advisory committee to Minister for Public Health includes animal welfare members (members appointed by Royal Academy of Sciences); annual report produced.
	Order in Council (1980) provisions in process of implementation		New controls being phased in currently; project leaders will be required (by 1986) to undergo training in laboratory animal science; detailed provisions on husbandry and care under consideration; a named person responsible for animal care in each institution (Art. 14); several now appointed; these are veterinarians, doctors and biologists who undergo training under the Chair of Laboratory Animal Science at Utrecht; neither they nor inspectors can stop experiments, but can impose restrictions; it is likely that institutional ethical committees will be recommended.
Norway	Welfare of Animals Act (1974); Regulations concerning Biological Experiments on Animals (1977)	Experimental Animals Board (EAB) appointed by Ministry of Agriculture	Licenses granted to approved institutes with qualified directors who accept responsibility for all personnel; this person must supply written consent to EAB for all experiments allowed at the institute; final decision rests with the Board; premises must comply with standards set

Country	Legislation	Authority	Details
Portugal	Legislation in line with Council of Europe Convention in preparation		by EAB; individual licenses issued (exceptionally) to suitably qualified persons; licenses refused if valid alternative available; inspection by police, members of EAB or persons it authorizes, e.g. county, district or state veterinarians, police, members of municipal animal welfare boards; explicit requirements for anaesthesia, analgesia and euthanasia; special exemptions needed for painful experiments without anaesthesia; all vertebrates and decapoda (crustaceans) covered; experiments unlikely to cause suffering exempted; special permission required for use of dogs, cats and non-human primates; annual return to EAB must be made by departmental head.
Spain	No legislation		
Sweden	Protection of Animals Act (1944) articles 12 & 13, amended 1979 and 1982	National Board of Agriculture (under Ministry of Agriculture); regional ethical boards and National Board for Laboratory Animals Laboratory Animals responsible to this body)	Individual licences subject to qualifications, purpose of experiment, etc.; purpose-bred animals must be used (exemptions can be granted); ethical review at local level; suffering not to exceed what is 'necessary'; euthanasia required if suffering continues; anaesthesia specified; pain relief required where anaesthesia not used; inspection by local veterinary officers under direction of municipal health boards; 'higher' and 'lower' vertebrates distinguished; no warm-blooded vertebrates to be used in teaching where other methods available; detailed records required of origin, purpose and numbers of dogs, cats, horses, ungulates and non-human primates. Legislation under revision.

Switzerland	Animal Protection Law (1978) and Animal Protection Ordinance (1981) part 7	Competent authority of each canton (cantonal committee of specialists to advise veterinary administration). Supervision by Federal Commission on Animal Experiments (a committee of experts including animal welfarists)	Authorization required for each experiment or series of experiments; experiments under direction of trained scientists; special permits required only for experiments causing pain, distress or disturbance of general condition; held by director of laboratory; inspection annually by permitting authority or by cantonal commission; anaesthesia required but exemptions allowed; detailed requirements on care and husbandry; some ethical assessment required where alternatives to animals not available; 'lower' species to be used where possible; only vertebrates covered; annual records required, kept for two years.
Turkey	No information available		
United Kingdom	See text		

Laboratory Animals: An Introduction for New Experimenters
Edited by A. A. Tuffery
© 1987 John Wiley & Sons Ltd

CHAPTER 3

The Design of Experiments

M. R. GAMBLE
Nottingham

INTRODUCTION

Research involving the use of animals can take place at various system levels. The highest level is the study of the interaction of a group of individuals. Next is the investigation of a single organism, followed by isolated organ studies and cellular research. The lowest levels involve the use of subcellular fractions and even biomolecules.

Until the knowledge of a particular process in the organism is complete, investigative research will move up and down between these system levels. It is only after investigation at all levels that the research retreats to the simplest level for reasons of convenience and directness (Weihe, 1985).

Research also progresses along a fairly common chronological sequence.

1. The first stage can be classified as an initial position finding or orientation in terms of primary morphology and gross physiology.
2. Work can then proceed to elucidate the basic mechanism involved, together with all the identifiable components and their interactions. This is usually a prolonged sequence as multiple variance may be evident according to the presence of different components or their concentration or the physiological state of the whole animal.
3. Pharmacological work can now proceed to investigate the effects of a wanted action on these basic mechanisms under as many different situations as possible.
4. Toxicological work can now proceed to elucidate the effects of an unwanted action on these basic mechanisms.
5. The target site, when selected, can now be used to obtain information (and primary diagnosis) of the action of any internal or external agonists as selected. This bio-assay work can be retained as a quality control function or be replaced by biochemical estimations.

It is clear that the design of any experiment requires a carefully prepared protocol. It may be that the stimulus for the work arises from the critical appraisal of the current knowledge on a particular subject which leaves certain gaps or questions which can only be answered by further study of the processes. Perhaps an observed response is noted which cannot be accounted for by the current knowledge of that system; a hypothesis can be found but experimentation is necessary to validate it or to corroborate previous work on the subject.

The following sections outline the major items (developed in more detail in later chapters) to consider in designing animal experiments. They can only act as signposts in any specific research area, however. There is no short cut for the research worker to achieve proficiency in investigative work other than by a period of attachment to an experienced research team. This period of discipline training should instil or bring out an ability to recognize a problem, analyse its various components, to formulate a working hypothesis to explain them and to design a test for it. The attempt should not be made if the researcher lacks total honesty, intellectual integrity, humanity for the subjects under his control and the ability for extended periods of hard work.

THE USE OF A CORRECT MODEL

A model is something worthy of imitation, and in this sense is used here to denote a series of reactions which are capable of generalized extrapolation, usually to man.

It should always be borne in mind that the use of common laboratory animals is not always the easiest or most accurate model to choose.

Careful thought should be given prior to designing the experiment to the relevance of using insentient material. The main limitation of insentient material is that it tends to oversimplify what is usually a very complex set of reactions.

Tissue culture

Laboratory animals can sometimes be replaced by cell or tissue culture, but this is usually after initial work has been done in the whole animal (i.e. a *post priori* rather than an *a priori* use). Mutagenic studies of substances on cell lines are useful prescreens for potentially toxic substances, which may well reduce the number of animals subsequently used in this type of work.

These *in vitro* methods are particularly useful in pharmacology, where new compounds may be difficult to make or where only a small amount of compound is available. By using small quantities the cost of compound development is reduced.

It should be realized that when cells are subcultured they tend to lose some of the specific properties they possessed when in the whole animal. Also any studies concerned with cellular metabolism can only be estimates of what the situation might have been in the body.

Physical or chemical methods

In certain cases the use of techniques involving gas chromatography and mass spectrometry can replace or reduce the numbers of animals used for specific types of work (e.g. vitamin D assays) and are also useful in identifying biochemical intermediaries in biological systems.

Computer modelling

It is sometimes of use to employ a computer where there are several biological variables operating simultaneously in a test system. Such projected models may rapidly identify possible negative combinations of variables which could save considerable time and reduce the amount of animal work necessary to elucidate the processes involved.

Type of animal

Having established that the use of animals is justified for the purpose of experiment it is important that the maximum amount of information is obtained from each animal used.

Careful thought should be given to the species and strain of animal to be used. Literature searches should give indications of the range of animal types for a given procedure and variants or mutants are constantly being produced which may be applicable. This choice also extends to invertebrate animals which are used widely for physiological and other types of experiments. Waynforth examines this problem of choice of animal in more detail in Chapter 11.

Quality of animal

As will be detailed later, the quality of any animal model chosen is important.

Micro-organisms are known to affect experimental results. This can be particularly important in immunological work where many bacteria are capable of modifying the immune response of an animal to an unrelated experimental antigen (Hunneyball, 1983).

Similarly, toxicologists require reproducible circumstances in which to study the activities of chemicals at varying dose levels over extended periods of time. Turnbull (1983) has detailed the effects of variations in microbiological status in toxicological investigations. The importance of health status investigation is stressed both at the start of any study and at regular intervals during the study.

The degree to which animals are placed under psychological and/or physical stress could also effect their ability to cope with both microbial and subsequent experimental challenges. The environment in which they are bred and held before use is therefore an important consideration in determining the quality of any animal (Clough, 1982; Gamble, 1982).

PLANNING THE DESIGN

Laboratory animal facilities

It is of little use obtaining good quality laboratory animals if there is nowhere suitable to keep them. As will be detailed in a later chapter, the ability to rear, use and obtain meaningful results from experimental animals depends to a large extent on the design and operation of the animal accommodation.

The European Convention for the Protection of Vertebrate Animals Used for Experimental or Other Scientific Purposes gives guidelines on accommodation and care of animals in its Appendix A. This lists the physical requirements of an animal facility and the recommended environmental conditions and husbandry procedures. In the UK the Animals (Scientific Procedures) Act 1986 has brought breeders and suppliers of laboratory animals as well as the scientific establishments under legislative control and guidelines are being produced by the Royal Society and the Universities Federation for Animal Welfare outlining the principles for the care and use of laboratory animals.

Careful consideration should be given to the space necessary to carry out the projected work. A common omission is to overlook the cages/racks necessary for cleaning procedures, equipment storage, diet storage, bedding storage, records and service areas.

It must be remembered that during an experiment the animals must be looked after on a daily basis. This includes weekends and holidays. Sufficient staff of the right calibre should be available for such attendance and clear instructions should be available at all times for the care and management of the animals. Maintenance or security staff should be made aware of areas housing laboratory animals so that plant failures, disturbances or other incidents can be rapidly acted upon and dealt with.

Caging and separation of animals

It has often been said that the ideal cage for a rabbit has a floor made of grass, walls made of hedges, a roof of sky and a floor area of half an acre. In practical terms the well-being of laboratory animals can be achieved by restricting their movements to areas smaller than this in cages or pens.

The type and design of cages will depend on the environmental conditions employed but will all have to conform to certain requirements regardless of type or size.

The caging should be escape-proof and the design should be such that infants cannot squeeze through small openings or that adults cannot tamper with any locking or latch mechanism. The material of which the cages are constructed should be such as to prevent damage by chewing, scratching or pulling. Cages should be resistant to corrosion by both urine and repeated washing.

Such materials are increasing in cost and there is a cogent argument for standardizing wherever possible and organizing the supply of cages centrally in any given facility. The ability to replace cages rapidly with similar ones is an obvious advantage.

However, some caging for certain experiments will have to be custom-made and sufficient time should be allowed for orders to be approved and placed and the equipment manufactured and delivered.

To increase the versatility of some research areas the use of cages on mobile rather than fixed wall racks is increasing. This enables easy changing of cages for cleaning and changes in types of caging for different species.

Veterinary support

The primary responsibility for the well-being of any experimental animals lies with the scientist whose charges they are. However, in the UK the Animals (Scientific Procedures) Act now requires a named veterinary surgeon and a named person responsible for the day-to-day care of animals to assume legal responsibility for the animals' health and welfare and to liaise closely with both scientist and the inspectors appointed under this Act.

The veterinary surgeon should also be consulted for preventative disease control methods and for diagnosis and treatment of disease or *postmortem* examinations as appropriate. He will also assist with surgical procedures where necessary and advise on methods of analgesia, anaesthesia and euthanasia.

Training of staff

Experimental animals can vary in their responses to experimental procedures if their handling and husbandry is not carried out carefully and with a degree of skill.

All staff, regardless of position or qualifications, should receive training in the sympathetic handling of a wide range of species, supplemented with regular practice.

Health of staff and animals

Chapter 7 will detail the hazards and safety aspects of animal work. In the design stage it is important to ensure that the provisions of any relevant legislations are observed, such as the Health and Safety at Work Act in the UK, and Occupational Safety and Health Act in the USA.

Measures should be taken to reduce the risk of laboratory animal allergic symptoms developing in the animal users and that the quality of animals bred or obtained is such that zoonotic diseases are avoided.

Chapter 4 will detail the precautions to be taken in the supply of animals and these should be augmented by the use of sentinel animals wherever possible.

Sentinel animals are high-grade animals, usually of the species being housed, whose bacterial, parasitic and virological burden is known. These animals act as living 'blotting paper' for microbes being generated in the animal room and by taking microbiological samples at periods from them it is possible to detect the acquisition of additional microbes at the subclinical stage which may affect the experimental procedures carried out.

Versatility

In planning the design of an experiment, provision should be made to change or alter various aspects of it that can be shown to be adversely affecting the results or are making them unacceptably variable. How critical the various components are depends largely on the type of work being conducted, but protocols and standard operating procedures should be routinely and frequently updated in the light of current knowledge and practice.

MANAGING THE DESIGN

Communication

It has frequently been said that most of the major and minor errors, oversights and mistakes in animal work arise as a breakdown in communication.

It cannot be stressed too highly that it is important to establish workable channels of communication, both formal and informal, with all personnel concerned with any experimental project. This applies from the head of department obtaining funds and space for the project to discussing with cage cleaners when clean cages can be delivered and when dirty cages can be received. The main areas for continued communication are outlined in the next three sections.

Literature searches/personal contacts

Fully computerized and regularly updated literature searches are now easily accessible to any researcher. Following up relevant publications and drawing implications from other people's work, however, is still as laborious and necessary as ever. Not every aspect of the experimental set-up is always reported in publication and it is by personal contact with other research groups that these often minor but possibly crucial omissions can be obtained.

Official and ethical supervision

As detailed in the previous chapter, it is the Home Office that administers the law relating to animal experimentation in the UK. In Canada the control of animal experiments is the responsibility of the Canadian Council on Animal Care and was designed to obviate the necessity for national legislation. Where there is formal legislation it is usually necessary to gain approval for a project or particular animal procedure from the appropriate administrative authority, for example the Home Office in the UK. Otherwise there usually exist systems of local or regional institutional ethical or animal care committees.

Availability of animals

Whether home-bred or commercially-bred animals are to be used, it is important to ascertain at an early stage in the management of any experiment that sufficient animals of a suitable age, weight and sex are available before that day.

By their very nature animals have a very short 'shelf-life' and breeding areas need extended periods of notice if quantities of older or inbred animals are required.

The transport of animals to the user areas does stress the animals and a period of recovery (usually not less than a week) is recommended in the area where they are to be used. The guidelines of the European Convention for the Protection of Vertebrate Animals recommend a period of acclimatization before use, even when the animals are seen to be in sound health.

Animals obtained from other countries have to conform to quarantine regulations in their holding and use and possibly disposal after use. Laboratory animals can be successfully transported with a minimum of stress if care is taken in deciding the most suitable method of transportation, the length of journey and whether food and moisture should be available (Bantin et al., 1984).

Statistical design

In assessing the numbers of animals necessary for a particular experiment calculations have to be made which will enable the results to be meaningful and amenable to statistical analysis. Biological variation does not make it possible in many cases for a simple 'yes' or 'no' to be the endpoint and numbers of animals have to be used and their combined results analysed. Quite frequently it is the statistician who finds that the degrees of accuracy which he can obtain from the experimental results are severely limited by poor experimental design.

Any experiment should enable the researcher to compare the effects of two or more 'treatments' on some attribute of the subject under consideration. The experiment should also give an indication of the magnitude of the apparent effect together with an estimate of the variability (over and above the uncontrolled or random variation).

To avoid bias the treatments should be allotted to the subjects at random (possibly from tables of random numbers). This will also give for similarly treated animals a 'within treatment' estimate of the uncontrolled variation that may be expected to affect the comparison between differently treated animals ('between treatments'). Tests of significance can then be carried out to calculate the chances of the differences observed between treatments being solely due to uncontrolled variation.

The fundamental principle is that all experimental groups should be treated exactly the same in all respects except for the treatment under test. Similarly all experimental procedures should be repeated, either in space or in time (replication).

Standard reference works on animal-related statistical design have been available for a long time (Bailey, 1959; Cochran and Cox, 1957; Cox, 1958; Fisher, 1966; Yates, 1937). A usefully straightforward introduction to the principles of statistical analysis together with all the basic variations in design has been written by Heath (1970).

Chanter (1982) has emphasized that to increase precision when dealing with heterogeneous systems such as animals the technique of blocking should be used. Simply, this involves splitting the experiment into subexperiments or blocks each one of which contains a replicate of each of the treatments. As long as each block is as homogeneous as possible different blocks can be subjected to different conditions, eg time of day, time of year, room temperature, batch of animals, batch of diet, animal technician, etc., without bias to the treatment comparison.

Precision is helped enormously by the uniformity of the experimental subjects. For at least the last decade Michael Festing (1975, 1979) has been appearing in print in the cause of the use of inbred and F1 hybrid strains in preference to outbred animals. The main deterrent (apart from their extra cost) has been the problem of availability of large numbers of them at any moment in time. Perhaps the use of blocking could overcome this inertia. Hunt (1980) has summarized other relevant aspects of the design of experiments.

It is important to seek the advice of a statistician at the design stage; as Chanter (1982) states, 'no amount of statistical wizardry can salvage information from a badly designed experiment; a statistician's most important contribution is often, therefore, to suggest ways of carrying out experiments so that they are in fact capable of providing answers to the questions originally posed.'

The 'animal guidelines' prepared by the Royal Society (Royal Society, in press) offer further advice on the use of specific statistical procedures in animal experimentation.

Controls

Linked to the previous sections, thought should be given to the necessity of control animals (and the allowance of space for them in the animal room). The type and form of any placebo compound to be administered should also be considered. Again 'dummy' handling should always be considered for certain stress-related reactions.

Interpretation of Results

To permit a clearer understanding of the possible mechanisms at work in any experimental situation clear, concise details should be given of the conditions under which the experiment was carried out. It has been a general failing that insufficient detail has been included in the reporting of experimental results in the scientific literature — particularly in the areas concerned with the animals and their environment (Clough, 1982; Gamble, 1979).

It is often the unexpected phenomena that arise which raise entirely new and quite often exciting problems for the researcher. Such discoveries cannot be anticipated and it will be the successful scientist who can realize the implications of his results, even though they were not those he was expecting. Laboratory animal science, conducted in a controlled and humane format, still offers great potential benefits to the future health and care of both man and animals.

REFERENCES

Bailey, N. T. J. (1959). *Statistical Methods in Biology*, English Universities Press, London.

Bantin, G. C., Deeny, A. A., Gregory, D. J. and Hewitt, R. A. (1984). Animals in transit, *Animal Technology*, **35**, 113–22.

Chanter, D. O. (1982). Protocols for routine toxicity studies: A statistician's view, *Arch. Toxicol. Suppl.*, **5**, 40–4.

Clough, G. (1982). Environmental effects on animals used in biomedical research, *Biol. Rev.*, **57**, 487–523.

Cochran, W. G. and Cox, G. M. (1957). *Experimental Designs* (2nd edn), John Wiley & Sons, New York and London.

Cox, D. R. (1958). *Planning of Experiments*, John Wiley & Sons, New York and London.

Festing, M. F. W. (1975). A case for using inbred strains of laboratory animals in

evaluating the safety of drugs, *Food and Cosmetics Toxicology*, **13**, 369–75.
Festing, M. F. W. (1979). *Inbred Strains in Biomedical Research*, Macmillan, London.
Fisher, R. A. (1966). *The Design of Experiments* (8th edn), Oliver and Boyd, Edinburgh and London.
Gamble, M. R. (1979). Effects of noise on laboratory animals (introduction), Ph.D. Thesis, University of London.
Gamble, M. R. (1982). Sound and its significance for laboratory animals, *Biol. Rev.*, **57**, 395–421.
Heath, O. V. S. (1970). *Investigation by Experiment*, Institute of Biology, Studies in Biology No. 23, Edward Arnold, London.
Hunneyball, I. (1983). The need of the immunologist. In F. J. C. Roe (ed.) *Microbiological Standardisation of Laboratory Animals*, Ellis Horwood, Chichester, pp. 66–87.
Hunt, P. (1980). Experimental choice. In RSPCA (eds) *The Reduction and Prevention of Suffering in Animal Experiments*, RSPCA, Horsham, West Sussex, pp. 63–75.
Royal Society/UFAW (1986). *Draft Guidelines on the Care of Laboratory Animals and their Use for Scientific Purposes* (in press).
Turnbull, G. (1983). The needs of the toxicologist. In F.J.C. Roe (ed.) *Microbiological Standardisation of Laboratory Animals*, Ellis Horwood, Chichester, pp. 9–230.
Weihe, W. H. (1985). Use and misuse of an imprecise concept: alternative methods in animal experiments, *Laboratory Animals*, **19**, 19–26.
Yates, F. (1983). *The Design and Analysis of Factorial Experiments*, Commonwealth Bureau of Soils, Farnham Royal, Bucks.

CHAPTER 4

The Supply of Laboratory Animals

H. DONNELLY
Laboratory Animal Science Unit, Royal Veterinary College, London

INTRODUCTION

Although many of the reagents and the apparatus used in biomedical research are easily available 'off the shelf' the animals used are not. They need to be bred, either by the user or a commercial breeder, or may have to be obtained by capture from the wild or from other sources and, depending on the user's requirements for age or body weight, their 'shelf life' may be very short.

It is unlikely that the experimenter will be able to obtain all the animals needed immediately, especially if there is a requirement for large numbers of one sex or for a narrow age or weight range. Instead, he or she will have to make their needs known some time in advance.

If the breeder is able to start producing the animals immediately then the user will have to wait for a 'production time', which is mainly determined by the animals' reproductive physiology. This includes the gestation period of the species, the period which the young spend with the mother before weaning, the time taken from weaning to when the animals are either of the age or weight required (Table 4.1) plus an allowance for delays and the time taken to organize the breeding facilities and set up matings.

Most breeders usually overproduce to make sure that they are able to meet standing orders, relying on occasional short-term demand to take up the surplus, and because of this they may at times be able to supply animals at short notice.

Not all animals may be available throughout the year either because they are seasonal breeders or have a period of hibernation or aestivation, and this is true not only of the more uncommon species which may have to be obtained from the wild but also of common domestic species such as the sheep or the ferret which are seasonal breeders.

The cost of the animals in relation to the total cost of the project will need to be considered, particularly the comparative costs of the various species which may be suitable, and the difference in price between animals bred 'in-house' and those bought from a commercial breeder.

There are also legal and ethical constraints on the use of animals in biomedical research; should one, for example, use stray or unwanted pets or animals from the wild?

The user therefore needs to be aware of not only potential sources of animals but also the limitations imposed on availability by the animals' biological characteristics, economic factors and legislation.

SOURCES

A wide variety of animals are used in research but the vast majority of experiments are performed on a relatively few species of small rodents and lagomorphs. For example, in the United Kingdom mice, rats, guinea pigs, other small rodents — such as hamsters and gerbils, and rabbits were used in almost 90 per cent of the experiments carried out under Home Office licence during 1984 (Home Office, 1985).

Virtually all of these animals are bred solely for laboratory use (LAC, 1974a), by the laboratory itself, by a centralized breeding unit attached to the institution or by commercial breeders.

Of the other species used a proportion have been bred for other purposes and 'diverted' into the laboratory, for example some of the more exotic pet animals or farm animals such as poultry, cattle, sheep, goats, pigs, horses and donkeys; some agricultural and veterinary institutes do, however, breed these species for research.

'Fancy' breeders may at times be a source of supply.

Some species are difficult, if not impossible, to breed in captivity and are only available from the wild. This is especially so for most primates and for many other species including amphibia and reptiles.

These animals may have to be captured by the user but, more usually, they will be available from a commercial supplier who specializes in providing animals for research.

In some countries, including the USA, stray dogs and cats which have been impounded may be used as experimental animals.

The use of strays and wild animals, particularly endangered species, raises not only legal but also ethical questions which the experimenter should consider carefully.

There has been an increasing awareness of the need to use animals for biomedical research which are defined microbiologically, genetically, nutritionally and environmentally, and the extent to which a potential source of supply is able to satisfy these various criteria must be taken into account.

In Great Britain the Laboratory Animal Breeders Association (LABA) has an Accreditation Scheme for breeders and suppliers of laboratory animals.

This scheme succeeded the MRC Laboratory Animals Centre's Accreditation and Recognition Schemes (LAC, 1974b) in 1982, and accreditation is awarded to LABA members who meet published standards for their facilities, husbandry and management and who follow a health monitoring programme which meets the requirements of the Scheme (LABA, 1982).

In the USA the Institute of Laboratory Animal Resources (ILAR) has published standards for the care of many species of laboratory animals (ILAR, 1980) and the American Association for Accreditation of Laboratory Animal Care (AAALAC) uses these standards as a guide in deciding whether to award accreditation to any institution which uses or breeds laboratory animals.

Many commercial breeders will supply details of the microbiological profile of their animals, their methods of genetic monitoring, dietary and environmental control, all of which may be used as a guide by intending purchasers.

No scheme can be an absolute guarantee of quality as any screening test, either microbiological or genetic, reflects only the status of the colony at the time it was carried out. The breeder's standards of husbandry and management do, however, offer guidance on the probability that the status of the animals will be maintained for long periods of time.

When animals bred for other purposes are used in research the standards of quality may or may not be very closely related to the needs of the laboratory, but animals coming from the wild are almost always of unknown age and background and will have a low degree of uniformity. They may also frequently be in need of treatment for infection and disease and be of a poor nutritional status. In the case of primates particularly there is a very real danger from zoonoses, for example tuberculosis, B-virus and salmonellosis, but these risks are not restricted solely to this group of animals.

In the United Kingdom a six-month period of quarantine is required when any mammals are imported to prevent the introduction of rabies (Rabies (Importation of Dogs, Cats and other Mammals) Order 1974).

REPRODUCTION AND BREEDING OF LABORATORY ANIMALS

Reproduction in vertebrates is characteristically a cyclical process and the periodicity of the cycle depends on the influence of external factors acting on the individual animal.

Most mammals show patterns of sexual behaviour only at certain times of year in response to an increase in the levels of gonadal hormones circulating in the bloodstream. In some species such as rodents, the 'long day breeders', this 'breeding season' occurs during spring and summer but in others, like the sheep or deer, the 'short day breeders', it occurs in the autumn.

The principal stimulus to the animal's reproductive system is the change in daylength which occurs in the spring and autumn and the effect operates through the interaction of the pineal body, the pituitary gland and the hypothalamus.

Because laboratory animals are kept in relatively stable conditions they are able to breed all year round although some species may show some seasonal effects. It is also possible, by artificially altering the photoperiod, to induce species such as the ferret and the sheep to breed out of season.

The length and frequency of the breeding season vary from one species to another. Mice and rats show no obvious breeding season, at least under domestication, and breed repeatedly throughout the year; the dog on the other hand has two breeding seasons in the year and the cat has two or three.

If, within the breeding season, the animal shows a succession of oestrous periods then it is described as seasonally polyoestrous but if there is no clearly defined breeding season then the species is described as being polyoestrous. Should there be only one period of oestrus during the breeding season, for example the dog, then the species is monoestrous.

In the female, after puberty, the ovary shows a cycle of activity which is controlled through its interaction with the anterior pituitary. The periodic release of ova is part of an internal endocrine rhythm which expresses itself in widespread changes in the physiology and behaviour of the female during the 'oestrous cycle'.

The cycle is divided into five stages, pro-oestrus, early oestrus, late oestrus, met-oestrus and di-oestrus, and these are usually identified by changes in the cells seen in a vaginal smear, a high proportion of cornified cells being present on the day of oestrus. The appearance of the external genitalia may also give some indication of the stage of the cycle.

The changes in the vaginal smear are similar for almost all small rodents and these and the underlying hormonal control of the cycle are described in many standard texts.

The release of ova, or ovulation, happens during the period of oestrus or 'heat'; this is when the female is sexually active and will allow the male to mate, even on occasions actively seeking him out.

With most mammals there is a sudden increase or 'surge' in the secretion of luteinizing hormone (LH) by the pituitary and this stimulates the rupture of the Graafian follicle and the release of the ovum.

With small laboratory rodents such as the rat, mouse and hamster this surge of LH occurs on the afternoon of the day of pro-oestrus, leading to ovulation twelve hours later in the early hours of the day of oestrus, thus fitting in with the nocturnal behaviour patterns of these animals.

In other species, such as the rabbit, ferret, mink and cat, the stimulus of mating is needed before ovulation occurs. Coitus provokes a reflex discharge of gonadotrophin-releasing hormone (GnRH) and hence of LH from the pituitary, and this leads to ovulation.

Although the laboratory rat, mouse and hamster ovulate spontaneously, the stimulus of copulation is needed to produce a fully functional corpus luteum, necessary for the establishment of pregnancy, otherwise the corpus luteum functions for only two or three days before regressing.

Even though the mating may have been sterile with no embryos being produced, an actively secreting corpus luteum can be established in these animals, resulting in an extension of the oestrous cycle known as pseudopregnancy.

The length of pseudopregnancy varies with species (Table 4.1) but progesterone is produced and the uterus undergoes the changes associated with pregnancy so that at first it is indistinguishable from a normal pregnancy.

If no embryos are present the corpus luteum will begin to regress but there is a great deal of variation in the time taken from one individual to another; in some cases pseudopregnancy lasts as long as a normal pregnancy.

Many small rodents, including the mouse, rat, hamster and guinea pig, have a 'postpartum oestrus' with ovulation occurring shortly after the birth of the young. With postpartum matings in the mouse fertility is less than that obtained with mating during the normal oestrous cycle and the implantation of any embryos produced may be delayed by lactation, especially if it is heavy. In the rat a litter larger than five (three in the mouse) will cause delayed implantation.

If female mice are housed singly in the absence of males they show lengthened oestrous cycles (5–7 days) with a high incidence of irregularity and spontaneous pseudopregnancy.

When the females are housed in groups the incidence of pseudopregnancy increases if the numbers are small (Lee–Boot effect); in large groups the females tend to become anoestrous.

Exposure of such females to a male has an immediate effect; individually housed mice begin to cycle regularly and anoestrous or pseudopregnant females begin a new cycle.

With females of some strains, whose cycles have been suppressed by grouping, there is a marked synchronization of their cycles and the incidence of oestrus and mating is significantly higher than expected on the third night after the introduction of the male (Whitten effect). Females will also mate on the first night of pairing if housed close to a male or treated with male urine two days beforehand.

The Whitten effect is not common to all strains of mouse but may be put to practical use to ensure that a large proportion of female mice ovulate or mate on any one night or especially when naturally ovulated eggs are required and the mice are refractory to injected hormones.

These effects are brought about by an androgen-dependent pheromone in the male's urine, as is another phenomenon also of practical importance — the Bruce effect, in which pregnancy or pseudopregnancy is blocked during the pre-implantation period by the presence or smell of a strange male.

Table 4.1 Data on reproduction for common laboratory animals

	Mouse	Rat	Syrian hamster	Chinese hamster	Rabbit	Guinea pig	Cat	Dog
Age at weaning (days)	18–21	18–23	21	21–25	6–8	2–3	4–8	6–8
Age at puberty (weeks)	4	7–9	6–8	7–14	16–24	m 8–10 f 4–5	m 36[a] f 20–28	m variable f 24–72
Age at mating (weeks)	6–10	9–14	6–12	10–12	24–35	12–16	m 36[a] f 28–36	40–72
Type of oestrous cycle	Poly	Poly	Poly	Poly	Poly	Poly	Poly	Mon
Length of cycle (days)	4–6	4–5	4	4	—	13–20	14–21	21–28
Ovulation	Spontaneous	Spontaneous	Spontaneous	Spontaneous	Induced 10–11 h pc	Spontaneous	Induced 24–38 h pc	Spontaneous

Gestation (days)	19–21	20–23	15–18	20–21	28–34	59–72	61–69	53–71
Pseudopregnancy (days)	12–21	12	8–10	—	14–16	Does not occur?	Uncommon 20–44	63
Next oestrus after parturition	P-partum	P-partum	P-partum	P-partum	4th week of lactation	P-partum	4th week of lactation	Next season
Litter size	6–12	6–15	5–9	4–8	5–10	3–6	1–8	[b]
Birth weight (g)	1–3	4–6	2	1.5–2.5	30–70	85–95	90–140	[c]
Weight at weaning (g)	10–12	35–50	35–40	6–8	0.8–1.5 kg	0.18–0.25 kg	0.7–1.0 kg	[c]
Adult weight (g) m	20–40	200–500	90–120	40–45	1.5–5.0 kg	0.75–1.5 kg	3.5–6.0 kg	[c]
f	25–40	250–350	95–130	40–45	1.5–6.0 kg	0.70–1.3 kg	2.2–3.0 kg	[c]
Useful breeding m	6–18	6–12	12–15	12–15	3–6 y	2–3 y	5–7 y	5–7 y
life (months) f	6–12	6–12	10–12	12–24	1–3 y	2–3 y	8–10 y	4–5 y

[a] Puberty depends on growth rate and season of birth; females are mature at 2.3–2.5 kg and males at 3.5 kg minimum.
[b] Small breeds 2–3; large breeds 4–12.
[c] Weight depends on breed.

Poly, polyoestrous; mon, Monoestrous; pc, postcoitus.

The period of the oestrous cycle is different for each species of laboratory animal (Table 4.1) but it is important to remember that in the natural state animals are either pregnant or anoestrous and the repeated unmated cycle is not the usual condition; cycles are instead usually the consequence of infertility or failure to mate.

The delay between fertilization and implantation is usually very short in laboratory rodents (4–6 days) but may be delayed by lactation if conception results from a postpartum mating.

Gestation or pregnancy is the time between fertilization and the birth of the young or parturition.

As the embryo develops the production of progesterone by the corpus luteum declines and the placental tissue takes over this function. The placenta also secretes oestrogens which inhibit further follicular development and ovulation.

The gestation period varies from one species to another (Table 4.1) and the young are also born in varying degrees of development, for example guinea pigs at birth are fully furred, have their eyes open and are quite mobile whereas rats and mice are born naked and blind.

In the female the latter half of gestation is associated with a major development of the alveoli in the mammary glands in response to oestrogen and progesterone secretion by the placenta.

In some species the first secretion produced, the colostrum, is rich in immunoglobulins and is important to the young in their acquisition of passive immunity.

After parturition the rate of milk secretion increases rapidly to a peak and then begins to decline. Suckling, through the reflex secretion of a hormone, is necessary for the maintenance of milk production and the duration of lactation depends on the state of development of the young at birth and their subsequent growth.

Weaning is the stage of development when the young animal changes from dependence on the mother's milk as a source of food and begins to use solid food and water. The age at which this happens again varies from species to species.

Although photoperiod is important in regulating the breeding cycle other factors such as temperature, suitable food and the availability of a nesting site also play a part.

For the female, sexual behaviour and willingness to mate are controlled by the oestrous cycle, which is open to the influence of seasonal and other factors. Males, on the other hand, do not usually show a breeding season and, given the correct circumstances, will mate at any time in the year.

BREEDING METHODS

The breeding method chosen for laboratory animals must take into account the behaviour and reproductive physiology of the species; for example, are adult females and males aggressive towards each other when together in the same cage, are the newborn and juvenile animals safe in the presence of the adult male, is oestrus regular or seasonal, is there a postpartum oestrus and are social factors important?

As well as these factors, the breeding method chosen will depend on the genetic characteristics required, the age at which the animals are needed and the numbers required over a given period.

Breeding methods are of two general types: those which produce inbred strains and those which produce random bred animals.

Inbred animals are produced by the mating of related individuals which have one or more ancestors in common. The parents of these individuals may or may not be themselves the result of such a mating but if they are related the young produced will be inbred.

The rate of inbreeding depends on the degree of relationship between the individuals and the closest which can be achieved, with mammals at least, is that of brother and sister: a full sib mating. The continuous mating of offspring to the younger parent or a single generation of parent × offspring mating is genetically the same as a full sib mating.

The degree of inbreeding is expressed as the coefficient of inbreeding F, which is the probability that the two genes at any locus are copies of one of the genes carried by the common ancestor.

When full sib matings are continued for 20 or more generations the animals become more and more homozygous for particular gene pairs and the coefficient of inbreeding increases from an arbitrary starting point of 0 to about 98.6 per cent, which is accepted as the minimum level required before most laboratory species can be designated inbred.

All individuals within an inbred strain should be genetically identical or 'isogenic' and this is achieved by ensuring that all individuals are derived from a common ancestral full sib breeding pair in the twentieth or a subsequent generation.

It is isogenicity rather than homozygosity that is the useful feature of inbred strains (Festing, 1979) but in practice total isogenicity cannot be obtained as it is impossible to produce a fully inbred strain and some degree of segregation remains.

During the first few generations of full sib matings there is a decline in reproductive performance, disease resistance and general 'fitness' which is termed 'inbreeding depression' and it may be attributed to the increase in the number of recessive deleterious genes which become homozygous.

When the strain becomes established no further inbreeding depression should occur, but when a new inbred strain is being derived allowance must be made for the possible extinction of a proportion of the lines during the first few generations.

The production, maintenance and genetic monitoring of inbred strains are specialized activities and are described in detail in several textbooks, including Festing (1979).

If two isogenic strains differ only at a single locus, the differential locus, they are called co-isogenic strains. These may arise naturally as a result of mutation within an inbred strain but an approximation to a co-isogenic strain can be produced by backcrossing a gene from a donor strain or stock into an inbred strain or by inbreeding with forced segregation of the locus in question. If a histocompatability locus is involved then the strain is called 'congenic resistant' because the animals will resist tissue and specific tumour transplants.

Random breeding is a method in which breeding stock is chosen without regard to parentage and mated without reference to relationships. The matings must, however, only be made between animals of the same stock, that is, they form a closed colony which has no introductions from outside.

In random breeding the intention is to maintain genetic variability and to avoid inbreeding depression, but random breeding does not in itself exclude the possibility of inbreeding, rather it prevents the subdivision of the colony into separate lines of descent.

If animals are mated at random then the larger the number of parents in each generation the more distant is the average relationship between the pairs and therefore the slower the rate of inbreeding. Very large numbers of parents are not needed to reduce inbreeding to a low level, for example with only ten pairs of parents the rate of inbreeding is 2.5 per cent per generation and eleven generations would have to be produced before the inbreeding coefficient was the same as that resulting from one generation of full sib mating, i.e. 25 per cent (Falconer, 1976). Although this degree of inbreeding may be acceptable in the short term, a rate of 1 per cent or less is desirable if the animals are to be maintained for a long time. In this case about 25 pairs or more per generation would be required.

The rate of inbreeding is at a minimum when there are equal numbers of males and females for a given number of parents but in some breeding systems each male may be mated with more than one female. If this is so then the rate of inbreeding depends more on the number of males than on the number of females.

It is possible to reduce the amount of inbreeding which inevitably occurs, especially in small colonies, by using a system of 'minimal inbreeding'. When breeding stock is chosen at random from all the available young animals each litter does not necessarily contribute equally to the next generation, whereas inbreeding can only be minimized by equalizing each litter's contribution. This may be achieved by ensuring that, as far as possible, each pair of parents in one generation contributes the same number of offspring to be parents of the next generation. If each pair contributes two offspring, either male or female, then the rate of inbreeding will be half that under random breeding with the same number of parents.

When animals are mated at random some pairs may well be full sibs whereas others are only distantly related, resulting in a variation of the coefficient of inbreeding within any generation and from one generation to another. If, however, the 'random' breeding system is modified to avoid the mating of close relations such as sibs and cousins, then the result is a greater uniformity of the animals and their characteristics but no reduction in the overall rate of inbreeding.

This modification is time-consuming to arrange in practice but it is possible to use systems of mating which are easy to apply and give uniformity of the degree of inbreeding within and between generations. Two of these systems are described by Falconer (1976).

Practical breeding systems are of two general sorts, those where the animals are permanently mated and those where the mating is only temporary.

Permanently mated groups may be set up as monogamous pairs with a single male and female or as polygamous groups of one or more males and several females. In either case the females are allowed to give birth in the presence of the male and the other members of the group. In the harem system a polygamous group of males and females is established but the pregnant females are removed from the group before parturition.

Hand mating is a system in which the male and female are left together only long enough for mating to occur.

Artificial insemination is possible with larger species such as the rabbit and embryo storage and transfer may be used on special occasions.

With some species only some of these methods are routinely used but with others several systems may be employed and the choice can have considerable influence on the number of animals produced and their cost.

Mice are usually permanently mated, with the ratio of males to females varying from 1:1 to 1:4. The number of young produced per female per week is lower with the higher number of females but the number produced per group per week is higher. When running costs are taken into account the larger ratio of females to males proves to be more economical.

Rats are normally bred either in monogamous pairs or in harems of one or two males to five or six females. The pregnant females are removed before parturition as the suckling mothers tend to be intolerant of each other unless there is sufficient nesting space.

Monogamous pairs have been recommended for small colonies as they are easier to manage but in large colonies the harem system may prove to be more efficient.

Syrian hamsters may be bred by using hand mating, harem or monogamous pair systems.

With hand mating, the female is placed with the male just after dark when the animals are naturally most active. If the female is in oestrus, mating will occur almost immediately but, if not, the female will fight off the male and may cause serious injury to him unless they are separated. If windowless rooms with a partially reversed lighting schedule are available hand mating may be carried out during the normal working day.

When harem mating is used the ratio of males to females is not important but is usually about one male to five females. The females are removed before parturition and the main disadvantage with this system is that fighting may occur when the female is returned to the group or when she is replaced by another female.

If hamsters are paired at weaning it is possible to establish monogamous pairs and this will allow the production of inbred strains.

Unless inbred strains are being maintained it is usually considered uneconomical to breed guinea pigs in monogamous pairs because they are costly of space and labour. The most commonly used method with guinea pigs is polygamous breeding in a group consisting of a boar and four to 20 sows. Unless accurate dating of birth is needed pregnant females are not usually removed from the group for parturition.

Rabbits are normally hand mated, the doe being brought to the buck's cage where mating should occur within a short period of time. The doe's response may vary from one buck to another and it is usual to try her with at least two bucks before assuming that she is anoestrous and will not mate.

Cats are normally bred in polygamous groups with one or more toms even though the majority of matings will be carried out by the dominant male.

Alternatively, the male may be kept near to the queens in an area which becomes his territory, the queens which are in oestrus being brought to him for mating.

Dogs are normally hand mated, the male and female being introduced to each other while restrained in case they show aggression. If the bitch is receptive the two are left together to go through their courtship behaviour and to mate.

COSTS

Although the cost of breeding and keeping good quality defined animals under experimental conditions is high, price is not a primary consideration with laboratory animals when seen in relation to the overall cost of a research project where the cost of the animals may be around 10 per cent of the total.

Nevertheless this is frequently an area where the research worker is tempted to economize by using cheaper animals only to find that it is a false economy and '. . . that cheap animals equal expensive and often bad research' (Festing, 1977).

A detailed analysis of the costing of heart valve replacement operations in either laboratory bred labradors, costing $81, or healthy pound dogs, costing nothing, has been published (Fletcher, Herr and Rogers, 1969). The survival of the labradors for more than five days after the operation was 93 per cent while that of the conditioned pound dogs was 73 per cent and it was concluded that the use of healthy purpose-bred animals resulted in a decrease in the overall cost of the experiment and in the number of animals used.

Assuming that one is not considering rare or extremely difficult to breed animals, then the relative costs of the common laboratory species are in proportion to their size.

The actual costs involved depend on the facilities available and so some generalizations must be made. In broad terms, the relative costs of producing common laboratory animals are: mouse — 1, rat — 4, guinea pig — 24, cat — 266, dog — 627 (Lane-Petter and Pearson, 1971).

The costs of maintaining the larger species are also greater, not only in monetary terms but in the space they occupy, for example an 18 m^2 experimental holding room will house 4800 mice or 840 rats or 300 guinea pigs or 75 rabbits or 40 cats or six dogs on short-term experiment (Lane-Petter and Pearson, 1971).

Taking these factors into account, Festing (1980) has produced an estimate of the economy of a number of species relative to the mouse, for example he calculated that the facilities for one rat take about five mouse units, a rabbit takes about 50 mouse units and a dog about 400 mouse units.

The use of larger species may also lead to an increase in the cost and difficulty of the experiment, for example in teratological or carcinogenic studies the amount of the test substance required will be much greater, that is, a 2 mg dose in a 25 g mouse becomes 4 g in a 50 kg pig. Apart from the proportional increase in the cost of what may well be an expensive test material, the handling of large quantities of potentially dangerous materials will add considerably to the cost and difficulty of the experiment.

In some cases the use of larger animals will be unavoidable, for example experimental surgery or in cases where a large amount or repeated sampling of body fluids or tissues is needed.

A frequently recurring question is whether or not it is more economical to buy animals from outside or to produce them by breeding 'in house'.

Usually a laboratory which breeds its own animals will have a smaller demand for any one species or group of animals than a breeder supplying a number of users. As a result it will be more difficult to provide large numbers of animals of a given age or weight on a given date.

Unless very large numbers of animals are to be produced it is unlikely that the user will be able to compete with commercial breeders on a cost basis. But some caution is necessary. If the demand is large but shows wide monthly fluctuations, then the overproduction needed to ensure a continuing supply incurs high holding costs which in turn greatly increase the cost of the animal at the time of use. In this case buying from commercial breeders may again be the best solution.

Apart from cost, there are other factors which must be taken into account when deciding whether to buy or to breed the animals required. Breeding animals will take up staff and facilities which could be better used for research but on the other hand these may already be available and it could be wasteful not to use them.

The type of research being carried out may itself require breeding, for example genetics and teratology. There may also be a need for quality control which it is felt cannot be met by commercial breeders, but usually small rodents may be obtained commercially at very high quality.

Other factors which must be considered are discussed in detail by Lane-Petter and Pearson (1971).

LEGISLATION

In Great Britain there is a considerable amount of legislation concerned with animals (Sandys-Wynsch, 1984). Apart from the 1986 Animals (Scientific Procedures) Act there are some 21 acts and orders which may be relevant to laboratory animals. None of these relates specifically to their acquisition or supply except the Dogs Act (1906), which does not allow stray dogs to be used for the purposes of vivisection.

Perhaps best known is the Rabies (Importation of Dogs, Cats and Other Mammals) Order 1974, which seeks to prevent the introduction of rabies into Great Britain by controlling the importation of some eleven orders of mammals and imposing strict quarantine requirements.

Under the Animal Health Act (1981) ministers are given powers to prohibit and regulate the importation of all mammals, fish and reptiles and other things by which diseases might enter Great Britain by sea or air.

As well as this legislation directed towards the control or eradication of disease there are laws concerned with the protection of endangered species and with animal welfare. Primates are the most widely used group of animals affected by such legislation.

In 1973 a Convention on International Trade in Endangered Species (CITES) was held in Washington in order to regulate the import, export and re-export of endangered species of wildlife and 33 nations including Australia, Canada, UK, USA and USSR have ratified the Convention.

In the UK, the Endangered Species (Import and Export) Act 1976 resulted from this Convention and the import or export of any live or dead animal described in the Act is an offence unless covered by a licence granted by the Secretary of State for the Environment.

Wild animals native to Great Britain are protected by a number of acts; seals and badgers have specific protection whereas other wild animals are covered by the Wildlife and Countryside Act 1981. It is an offence under this act for anyone to intentionally kill, injure or take any of 39 different kinds of wild animals named but licences may be issued either for specified purposes or to cover actions which would otherwise be offences.

The regulations controlling the movement of animals within Great Britain are concerned with the spread of diseases of animals and the welfare of the animals whilst travelling.

The Transit of Animals (Road and Rail) Order 1975 applies to cattle, sheep, swine, goats and horses and the Transit of Animals (General) Order 1973 applies to all animals, fish, reptiles and crustaceans and other cold-blooded creatures of any species and poultry, except those animals covered by the 1975 Order.

The 1973 Order applies to transport by road, rail, sea and air and it seeks to prevent injury or suffering during loading, unloading and carriage. It also deals with provisions for feeding, watering and accommodation and provides for unfit and pregnant animals as well as those injured during transit.

The Breeding of Dogs Act 1973 requires that all breeding establishments are licensed by the local authority and makes provision for their inspection.

In 1971 the Council of Europe appointed a Committee of Experts for the Protection of Animals, and the Draft Convention which the committee produced requires that most species of laboratory animals should be acquired from breeding establishments registered by the appropriate authority. This Convention, and the UK Government response to it, is described in Chapter 2.

REFERENCES

The following list contains cited references and a selection of other important sources of information relevant to the topics discussed.

Austin, C. R. and Short, R. V. (eds) (1973). *Reproduction in Mammals: 4 Reproductive Patterns*, Cambridge University Press, Cambridge.

Austin, C. R. and Short, R. V. (eds) (1984). *Reproduction in Mammals: 3 Hormonal Control of Reproduction*, 2nd edn, Cambridge University Press, Cambridge.

Bacharach, A. L., Cuthbertson, W. F. J. and Flynn, G. W. (1958). *The Economics of Laboratory Animal Breeding: Rats and Mice*, LAC Collected Papers, 7, 31–43.

Brooksby, J. B. (1958). *Economics of an Animal Division: Problems of Equating Supply and Demand*, LAC Collected Papers, 7, 19–29.

Carter, T. C. (1957). Breeding methods — II Economic considerations. In A. N. Warden and W. Lane-Petter (eds) *UFAW Handbook on the Care and Management of Laboratory Animals*, 2nd edn, UFAW, Potters Bar, pp. 108–16.

Clough, G. and Festing, M. F. W. (1978). Some new animal models for biomedical research. In T. Nevalainen and K. Pelkonen (eds) *Symposium on Design of Experiments and Quality of Laboratory Animals*, Papers and Abstracts, University of Kuopio, pp. 72–85.

Clough, G. and Gamble, M. R. (1976). *Laboratory Animal Houses, a Guide to the Design and Planning of Animal Facilities*, Manual Series No. 4, Medical Research Council Laboratory Animals Centre, Carshalton, Surrey.

Council of Europe (1982). Ad Hoc Committee of Experts for the Protection of Animals (CAHPA), Appendix III, Draft European Convention for the Protection of

Vertebrate Animals used for Experimental and Other Scientific Purposes.
Eaton, P. (1979). Financial control of laboratory animal facilities: partial cost recovery in an academic institution, *Lab. Animals*, **13**, 153–8.
Falconer, D. S. (1976). Genetic aspects of breeding methods. In UFAW (eds) *UFAW Handbook on the Care and Management of Laboratory Animals*, 5th edn, Churchill Livingstone, Edinburgh, London and New York, pp. 7–26.
Festing, M. F. W. (1977). Bad animals mean bad science, *New Scientist*, 20 Jan., 130–1.
Festing, M. F. W. (1979). *Inbred Strains in Biomedical Research*, Macmillan, London.
Festing, M. F. W. (1980). The choice of animals for research. In H. V. Wyatt (ed.) *Handbook for the Animal Licence Holder*, Institute of Biology, London.
Fletcher, W. S., Herr, R. H. and Rogers, A. L. (1969). Survival of purchased labrador retrievers versus pound dogs undergoing experimental heart valve replacement, *Lab. Anim. Care*, **19**, 506–8.
Heine, W. (1965). Problems of large scale production, *Fd. Cosmet. Toxicol.*, **3**, 223–8.
Hill, B. F. (1965). The economics of laboratory animal production, *Fd. Cosmet. Toxicol.*, **3**, 217–21.
Home Office (1985). *Statistics of Experiments on Living Animals, Great Britain 1984*, Cmnd 9574, HMSO, London.
ILAR (1980). *Guide for the Care and Use of Laboratory Animals*, United States Department of Health, Education and Welfare, National Institutes of Health (NIH) Publication No. 80–23.
LABA (1982). *The Laboratory Animal Breeders Association Accreditation Scheme Manual*, Laboratory Animal Breeders Association of Great Britain Ltd, Margate, Kent.
LAC (1974a). *Survey of the Numbers and Types of Laboratory Animals Used in the United Kingdom in 1972*, Medical Research Council Laboratory Animals Centre Manual Series No. 3.
LAC (1974b). *The Accreditation and Recognition Schemes for Suppliers of Laboratory Animals*, Medical Research Council Laboratory Animals Centre Manual Series No. 1.
Lane-Petter, W. and Pearson, A. E. G. (1971). *The Laboratory Animal — Principles and Practice*, Academic Press, London and New York.
NIH (1979). *Cost Analysis and Rate Setting Manual for Animal Resource Facilities*, United States Department of Health, Education and Welfare, National Institutes of Health (NIH) Publication No. 80–2066.
de Ome, K. B. and Barnawell, E. B. (1958). The economics of animal colony operation, *Proc. Animal Care Panel*, **8**, 113–27.
Parrott, R. F. and Festing, M. F. W. (1971). *Standardised Laboratory Animals*, LAC Manual Series No. 2, MRC Laboratory Animals Centre, Medical Research Council Laboratories, Carshalton, Surrey.
Paterson, J. S. (1965). Laboratory animals: the equation of supply and demand, *Fd. Cosmet. Toxicol.*, **3**, 25–35.
Perry, J. S. (1971). *The Ovarian Cycle of Mammals*, Oliver & Boyd, Edinburgh.
Poiley, S. M. (1960). A systematic method of breeder rotation for non-inbred laboratory animal colonies, *Proc. Animal Care Panel*, **10**(4), 159–66.
Poiley, S. M. (1972). Growth tables for 66 strains and stocks of laboratory animals, *Lab. Anim. Sci.*, **22**(5), 759–79.
Sandys-Wynsch, G. (1984). *Animal Law*, 2nd edn., Shaw & Sons, London.
Walker, A. I. T. and Stevenson, D. E. (1967). The cost of building and running laboratory animal units, *Lab. Anim.*, **1**, 105–109.

Laboratory Animals: An Introduction for New Experimenters
Edited by A. A. Tuffery
© 1987 John Wiley & Sons Ltd

CHAPTER 5

Quality in Laboratory Animals

G. CLOUGH
Alanann Consultancy Services, York

INTRODUCTION

Choice of animals

In a recent White Paper (Home Office, 1983) setting out the Government's proposals for new legislation to replace the Cruelty to Animals Act 1876, sponsors of applicants for a licence to use animals in scientific procedures are asked to ensure that: '. . . the type of animals which it is proposed to use is appropriate.'

This disarmingly simple statement summarizes the theme of this chapter most succinctly, the key words being 'type' and 'appropriate'.

How should the choice of animal be made? How do you know which 'type' of animal is best and whether or not it is 'appropriate'? Certain aspects of the choice are prerequisite: clearly only a female can be used as a source of ova; man-sized animals such as a pig or dog make a better 'model' of certain human surgical procedures than would, say, a mouse or a rat. The size of animal is also relevant to the scale of an experiment: whereas a pilot study to raise specific antisera might be carried out using guinea pigs or rabbits, subsequently, when larger quantities may be needed, it is obviously advantageous to raise antisera in species such as sheep, goats or horses. Even for the pilot study, however, it is very relevant to appreciate that obtaining repeated blood samples from a rabbit, with its large and easily accessible ear veins, is very much easier than from a guinea pig, which has very few superficial blood vessels.

Again, although the size of a pig may make surgical procedures much easier than they would be in a mouse, whereas mice are available which are congenitally immunodeficient to the extent that they will not reject tissue transplants even from other species — including man — this is not true in the case of the pig.

Yet again, if your interest is in geriatrics, then whereas an eighteen-month-old mouse is definitely 'old', a similarly aged pig would be just mature and a Rhesus monkey would still be a 'baby'.

Cost is often a major factor in making the choice, but this may not be as straightforward a matter as it may at first appear. It seems obvious that the capital cost of a pig is very much higher than that of a mouse; certain strains of mouse, however, such as the obese nude which is difficult to breed are surprisingly expensive. One must also bear in mind the maintenance costs and facilities required: pigs certainly need more space (and food, anaesthetic, etc.) than mice, but again certain mice, including the strain mentioned above (which, being immunodeficient is consequently highly susceptible to disease), will need very special, and thus very expensive, housing conditions if they are to be kept disease-free for any length of time.

It is clear then that the choice of animal to be used in any given research project depends upon many factors including size, species/strain, age, sex and suitability for the nature of the investigative techniques to be used as well as the cost (both capital and maintenance) and the facilities available for housing.

Having taken all these factors into account, however, a further important consideration is the quality of the animals which are to be used.

'Quality' of animals

It is well known that modern methods of investigation frequently involve the use of sophisticated and highly sensitive equipment which is carefully calibrated and meticulously maintained to ensure a standard level of performance which guarantees — as far as is possible — reliable and reproducible results. Very often the major function of this high-quality equipment is to measure the biological responses of the animals on which the research is based. It seems only logical, therefore, that the animals themselves should be of high quality and that they should be similarly maintained in conditions that permit — as far as is possible — reliable and reproducible results to be obtained.

Unfortunately, however, laboratory animals are still too often regarded as standard items which can be purchased and stored like any other 'reagents'. This is, of course, definitely not the case and never will be because of the large number and complexity of delicately balanced biological systems that each animal represents. The very adaptability conferred upon the animals by the integration of these systems to achieve stability of, for example, deep body temperature (T_b) can simultaneously be the very cause of variability in response to an experimental procedure. The following simple example will clarify this point.

Let us suppose a long-term experiment involves rats being fed a diet containing a test substance which is toxic and that the room in which the animals are housed has an inadequate system of temperature control. During the course of the experiment it becomes clear that more animals show symptoms of toxicity during the winter months. It could be assumed, therefore, that they are exhibiting seasonal variation in sensitivity to the test substance; a more likely explanation, however, is that the average room temperature during the wintertime is several degrees cooler than that of the summer; during this period the rats eat more food merely to maintain their T_b and thus take in a relatively higher dose of the toxic test material with their food.

Again, there are certain aspects of 'quality' which are more obvious than others. Nobody, for example, would start an experiment using animals which were known to be diseased, malnourished, or of a genetic make-up which rendered them unsuitable for the investigation to be undertaken; for example the mouse mutant dy, individuals of which exhibit muscular dystrophy; or strain AKR, up to 90 per cent of which develop leukaemia during the first 18 months of life (Festing, 1979).

Having made that clearly reasonable and acceptable statement, however, how in practical terms can one know whether an animal is diseased, malnourished or genetically suitable for the purpose for which it is to be used? Whereas it is (usually!) a straightforward matter to check on those aspects of the choice process mentioned earlier (species, size, sex, cost, space requirements, etc., of the animals to be used), assessing their microbiological and nutritional status and their genetic make-up is not so simple; for example, there are now over 1000 recognized strains of mice and rats used in biomedical research (Festing, 1982), a large number of which, being albinos, all look alike!

Defined animals

For reasons such as those just mentioned it is becoming increasingly common to apply certain methods of quality control to animals at all stages of their production and maintenance. In practical terms these are designed to minimize disease, to maximize genetic purity and to reduce those variables of environment and nutrition which are known to be important in this respect. Such techniques go at least some way towards 'standardizing' animals which, when produced under such conditions, are generally referred to as 'defined animals'. The use of this sort of animal can lead to an increase in experimental accuracy and a reduction in the number of animals needed to yield statistically significant results; this in turn can lead to savings in time, effort and money and should facilitate the comparison of results obtained in one laboratory with those from another.

The rest of this chapter will be devoted to further illustrating the needs for quality control and describing some of the methods in current use.

THE NEED FOR QUALITY CONTROL

Genetic

As described in Chapter 4, a large proportion of all the animals are outbred; this means they are produced by random mating in such a way as to maximize genetic variability. Although animals from the same outbred stock will superficially appear to be uniform (virtually all are albino), genetically they will be variable. Such stocks sometimes have generic names — for example 'Wistar' rats, 'Swiss' mice, 'Dutch' rabbits — also suggesting uniformity, but there is little likelihood that 'Wistar' rats from different colonies will have much in common except perhaps their ancestral origin. The main advantage of outbred animals is that they breed prolifically and are therefore cheap; a major disadvantage is that few are available internationally, so that work carried out on such stock in one country is unlikely to be repeatable elsewhere.

As also mentioned in the previous chapter, one can achieve genetic uniformity by creating inbred strains. The effect of an inbreeding programme is to eliminate virtually all genetic variation within the colony; in practical terms, all the individuals become homozygous and genetically identical, with great uniformity. Because all individuals are genetically identical, determining the genotype of any individual will characterize the whole strain.

Genetic uniformity also confers long-term genetic stability and makes it possible to be certain that daughter and parent colonies will be identical also; because of this, many inbred strains are internationally distributed so that work involving such strains is easily repeatable in other countries. On the other hand, one disadvantage of inbreeding is the resulting inbreeding depression which commonly occurs; generally speaking the animals become less healthy, more susceptible to disease and their reproductive capacity is reduced. This often makes inbred strains troublesome and more costly to produce. In order to produce a more vigorous type of animal, F1 hybrids can be used. An F1 hybrid is the result of a first-generation cross between two inbred strains; such animals have most of the advantages of inbred strains (in that all individuals are genetically identical) but with the added advantage that F1 hybrids tend to be very vigorous. It must be remembered, however, that they will not breed true and so must be produced each time they are required by crossing the two relevant inbred strains.

The species most commonly used in biomedical research are the mouse, rat, guinea pig and rabbit (Clough, 1987) and historically speaking there has been a strong tendency to use albino varieties. This probably reflects the aim of early research workers towards the 'standardization' of their animals as, superficially at least, albinism does indicate a degree of uniformity. Even once the choice of species has been made, however, great care must still be taken to ensure that the most suitable genetic type within that species is used.

Whereas in farm animals, cats, dogs and primates it is difficult to pay much attention to the genetic status, in the case of the common species — including mice, rats, hamsters, guinea pigs (and to a lesser extent rabbits and poultry) — there is a wide variety of stocks, strains, mutants, hybrids and other genetically defined types available which may or may not be suitable for certain research applications. Some mouse mutants, for example, provide medically interesting 'models' of disease such as anaemia, diabetes, muscular dystrophy, obesity, sensitivity or resistance to tumours and other metabolic disorders (for details see Festing, 1979; Green, 1981). Clearly, the designation of such a mutant as a 'model' of a similar disease in man does not indicate that the aetiology and symptoms of the disease are necessarily the same in the two species; nevertheless, a study of the disease in laboratory animals often leads to an insight into the disease in man or other animals. Other useful characteristics arising from single-gene mutations — though not usually classed as 'models' of disease — include various types of hairlessness (useful where substances are to be tested by application to the skin) and developmental abnormalities such as absence of tail, spleen or thymus. Indeed, absence of the thymus in the mouse and rat mutants 'nude', which leads to a deficiency of the immune system such that homozygous nude animals are unable to reject xenografts (that is, tissue transplants from other species, including man), has made them one of the most exciting 'models' of recent years; in such animals, for example, it has been possible to grow human tumours on which to try out potentially chemotherapeutic agents against cancer directly on a human rather than a murine tumour (Rygaard and Povlsen, 1974).

Inbred strains, which as we have seen were initially developed as a means of controlling genetic variation (in effect by reducing genetic 'noise' and thus increasing experimental precision), are now widely used in research as many of them have very useful characteristics. These cover factors affecting, for example, anatomy, behaviour, lifespan, susceptibility to drugs, immunology and other aspects of physiology and biochemistry which are of great importance in various areas of research.

Unfortunately, very many of the mice with different genotypes all exhibit the same phenotype — in that they all carry that all-too-misleading indicator of uniformity, albinism! It is clearly very important, therefore, not only that the name of the strain be clearly reported in publications, but also that some effort is made to ensure that the animals being used are indeed of the strain quoted. The consequences of inadvertently using the 'wrong' strain can be disastrous; for example, a contract research organization was screening a potential antileukaemia drug using commercially purchased AKR mice (in which, as already reported, up to 90 per cent of individuals develop leukaemia during the first 18 months of life). A year later none of the control mice had died although by then more than half should have developed leukaemia. Subsequent investigations revealed that the mice were not in fact AKR and were not even inbred! In this particular example the estimated loss was about $200,000 and there are a number of reports of other work being frustrated by such genetic 'mistakes' (Festing, 1982; Hoffman, 1983) some of which have given rise to legal action (Fox, 1983). A useful discussion of the subject can be found in 'Genetic control of laboratory animals, possibilities and limitations' (Spiegel *et al.*, 1980).

Health status

Like people, animals can suffer from a wide range of different diseases. These can vary from the one extreme of being lethal to the other where the disease is subclinical or latent; that is, the animals are infected but show no clinical symptoms and so appear perfectly 'healthy'.

Clearly, an acute outbreak of a disease which kills off a large proportion of animals will seriously interfere with any research programme; because of this, everyone would unreservedly agree that the most stringent precautions should be taken to prevent such diseases from being introduced into animal colonies.

It is much less widely appreciated, however, that subclinical or latent diseases (whose effects, as has already been pointed out, often go unnoticed) can have significant effects on the course of an experiment. At best this can make the experiment less efficient (so that larger numbers of animals have to be used to reach a satisfactory result) and at worst give rise to completely misleading results. Thus, in a case where the control animals are not subjected to the same degree of stress as those undergoing the full experimental procedure, the apparently clear-cut difference which might arise between the experimental and control groups can be due merely to the activation of latent disease in the experimental group rather than to the experimental 'treatment' itself.

A classical case of this sort of confusion arose in the field of inhalation toxicology using rats as the experimental model. In such experiments, unless disease-free animals are used, any lesions which might be caused by the test substance can be completely obscured by lesions existing in the lungs due to respiratory disease.

There are many well-documented examples showing that inapparent infections of this sort have been responsible for numerous problems; not only interference with experimental results, but also human infection, contamination of biological reagents, changes in enzyme levels, both long- and short-term changes in cell and organ function, longevity, immunosupression and immunoenhancement, pathological lesions and changes in sensitivity to various external stimuli. (For useful discussion see Parker (1980) and Hsu *et al.* (1980).)

It is for reasons such as these that over the past 20–25 years there has been an increasing tendency to make use of animals of a known or stated health status. This has led to the introduction of a variety of descriptive terms, many of which are far from precise. The 'normal' animal living under ordinary laboratory conditions with no attempt being made to limit the entry of undesirable micro-organisms is generally referred to as conventional. However, terms such as disease-free, minimal disease, barrier maintained and specified pathogen free (or SPF) can mean different things to different people; hence a more precise terminology would be very helpful. Probably the best attempt so far made to establish such a terminology was the category, or star rating system introduced by the UK Medical Research Council's (MRC) former Laboratory Animal Centre in relation to the accreditation scheme it operated from 1950 to 1982 (LAC, 1974; Townsend, 1969).

In that system, animals were classified into five categories, each category being allocated a star rating (1*–5*); rather like hotels, the more stars the better! The lowest acceptable standard (1*) required that the animals must be free from zoonoses (diseases transmissable to man). This ensured that they would be safe even for schoolchildren or students to handle without risk of acquiring disease. The highest standard (5*) had to be free from any other form of life; such animals are generally known as germfree and they can only be maintained by the use of very specialized, closed-system techniques. The term axenic (literally 'without strangers') is also used for these animals. If a single strain of microorganism is introduced to such animals, they are said to be monoassociated, whilst the introduction of several kinds of organisms results in a polyassociated system. As long as the organisms involved are known, then the system is said to be 'gnotobiotic' (from the Greek *gnosis*, knowledge and *bios*, life).

Clearly, germfree animals are not 'normal' and they are only suitable for more specialized experiments. Category 1* animals should be suitable for relatively crude experimental work, teaching and some types of biological assay though they are likely to be a health hazard to other laboratory animals as they are not required to be free from any of the serious epidemic diseases of the species. For longer term experiments, the equivalent of the higher categories (particularly 4*, probably equivalent to those commonly known as 'SPF') will be much more suitable as these must be free not only of the common epidemic diseases, but also demonstrably free from all the known disease-causing organisms (pathogens) of the species concerned.

Such animals are usually Caesarian derived and are maintained under special conditions which minimize the likelihood of them being exposed to unwanted pathogens. As such organisms (bacteria, viruses, protozoa, fungi, parasites, etc.) can be introduced by people, contaminated materials (food, cages, bedding, water, air, etc.), wild rodents, birds, insects and other arthropods, and even via apparently healthy animals brought into the building, quite complex procedures must be adopted to ensure that animals remain SPF over the long term.

Unfortunately, the MRC grading system was discontinued in 1982 but another scheme, partly derived from it, came into operation immediately. This is run by the Laboratory Animal Breeders Association (LABA) and is generally referred to as LABAAS — the Laboratory Animal Breeders Association Accredaitation Scheme. Under this new scheme there are three grades of accreditation:

1. Barrier breeder
2. Non-barrier breeder and
3. Supplier.

Additionally, the scheme has three schedules, A, B and C, which define:

(A) the minimum requirements for management, husbandry and housing
(B) the minimum requirements for health monitoring and
(C) lists those infections which it is mandatory for accredited breeders and suppliers to report both to their customers and the Secretary General of the scheme.

Details of the LABAAS and how it works have been published (Bantin and Smith, 1985). Two disadvantages of it are that it is open to LABA members only and the relevance of the 'standards' it sets concerning animal health are not very easily understood by most animal users.

Environment

In that section of the Introduction dealing with 'Quality' it has already been noted that the reliability of results largely depends upon the animals showing as standard a response as possible to any experimental treatment to which they are subjected.

It should always be remembered, however, that the majority of species used in biomedical research are small, homoiothermic mammals which have sensitive and strong homeostatic mechanisms. This means that by their very nature, such animals respond to changes in the external environment by altering their physiology and/or behaviour so as to 'compensate' for those changes and so maintain the stability of their internal environment. Such compensatory changes in the animal — which are generally mediated through the neuroendocrine system — are potentially capable of influencing experimental results and there is a wealth of published information showing that this is the case (for review see Clough, 1982). Some of those factors reported in the literature as being relevant include temperature, relative humidity, air movement, light, sound, vibration, air pressure, gravity, electrical and magnetic fields of force, ionizing radiation, odours, dust, chemicals and other pollutants.

For anyone embarking upon research using animals, therefore, it is clearly important to attempt to control at least those environmental factors which are most likely to influence the work it is intended to carry out. The following brief survey of those factors which are thought to have the most relevance in a variety of fields of investigation will serve to demonstrate this point (for references see Clough, 1982).

Temperature

We have already noted that most of the animals used are homoiotherms; thus, changes in ambient temperature or any other factor affecting the animal's thermoregulatory system will result in compensatory changes in an attempt to maintain the deep body temperature (Tb) as close as possible to 'normal'. As such changes frequently involve disturbances in metabolic rate, peripheral circulation, activity and behaviour, it is perhaps not surprising that there are many published records of temperature affecting experimental results; the following are just a few examples:

1. Changes in food and water intake
   — thus affecting the 'dose' of any test substances that may be incorporated into the food or water.
2. Changes in drug activity
   — in some cases increasing and in others decreasing their toxicity. Even small fluctuations (4°C) in ambient temperature during the course of an experiment can cause large (tenfold) variations in toxicity, and even larger variations (several thousandfold) can be caused by an animal's previous thermal experience.
3. Survival of micro-organisms in the environment and hence disease control in animals.
4. Changes in Tb
   — whether physically induced or caused by infection can also interfere with endogenous functions such as lactation, pregnancy and spermatogenesis, as well as causing varying degrees of teratogenesis; the latter can be particularly insidious when the changes involved are relatively minor, as has been found to occur in the developing nervous system of the foetal guinea pig.
5. Strange personnel, stormy weather and mere handling of animals can cause significant alterations in Tb which can confuse studies involving pyrogenic substances.

Relative humidity (RH)

Because the commonly used laboratory species are unable to sweat to any significant degree, they respond to thermal stress by a large rise in their respiratory rate as their only means of increasing evaporative heat loss. There is an obvious relationship between the RH of the inspired air and the animal's ability to lose heat by this method. The following are examples of some published affects of RH on animals:

1. Five per cent increase in food consumption of rats kept at 35 per cent RH as compared to others kept at 70 per cent (both groups at 21°C); again this can affect the 'dose' of any test substance incorporated into the food or water.
2. Variation in the absorption rate of test substances applied to the skin (particularly if shaved or hairless animals are used), either due to changes in the viscosity of the applied substance or to RH/temperature related changes brought about in the animal's peripheral circulation.
3. Increased incidence of disease, either directly (e.g. 'ringtail' in rats, which is encouraged by low RH) or indirectly by the beneficial effects of the ambient RH on the survival of micro-organisms.
4. High RH encourages the production of ammonia in animal cages. This, as well as being responsible for pathological changes in the respiratory tract which are indistinguishable from those caused by other 'irritants' (experimental or otherwise), also facilitates respiratory disease by depressing the defence systems of the respiratory tract.

Air quality

Of necessity, animals obtain the air they breathe from the cage. A typical mouse cage has a volume of some six litres; five mice living in such a cage would 'respire' that amount of air in about 45 minutes. Clearly the quality of this air can be critical, not only in terms of its temperature and RH, but also in that it can introduce contaminants both in the form of organisms and polluting substances. Examples include:

1. Aromatic substances (some potentially carcinogenic) arising from sawdust bedding affecting liver function.
2. Pesticides (again from sawdust made from treated timber) altering the immune response of rodents.
3. Introduction (through the air-conditioning system) of an imbalance in positively/negatively charged ions which can affect the activity and behaviour of rodents, as well as facilitating certain types of infection.

Air movement

The speed of air movement determines whether or not there is a 'draught'. The fact that air at 21°C, moving at the slow speed of 30 cm/sec (about six times slower than normal walking speed) has a cooling effect equivalent to a reduction of almost 7°C emphasizes one factor of moving air. This is clearly important in relation to the animal's T_b, and hence once again to food and water consumption.

Light

Most laboratory animals are either crepuscular (active at dawn and dusk) or nocturnal in their habits; their eyes are therefore adapted to dim light conditions and very few of them have any colour vision. The photoreceptors of the mammalian eye undergo a continual renewal process which is dependent upon frequent periods of darkness to prevent degeneration of the retina. Light-induced retinal damage is not uncommon in albino rodents under laboratory conditions and its effects are increased when recovery periods in darkness are too short; hence both light intensity and photoperiod are relevant to the proper functioning of the eyes. Light, in its various aspects of intensity, colour and periodicity, has been shown to be important in:

— stimulating and synchronizing breeding cycles
— gastrointestinal mobility
— wheel-running activity in mice
— growth rates and age of maturation in rats
— pathological changes in the retina
— interactions with chemical carcinogens in the mouse

It is important to note that even very short periods of light (as little as one second — or less in some cases) occurring during the dark phase of the photoperiodic cycle can have dramatic effects on the reproductive performance of some animals (hamster) and the digestive function of others (rabbit).

Sound

The significance of sound for laboratory animals was reviewed extensively by Gamble (1982) and the importance of this particular factor in relation to experimental results was stressed by Clough (1982). For present purposes it is necessary to note that the ears of most animals respond to a much wider range of frequencies (pitch) than do those of man; for example, mice cannot hear the Greenwich meantime 'pips' broadcast as a time check on the radio because their frequency (1000 Hz — or 1 kHz) is too low. On the other hand, many mice can hear up to 80 kHz or higher, whereas the upper cut-off frequency for most people is around 18–20 kHz. All 'sounds' above that frequency — because they are too high pitched for us to hear — we refer to as 'ultrasound'. A well-known example of this is the 'silent' dog whistle which, with a frequency of about 25 kHz, is inaudible to us but very loud to a dog. Due to the complexity of the subject, it is not feasible to give details here. Suffice it to say that whether or not a sound is 'damaging' depends not only on its intensity (loudness), frequency and duration, but also on the hearing ability of the species concerned, the age and physiological state of the animal at the time of exposure and possibly what other sounds it has been exposed to during its lifetime.

Thus, the would-be research worker should make himself aware of what sounds his animals can hear and should also remember that the body can 'respond' (not necessarily obviously) to those sounds during sleep, under deep anaesthesia and, indeed, even after removal of the cerebral hemispheres! The relevance of sound in relation to experimental animal work is very extensive and the following effects are merely a few examples:

- physical damage to the ear (particularly the cochlea)
- hypertension
- changes in body weight
- changes in immune response
- changes in tumour resistance
- effects on reproductive function (increase/decrease in fertility; termination of pregnancy; teratological effects)
- cannibalism
- audiogenic seizures
- audioconditioning related to atypical responses to drugs administered at some future date
- changes in blood chemistry and cellular distribution

If the different published 'guidelines' for the environmental requirements of laboratory animals are examined, it soon becomes clear that there is considerable variation between them (Clough, 1984). This may be because they are based largely on practical experience related to the different parts of the world from which they originate. There is an increasing tendency for such differences to be lessened as the effects of international regulatory authorities, such as the American Food and Drug Administration and the Council of Europe, begin to take effect. Nevertheless, there still are significant differences between various national guidelines. From the above considerations, it is clear that in order to enable other investigators to repeat experiments or carry out comparative studies, the environmental conditions pertaining during an experiment should be adequately described in any publications.

Nutrition

This chapter would be incomplete if it was not stated that the food given to laboratory animals, particularly in relation to its nutrient specifications, can have significant effects on their response to many other environmental factors and to any experimental treatments to which they may be subjected. Nutrition is such an important subject, however, that it is treated in detail in another chapter (see Chapter 12). It is mentioned here merely to emphasize to the reader its importance as one of the factors contributing to the quality of experimental animals.

From reading this far it will be clear that there are many aspects to be considered in selecting the right animal for any research project. Let us now briefly examine some of the methods which can be used to try and ensure that the animals chosen are 'appropriate' for the proposed work.

METHODS OF QUALITY CONTROL (QC)

Genetic monitoring

There is clearly no problem in determining the species of animal in use. However, particularly in those species with large numbers of phenotypically similar genetic types, there definitely is a need to ensure that the animals being used are genetically authentic. In most cases, the implementation of routine genetic QC programmes in the breeding colonies will be adequate, but in some cases individual batches of animals must be authenticated; this may become critical if legislation is involved as was described earlier.

Methods of genetic monitoring include those given below. References are given where appropriate so that the reader can obtain more information about those techniques which may be useful.

1. Breeding methods
   Whenever possible, coat colour markers are used. Strains with the same coat colour should always be kept physically separate. Meticulous records of all matings and animal movements must be maintained. Breeding programmes must be under the supervision of trained and competent staff. (Festing, 1979)
2. Biochemical markers
   There are many biochemical markers available which characterize individual strains and these provide an extremely sensitive monitoring method. The markers can be detected by various types of electrophoresis (Krog, 1976; Groen, 1977; Hoffman et al., 1980). A comprehensive list of known markers in inbred mouse strains is given in Green (1981) and a few markers are also known in the rat (Brdicka, 1980).
3. *In vivo* histocompatibility testing
   This includes such techniques as skin grafting, lymphoid tissue transplantation and tumour transplantation between individuals. The acceptance of a skin graft shows that a strain is fully inbred and also provides a sensitive method for detecting recent genetic contamination (Festing, 1979).
4. *In vitro* histocompatibility testing
   Serological methods can be used and are particularly valuable in authenticating individual strains that differ at the major histocompatibility complex; although conventional antisera needed for these tests are not readily available, it is possible to use polyvalent, strain-specific alloantisera

which can be prepared without the need for specialized strains as described by Festing and Totman (1980). Monoclonal alloantisera which are particularly valuable for this type of screening are gradually becoming available commercially.

5. Mandible analysis

    This is a morphological method based on the shape of the right mandible (Festing, 1973). It is especially valuable for the routine monitoring of mouse and rat colonies which have already been authenticated by other methods. This is because it is quick, economical and requires little expertise. It is also of special value for authenticating outbred stocks which have not yet been characterized biochemically (Festing and Lovell, 1980).

6. Embryo cryopreservation

    This technique (Whittingham, 1974) involves the storing of frozen embryos with regular restocking of the foundation colonies from the frozen stocks after 10–20 generations; this is equivalent to 5–10 years. By this technique, an almost steady state is achieved since the reconstituted colonies all display exactly the same genetic constitution as at the time of the original freeze-preservation. Although this technique is expensive in laboratory and highly trained personnel requirements, it is, nevertheless, the only technique currently available which ensures genetic constancy as well as integrity (Hedrich, 1980).

Health status monitoring

There are several aspects to any animal health monitoring programme and they should include as a minimum:

1. Clinical examination and treatment

    The treatment of individual small rodents is rarely practicable or desirable. Regular clinical checks can, nevertheless, be invaluable in the early detection of disease and the promotion of animal welfare. It is certainly essential in the case of the larger, non-rodent species in which treatment can be very successful.

2. Laboratory investigations (of 'normal' and apparently normal as well as 'sick' animals)

    These investigations play a most important part in the diagnosis of disease. Clearly, postmortem investigations must be carried out on all animals that die or are killed. Additionally, however, samples of animals should be routinely screened for the presence of parasites, bacteria, viruses and pathological lesions. A wide range of techniques including bacteriology, histopathology, haematology and clinical chemistry can be used as aids to diagnosis. It must be appreciated that the results of such investigations are always retrospective; that is, the findings indicate a situation that has existed

for some time. This fact emphasizes the need for such investigations to be made on a routine basis. Only by the use of such techniques can 'early warnings' of changes in the health status of a colony or group of animals be detected. This is essential if there is to be any hope of preventing a disease from spreading either to other colonies, to other animal houses, to other sites, or even to members of staff. Special attention must be paid to the monitoring of all breeding colonies, rodent holding rooms and any animals arriving on site from outside sources.
3. Managemental procedures
These should be aimed at minimizing the likelihood of undesirable organisms gaining access to the animals. They will include procedures such as:

— the housing of incoming animals in a special quarantine area where they will be screened to ensure freedom from disease before mixing with existing stocks
— the housing of sick animals in isolation or their disposal by approved methods
— the routine care of healthy before sick animals in the daily routine
— careful control of staff and visitors
— decontamination of all materials entering units; in the case of SPF animals, this will include filtration of the air, the creation of positive pressure within the unit, the sterilization of food, bedding and equipment (by the use of devices such as autoclaves, fumigation chambers, gamma irradiation, etc.), as well as a full change of clothing (and possibly compulsory showering) for staff on entry.

For further reading on this topic consult the *UFAW Handbook*; Needham (1979); Spiegel *et al.* (1980).

Environmental monitoring

Whichever standards are followed for the establishment of environmental control it will be necessary to make either continuous or periodic checks to ensure that those standards are being met. As a guide, the following general recommendations will prove suitable for the majority of animal facilities:

Temperature
Although various sources (for review see Clough, 1984; 1986) quote 'optimum' temperature ranges, with the exception of some primates and other special cases (such as hairless animals), there is little scientific evidence to show that the common laboratory species really require different temperatures. All the species are adaptable and it is unlikely that an overall room temperature of 20±3°C will create any major problems for the staff or the animals; however, impending legislation may require stricter control than this in due course.

Relative humidity
Under the temperature conditions suggested, there should be no adverse effects if the RH is maintained at 55±15 per cent. The most likely exceptions are recently imported primates.

Light
A regimen providing a 12 hr dark:12 hr light photoperiod, utilizing 'daylight' colour fluorescent tubular lights and providing an intensity of 350–400 lux at 1 m from the floor, should be adequate for most purposes. Particular care should be taken, however, to provide shade for albino rodents in the top row of cages (in which the light intensity should not exceed 60 lux) and to ensure that the room light is not turned on, even for very short periods, during the normally dark part of the light cycle. If it is necessary to examine or observe animals during their 'night', then the use of a dark red light will ensure that they are not affected, as dark red is invisible to most of the common species.

Sound
Due to the variations in the hearing ability of different species and the potentially endless combinations of temporal and spectral variations of sounds that can occur in animal houses, it is impossible to give specific recommendations for permissible noise levels. From past experience, however, if the background sound level in the empty animal room is not allowed to rise higher than about 50 dBA, then the resulting overall sound level achieved when the room is in use is unlikely to give rise to any great problems for either animals or personnel.

Ventilation
Because of the infinite variation in room layouts, furnishings and animal stocking densities, it is not possible to give any generally useful recommendations on this topic. In most cases it will be necessary to seek professional advice if satisfactory control of this aspect of the environment is to be achieved.

There are many commercial sources of instruments suitable for environmental monitoring (see survey by Johnstone and Scholes, 1976). There are also companies whose business it is to check those aspects of the environment which require more sophisticated instrumentation than is usually available in the average animal house; this generally applies to such aspects as sound levels, ventilation efficiency and balancing of the ventilation system. Generally speaking, frequent or continuous checks should be made of the temperature, relative humidity and photoperiod in every animal room, with less frequent checks being made of light intensity (including that in the cages) and sound levels (including possible sources of ultrasound). Routine maintenance checks (with occasional specialist advice when necessary) should ensure that the air conditioning system is working according to specification. Particular attention should be paid to the condition of filters and also to ensure that the air distribution system is functioning effectively. The best way to do this is by use of a ventilation rate meter which relies upon the use of a tracer gas decay method.

Nutrition

There are no readily available guidelines for the monitoring of laboratory animal diets. Clearly, the quality of the finished product will depend largely upon the manufacturer and the standards to which he works. The premises should be as free as is practicable from rodent and bird life. Faecal contamination should be kept as low as possible from the harvesting of the raw materials onwards, but this becomes particularly important once the diet has been pelleted.

Factors likely to damage the diet include exposure to light, air, heat, chemical fumigants, radiation and attack by micro-organisms and vermin. Storage conditions should be controlled, being vermin-free, clean, dry, cool and at a low RH. Most manufacturers now date stamp their diet packaging and some will guarantee the storage life of their products, provided their specified storage conditions are met (see also Chapter 12).

CONCLUSIONS

In conclusion, let us return to the quotation at the beginning of this chapter and see if we are any closer to defining what is '. . . the most appropriate type . . .' of animal for any research programme. One brief definition could be:

> That animal which will provide the most useful information in answer to the question posed by the researcher, in the shortest time, using the minimum number and with a simultaneous guarantee of minimal complication of the results by genetic, environmental or other extraneous influences.

Considering this definition in relation to the theme of the chapter as a whole, it seems likely that this may be an unachievable ideal. Nevertheless, a sound understanding of the problems involved, together with the implementation of a good system of quality control, can go at least some way towards achieving this goal.

Care must always be exercised, however, to avoid acquiring a false sense of security; only constant revision and updating of the QC programmes will lead to reasonable success.

REFERENCES

Bantin, G. C. and Smith, M. W. (1985). The Laboratory Animal Breeders Association Accreditation Scheme for commercially bred laboratory animals within the United Kingdom, *Anim. Tech.*, **36**, 1–6.

Brdicka, R. (1980). Present status in searching for biochemical markers among laboratory inbred rat strains (*Rattus norvegicus*), *Folia Biol.* (Prague), **26** 130–9.

Clough, G. (1982). Environmental effects on animals used in biomedical research, *Biol. Rev.*, **57**, 487–523.

Clough, G. (1984). Environmental factors in relation to the comfort and well-being of laboratory rats and mice, Proc. Symp. LASA/UFAW, 1983, *Standards in Laboratory Animal Management*, The Universities Federation for Animal Welfare, Potters Bar.

Clough, G. (1987). The animal house: design, equipment and environmental control. In *The UFAW Handbook on the Care and Management of Laboratory Animals*, 6th edn, Longmans, London and New York.

Festing, M. F. W. (1973). Mouse strain identification by mandible analysis. In A. Spiegel (ed.) *The Laboratory Animal in Drug Testing*, Proc. Vth ICLA Symp., Hanover, 1973.

Festing, M. F. W. (1978). Genetic variation and adaptation in laboratory animals. In W. H. Weihe (ed.) *Das Tier im Experiment*, Verlag Hans Huber, Berne, Stuttgart and Vienna.

Festing, M. F. W. (1979). *Inbred Strains in Biomedical Research*, Macmillan, London and Basingstoke.

Festing, M. F. W. (1982). Genetic contamination of laboratory animal colonies: an increasingly serious problem, *ILAR News*, **XXV**, 6–10.

Festing, M. F. W. and Lovell, D. P. (1980). Routine genetic monitoring of commercial and other mouse colonies in the UK using mandible shape: Five years of experience, 7th ICLAS Symp., Utrecht, 1979, *Animal Quality and Models in Biomedical Research*, Gustav Fischer Verlag, Stuttgart and New York, pp. 341–8.

Festing, M. F. W. and Totman, P. (1980). Polyvalent strain-specific alloantisera as tools for routine genetic quality control of inbred and congenic strains of rats and mice, *Lab. Animals*, **14**, 173–7.

Fox, J. L. (1983). News and Comment: Scientist sues over genetically impure mice, *Science*, **221**, 625–8.

Gamble, M. R. (1982). Sound and its significance for laboratory animals, *Biol. Rev.*, **57**, 395–421.

Green, M. C. (1981). *Genetic Variants and Strains of the Laboratory Mouse*, Gustav Fischer Verlag, Stuttgart.

Groen, A. (1977). Identification and genetic monitoring of mouse inbred strains using biochemical polymorphisms, *Lab. Animals*, **11**, 209–14.

Hedrich, H. J. (1980). Aiming at genetic constancy of inbred strains via genetic monitoring and cryopreservation, 7th ICLAS Symp., Utrecht, 1979, *Animal Quality and Models in Biomedical Research*, Gustav Fischer Verlag, Stuttgart and New York, 1980, pp. 329–31.

Hoffman, H. A. (1983). Profile of a genetic contamination: BALB/c-nu mice, *ILAR News*, **XXVII** (1), 10–11.

Hoffman, H. A., Smith, K. T., Crowell, J. S., Nomura, T. and Tomita, T. (1980). Genetic quality control of laboratory animals with emphasis on genetic monitoring, 7th ICLAS Symp., Utrecht, 1979, *Animal Quality and Models in Biomedical Research*, Gustav Fischer Verlag, Stuttgart and New York, pp. 307–18.

Home Office (1983). *Scientific Procedures on Living Animals*, Cmnd 8883, HMSO London.

Hsu, C. K., New, A. E. and Mayo, J. G. (1980). Quality assurance of rodent models, 7th ICLAS Symp., Utrecht, 1979, *Animal Quality and Models in Biomedical Research*, Gustav Fischer Verlag, Stuttgart and New York, pp. 17–28.
Johnstone, W. W. and Scholes, P. F. (1976). Measuring the environment. In T. McSheehy (ed.) *Control of the Animal House Environment*, Laboratory Animal Handbooks 7, Laboratory Animals Ltd, London, pp. 113–28.
Krog, H. H. (1976). Identification of inbred strains of mice, *Mus musculus*. 1. Genetic control of inbred strains of mice using starch gel electrophoresis, *Biochem. Genet.*, **14**, 319–26.
LAC (1974). *The Accreditation and Recognition Schemes for Suppliers of Laboratory Animals*, 2nd edn, Manual Series No. 1, Medical Research Council Laboratory Animals Centre, Carshalton, Surrey (out of print).
Needham, J. R. (1979). *Handbook of Microbiological Investigations for Laboratory Animal Health*, Academic Press, London.
Parker, J. C. (1980). The possibilities and limitations of virus control in laboratory animals, 7th ICLAS Symp., Utrecht, 1979, *Animal Quality and Models in Biomedical Research*, Gustav Fischer Verlag, Stuttgart and New York, pp. 161–72.
Rygaard, J. and Povlsen, C. O. (1974). *Proceedings of the First International Workshop on Nude Mice*, Gustav Fischer Verlag, Stuttgart.
Spiegel, A., Erichsen, S. and Solleveld, H.A. (eds) (1980). 7th ICLAS Symp., Utrecht, 1979, *Animal Quality and Models in Biomedical Research*, Gustav Fischer Verlag, Stuttgart and New York.
Staats, J. (1976). Standardized nomenclature for inbred strains of mice, 6th listing, *Cancer Research*, **36**, 4333–77.
Townsend G. H. (1969). The grading of commercially-bred laboratory animals, *Vet. Rec.*, **85**, 325–6.
UFAW (1972). *The UFAW Handbook on the Care and Management of Laboratory Animals*, 4th edn, Churchill Livingstone, Edinburgh and London.
UFAW (1976). *The UFAW Handbook on the Care and Management of Laboratory Animals*, 5th edn, Churchill Livingstone, Edinburgh and London.
UFAW (1987). *The UFAW Handbook on the Care and Management of Laboratory Animals*. 6th edn, Longmans, London and New York.
Whittingham, D. G. (1974). Embryobanks in the future of developmental genetics, *Genetics*, **78**, 395–402.

Laboratory Animals: An Introduction for New Experimenters
Edited by A. A. Tuffery
© 1987 John Wiley & Sons Ltd

CHAPTER 6

Principles of Animal Husbandry

MARIE S. WILSON
Smith Kline & French, Welwyn, Herts

INTRODUCTION

In this chapter the term 'research worker' will be used for all those, graduate or not, undertaking scientific procedures on live animals. 'Scientific procedures' includes research, experimentation, assay, demonstration and teaching. In the UK such people will generally require a Home Office licence.

It is unlikely that research workers will be required to involve themselves in the more routine and menial aspects of the day-to-day care of their experimental animals. An exception might be the deliberate choice to do so as part of the experimental regime or where working in a small unit with limited resources makes this a necessity. More usually, these tasks will be undertaken by trained animal technicians with their own expertise and knowledge of animal requirements. The routines will to some extent be dependent upon the facilities available, the species concerned and the methods of husbandry adopted. Therefore it will be of limited value to go into these in depth. However, this does not mean that research workers can absolve themselves from this aspect of the work. There are two sound reasons why a basic understanding of these requirements is necessary.

First, in some countries legislation or the policy of grant-awarding bodies lays clear responsibility for proper maintenance of animals on the research worker. In the UK this is through the Animals (Scientific Procedures) Act 1986, where responsibility is not limited only to experimental procedures but extends to all aspects of the animals' care and welfare. Some of this responsibility will be shared with or exercised through managerial, technical and veterinary associates. Delegation should only be entrusted to reliable people, as it is the delegator who is finally accountable. Irrespective of formal requirements, good animal husbandry always requires daily attention to the animals through interested, appreciative eyes.

Secondly, animals are extremely sensitive to the routines and influences to which they are subjected during maintenance and when undergoing investigation. Therefore it is important that the researcher is familiar with these factors and the effect they may have on the interpretation of experimental findings (Gärtner et al., 1980; Sadjak et al., 1983; Schade, Petzoldt and Friedrich, 1983).

With these two points in mind, the main principles of husbandry, including humane consideration of the animals, economy and a high standard of scientific research, will be discussed under the following headings: caging; hygiene and disease control; watering; feeding; environment.

The extensive literature available in many books and journals is indicated by the Bibliography at the end of the book. Specific references in the text are listed at the end of this chapter.

CAGING

Materials and design

The purchase of caging will be one of the most significantly expensive items and may well be subjected to financial constraints. Therefore it is essential to ensure that it will meet the needs of the animal, those of the worker and experimental requirements. In some cases the research worker may inherit caging with little choice, but even so there may be occasions when this should be resisted.

The main function of a cage is to confine the animal but, within these confines, to provide for adequate comfort and mobility. Animals are experts at escaping and their ability and persistence should never be underestimated. Much attention is given to the ingenuity of monkeys to undo nuts or cage clasps, but even the mouse, given the opportunity, can be just as inventive. Equally, hamsters can easily chew their way to the outside from a plastic box with a slightly roughened corner. Ill-fitting cage lids provide ideal escape routes for many species. As well as the inconvenience of recapturing the animals and the possibility of them escaping from the room, it can prematurely end an experiment if several animals escape from different sources and cannot be individually identified. Escapees may also be attacked by inmates of other cages and suffer damage to feet and tails.

Durability is essential. Not only in strength to hold the animals but also to withstand the rigours of handling and cleaning. As part of essential hygiene routines it may be necessary to autoclave cages regularly — subjecting them to temperatures of 121°C (for 20 min) and above.

The most frequently used materials for cage construction are metal and plastic. Glass has obvious disadvantages but may be appropriate for lower vertebrates. The commercially available off-the-shelf moulded plastic boxes are usually made of polycarbonate (macrolon), polypropylene or polystyrene (the latter cannot be autoclaved). They come in a range of sizes and designs and have the advantage of being readily available and of giving standardization to the caging system.

Metal cages are either galvanized steel, stainless steel or less commonly aluminium. Galvanized cages (mild steel protected by a zinc metal coating) are prone to the effects of urine and detergent acids and rust when the zinc coat breaks down. Stainless steel, although more expensive, is undoubtedly the better buy in the long run. It is not subject to corrosion and will give years of continuous use: it is the metal of choice.

Cages are also available with either a solid or a grid floor, the choice a matter of preference and experimental requirement. Grid floors are either formed by a sheet of metal with regularly stamped holes of an appropriate size and shape for the species concerned to allow urine and faeces to drop through, or are made up of varying thicknesses (again species-dependent) of wire going horizontally from front to back or from side to side or both. Floors are also now available in a strengthened plastic fibreglass material, suitable for the larger species. The type of grid floor and size needs careful consideration and should be chosen with a particular species in mind. This is especially important when breeding to ensure that the young do not fall through or become trapped. A wrong choice can mean uncomfortable conditions for the animal or even injury to feet or legs. Most species appear content to live on the correct grid but problems can occur in long-term studies with the development of sore feet, probably akin to pressure sores in man. Obesity and infrequent cleaning of the grids can predispose the animal to this condition.

A major difference is that animals housed on grid floors never come into contact with their bedding, therefore grid-floored cages are particularly useful if an animal needs to be fasted. Rats in solid cages with their food removed will often substitute bedding material instead. Faeces and urine fall through the grid onto a tray covered in some absorbent material — usually sawdust, paper or wood chips — or onto a tray with a manual or automatic flushing system. Although this system of housing is often more convenient, it can mean that the animals are not handled as frequently as those in solid-bottomed cages. Here the animals are in direct contact with their bedding and need either to have this replaced at regular intervals or to be moved to a clean cage each time.

If animals are supplied with bedding material it must of course be non-toxic and of no significant nutritional value. Local agriculture can often supply acceptable bedding such as corn cob preparations. Absorbent paper might be an advantage to use with postoperative animals where fine sawdust particles could enter or stick to a new wound and where a grid floor could press and cause undue discomfort on the operative site.

Nesting material (with or without a nestbox) in the form of hay, straw, woodwool, shredded paper or cotton wool is sometimes supplied for pregnant and lactating females, or perhaps for animals requiring additional warmth, for example hairless mice, postoperative animals or those that would normally sleep in a nest. Again thought must be given to the material of choice. Young animals could accidentally become entangled or strangled, and hamsters will fill their cheek pouches as part of their natural behaviour.

Cage size and choice

How the animals will be grouped depends on cage size, the species concerned and experimental requirements. Gregarious animals such as rats and mice show distinct signs of withdrawal if housed singly, especially if they are not able to sense the presence of other rats and mice by smell. A frequent sight in a rat cage is that of all the occupants piled on top of one another. If you were to stay and watch, periodically an eruption ensures that the rat at the bottom moves to the top. Although this appears to be a natural choice of sleeping for the rats, this is no excuse to overcrowd the animals, who should have sufficient room to disperse if they so wish.

Too high a stocking density can affect behaviour, causing more aggression, possible suffocation and a decrease in fertility, and a higher mortality rate amongst the young in breeding colonies.

Some species, however, must be housed singly. Hamsters are rarely sociable to one another, even at mating, and will often fight savagely unless they have been housed together from weaning. Even then fighting can occur as the animals become sexually mature. Equally, rabbits should be housed separately — not only male rabbits but females also. Does may disrupt reproductive studies, as by mimicking mounting behaviour they can stimulate ovulation and cause pseudopregnancy. Monkeys often have to be housed alone for their own safety. Even so, social interaction is still important for the well-being of these animals and it is essential that they should be in easy view of one another.

Guidance on the floor area required by stock, experimental or breeding animals of different species is available (CCAC, 1980; 1984; ILAR, 1985; Lantbruksstyrelsens författningssamling, 1982). In the UK guidelines are currently being compiled under the auspices of the Royal Society and are to be published by the Universities Federation for Animal Welfare (UFAW). Should these receive general support and prove acceptable in implementation of existing or new legislation they could well be used as a point of reference by Home Office inspectors and others monitoring proper care and maintenance. The figures were based on, and in many cases improve, those laid down by the Strasbourg Council of Europe European Convention for the Protection of Vertebrate Animals Used for Experimental and Other Scientific Purposes, which sets standards throughout Europe. But these are only guides. Feelings, judgement, observations about the animals, an interest in the work and commonsense should also be gauges when choosing cage sizes.

Further information about cage sizes can be obtained by attending meetings and visiting trade exhibitions: in the UK organized by the Institute of Animal Technicians (IAT) and the Laboratory Animal Science Association (LASA), in North America by the American Association of Laboratory Animal Science (AALAS) and the Canadian Association of Laboratory Animal Science (CALAS), in Germany by the Gesellschaft für Versuchstierkunde/Society for Laboratory Animal Science (GV–SOLAS) and in the Nordic countries by the Scandinavian Laboratory Animal Science Association (ScandLAS). A comprehensive list of cage manufacturers in western Europe is published in *Buyer's Guide*, produced by Laboratory Animals Ltd, and in North America by the publication *Lab Animal*.

Much thought has gone into the design of animal cages over the years. One important feature must be the ease and safety with which the animal can be brought out from and returned to its cage. Most cages have a door either in the front or the lid, or the complete lid can be removed. Do not be tempted to try and retrieve an animal from a small aperture, which can mean the handler going in 'blind', resulting in the animal becoming frightened and possibly hurt and the handler being scratched or bitten.

Smaller animals are usually housed in cages that provide room for movement. Larger species frequently need access to exercise facilities. This can also be particularly important with postoperative animals to help a smooth recovery. Some species may require perches in their cages. In the case of marmosets they actually chew the perches as part of their normal behaviour and therefore it is essential to choose a wood that is non-toxic and reasonably resilient.

It may be necessary to restrain the animal at some stage of an experimental procedure. When using non-human primates, it is an essential feature of the cage to have a 'crush' back (Baker and Morris, 1980). This causes the monkey to move to the front of the cage, enabling a limb to be eased out through the cage front for easier manipulation, for example taking a blood sample or injecting a sedative prior to removing it from the cage.

Special restraining devices are commercially available for laboratory animals in a variety of designs, or can be made up to suit particular needs (Hearn, 1977; Lennox and Taylor, 1983; Reigle and Bukva, 1984). Materials range from adjustable perspex cylinders to wrapping animals in cloth. It has been found that in some instances rats seem to prefer the latter method (Danscher, 1972; Owen, Tasker and Nakatsu, 1984). Badly designed restraint apparatus could injure the animal being restrained: rabbits are particularly susceptible to back injury. Sometimes, however well thought out, it is difficult to outwit an animal. Rats seem to possess the suppleness to turn completely round, however tight the space may be.

In the UK experimental procedures involving prolonged periods of restraint have to be described in detail to the Home Office. Tethering devices operated on a swivel are available for chronically implanted and instrumented animals which allow them unimpeded movement within the cage. Metabolism cages for the separate collection of urine and faeces may also need to be used. Again these are available commercially in a selection of designs and materials or can be constructed in-house (Rucklidge and McKenzie, 1980). Animals should be left in these cages only for periods sufficient to collect the samples.

On finishing a task with a cage of animals it is worth checking that:

1. The number of animals expected is still in the box.
2. Cage lids are refitted correctly.
3. No animals are entrapped in the cage (mice do have long tails!).
4. Cages are securely seated on the shelf and are not likely to topple over onto the floor. Full water bottles can destabilize the cage.
5. Water bottles removed during manipulation are returned and available to the animals.
6. All animals look okay on a final check, for example if animals have been bled all signs of bleeding have stopped and any devices used for this purpose, such as the poor practice of putting paper clips on rabbits' ears, have been removed. The animals should be calm and contented.

Never return sick or unconscious animals to a cage containing fit, conscious animals. Rats are cannibalistic and often all that will be found of the returned animal next morning will be a piece of skin and tail. Finally, all cages should (and in the UK must) bear an identification of the animals they contain — referring to research workers' name, nature of experiment, dates for the beginning and end of the experiment, species and individual animal identification. If cage cards are used, they must be tamperproof and waterproof so as not to obscure information. It is unwise to write experimental details on the cards. This is not an accepted practice and you run the risk of the card being lost by the animal destroying or eating the whole card or at least the very portion that contains all the vital information.

It is customary to fit cages into mobile racks, either on shelves or suspended. These can be moved easily from room to room as required. Alternatively, wall-mounted shelving can be installed on either side of the room. In the interests of hygiene, this must be easy to dismantle for cleaning. Equally, shelving should not be mounted so high up the wall that the animals cannot easily be observed. It should be possible to walk into a room and see all the animals at a glance. If it is necessary to look for step ladders there is always a danger (apart from the safety aspect) that it becomes too much trouble to check the row of animals above immediate view.

HYGIENE AND DISEASE CONTROL

Various systems of husbandry have been devised to regulate the exposure of animals to infection. Whatever system is adopted, it can be effective only if all those working in and employing the animal facilities use their commonsense and comply with whatever rules or codes of practice are relevant. Non-compliance or short cuts will inevitably result, sooner or later, in breakdown, with the penalty of financial and often irretrievable experimental loss.

The hygiene measures taken and the degree of control sought will depend upon the system adopted. Examples are given below.

Conventional unit: Animals housed in an area where no special precautions are required. Although strict attention is still given to cleanliness and hygiene, the research worker will usually be free to move animals to and fro within designated areas. However there may be some restrictions, for example on entry to the unit, putting on overshoes or walking across a special microbiological mat with a sticky surface to reduce the level of bacterial and fungal contamination. Some form of protective clothing such as a gown may be required for certain rooms.

Barrier unit: At its simplest, this can mean stepping over a physical barrier or through a liquid dip. Full barrier facilities require much more disciplined entry procedures, with showering and a complete change into sterile clothing. Supplies (food, bedding, etc.) and services (air, water) into the unit are strictly and routinely checked for sterility, and a system of clean and dirty corridors is enforced. Animals maintained in this way are referred to as SPF or specified pathogen free. This means that they are free from stipulated, but not all, organisms. Often entry by personnel is restricted. Where an experiment, for example a toxicological study, is conducted within a barrier system, all personnel — not just the animal staff — must follow the procedure laid down. Experiments can be destroyed by infringement, whatever the grade or rank of the infringer.

Isolators: This is a barrier taken to its maximum. Animals are maintained (and bred) inside a sterile flexible clear plastic bubble or the equivalent. All supplies and services going inside are sterile and the animals' needs are tended to by staff outside the bubble via gloves or a half suit protruding into the isolator. The animals are free of all known organisms and are referred to as germfree (axenic). Animals that have deliberately been given a known 'cocktail' of organisms are defined as gnotobiotic and are also kept in an isolator.

There can be no doubt that one of the biggest risks of infection is the animals themselves. Ignorance or insistence by a research worker on obtaining animals from an inappropriate source can wreak havoc in a unit, with widespread involvement of fellow workers' animals which were not initially involved.

Before bringing in any animals, the advice of the head of the animal facilities or the senior technical staff should be sought. They will have accumulated considerable knowledge of which species and strains of acceptable quality are available and will know what sources should be avoided at all costs. A comprehensive worldwide guide to the sources of animals and their health status can be found in the *International Index of Laboratory Animals* (Festing, 1980).

A wide selection of species and strains is now available from reputable commercial breeders and suppliers. In the UK many of these are now registered under the Laboratory Animal Breeders Association Accreditation Scheme (LABAAS). The scheme ensures that any breeder or supplier applying for accreditation has undergone a thorough inspection of his premises (and periodic reinspection thereafter) to ensure that his facilities, husbandry, management and health monitoring comply with the prerequisites laid down, depending on the category of accreditation applied for. The breeder or supplier must ensure that his animals undergo microbiological and serological screening for specified organisms depending on the class at the intervals stipulated (although this may be more frequent by choice). The details and results of the health monitoring scheme must be made available to the customers on request and there is a list of infections which must be reported to them. Outside the UK, the procurement of animals is variously addressed in national guidelines, including where still practised the procedure for the acquisition of pound animals (collected strays). In the UK, all laboratory animal supply will be regulated by the 1986 Act.

If a particular breed or strain is not available locally it may be necessary to import from a reputable source abroad. In the UK these animals will automatically be subject to The Rabies (Importation of Dogs, Cats and Other Mammals) Order 1974 (As Amended), the regulatory authority being the Ministry of Agriculture, Fisheries and Food (MAFF). Regulations will vary from country to country but the aim of confining imported disease — especially rabies — will be common, so that it may be useful to examine in detail the situation in the UK.

Most animals must be kept in quarantine for six months even though they may have been bred for experimental use in a high-grade facility; very strict adherence to the rules must be observed. Animals can only be housed in accommodation approved by the Ministry. This is often a room or rooms within the facilities where the animals will ultimately be used, but they will be subject to periodic inspection by MAFF veterinary inspectors to ensure compliance with the Act. A barrier must be erected to prevent the escape of animals and this is often elaborate and troublesome. Each batch of animals arriving must be separately housed; if batches are mixed in the same room the quarantine period for all the animals commences again from when the last batch of animals came in. Monthly returns have to be made of a head count of all the animals (including any bred on the premises), and the circumstances of any deaths noted. Any suspicion of rabies must be reported immediately. No animals may be moved from the approved premises without prior consent of the MAFF inspector. If this is agreed, any special precautions laid down must be followed. This places obvious limitations on the research worker, who has the choice of waiting for the quarantine period to end, working inside the premises or adhering to the agreed procedure for moving the animals and handling their tissues or body fluids.

Research workers may be tempted to bring in animals from other units because they have contacts who are willing to supply the animals or because they wish to follow through a line of research involving a particular strain.

Where this is proposed full details of the animals must be given to the person in charge of the animal facilities so that they can be evaluated. Often this can and must end in refusal to accept the animals despite their interest from a research point of view. The very best that can be offered is a rederivation programme ('cleaning up'): delivering the young by hysterectomy or Caesarean section and rearing them in a clean environment (isolator or barriered area) by hand or using established foster mothers (Owens and Berg, 1981). This will inevitably result in delay before sufficient numbers of animals are available for use.

Diseased animals are most certainly not suitable subjects on which to base experimental work. Often there will be the frustration of animals dying before the end of the experiment or becoming so sick that they have to be destroyed. Curing sick animals is in the long term disappointing. It is often a better bet to put the animal down and replace it. Even if the animals survive, the results obtained must be regarded with suspicion. Only by prompt, decisive action and cooperation from all parties can disease control be effective. In many cases the junior animal technician is the first to notice if an animal is sick. This may be because of clinical signs but is often just an indefinable 'feeling' that the animal is not its normal self — often subtle changes in stance and behaviour. Sometimes this is a false alarm, but it can be an early warning of illness and should never be ignored. Any symptoms that cannot be accounted for as part of the procedure must be treated with suspicion until proven otherwise. If there is a veterinary surgeon who is responsible for the health and welfare of the animals they or other appropriate specialists should be consulted without delay to undertake the necessary examinations, microbiological testing, clinical chemistry, haematology and necropsies, etc. If these services are not available in-house, steps should be taken in consultation with the senior personnel in the animal unit to arrange for them from outside sources. If it is deemed necessary to isolate specific groups of animals or a whole room or rooms, research workers must be prepared to abide by the restrictions even if this means suspending work until the situation is clarified.

If bringing in animals of unknown health status is unavoidable, sensible precautions must be planned before the animals' arrival. Isolation of the animals in a purpose-built area would be the ideal solution. If this cannot be done, at the very least the animals should be housed as far away as possible with no access by any members of staff in contact with the other animals on site. A worthwhile additional precaution would be to house the animals in a walk-in plastic tent of negative air pressure relative to that in the room, so that the flow of air is always into the tent. An alternative system is a negative-pressure isolator. Animals may also be housed either in individual cages with a disposable filter cover incorporated in the lid or a number of cages can be supported in a laminar air flow cabinet. Both systems will help to contain the possible spread of airborne infections.

A mixture of attention to cleansing and sterilization routines and commonsense by all users of the animal facilities will go a long way to control the ingress of disease and to contain it if it should get in. Much of the specialized cleansing of the equipment and premises will be undertaken by the technicians and animal care personnel. However, attention to personal hygiene and compliance with the procedures laid down by all those involved will do much to ensure that these efforts are not in vain.

WATERING

Water must be available to all animals at all times unless it becomes necessary to restrict it as part of the experiment. This should be a carefully monitored rarity and in the UK requires to be mentioned in detail to the Home Office. Sometimes, as part of the pre- and postoperative care of the animals it may be necessary to restrict their water intake for short periods for their own sake.

Water is either taken straight from the tap or may undergo some form of sterilization such as filtration or chlorination before it reaches the animals. It is usually presented in open bowls or troughs (with the problems of wastage and fouling), inverted glass or plastic bottles or automatic water systems with a drinking valve protruding into the cage. Bottles come in a selection of sizes, and the animal obtains water through a purpose-built spout of hard plastic, glass or metal. Bottles are usually suspended in a frame attached to the outside of the cage, or as an integral part of the cage lid. Most species will drink readily from these, even those that usually lap, but care should be taken to ensure that young or sick animals can reach the spout. The spout should not be too near the floor of the cage; mice and rats will often back their bedding up against the spout and cause spontaneous emptying of the bottle into the cage which, with a full bottle, can have disastrous effects on the occupants.

Various automatic watering systems are now on the market. This is certainly a convenient method but does need constant attention to ensure that valves are working properly and the system remains clean. There is the disadvantage that if a valve should leak unnoticed into a solid-bottomed cage, the occupants can be drowned. It is therefore more safely used in cages where the water can escape even though the animals may get wet: it is possible to buy automatic systems where the valve is accessible but outside the cage. Guinea pigs will constantly play with the valves and should most certainly be housed in a cage such as this. Automatic systems also have the disadvantage that substances cannot easily be added to the water supply, nor can fluid consumption be monitored. Measurement of individual water consumption is anyway not possible in animals maintained as a group. However, treated sensibly, this is an extremely efficient and labour-saving way of watering animals. It does mean that when replacing cages on a rack or shelf care must be taken to ensure that the valve is correctly repositioned.

Observation of the fluid intake is a useful indication of the animal's condition. Often an animal that is not drinking will not eat. If the water bottle remains full there are several possibilities:

The animal cannot reach it
The water spout is blocked
The animal is sick or dead.

Similarly, if the bottle is unexpectedly empty it may mean that:

The animal is unusually thirsty, which in turn might be indicative of ill-health
The environmental control has failed (too hot)
That it has simply spontaneously run out, in which case the cage and bottle will need changing.

Although these conclusions may seem obvious, the important thing is that because they noted that the water level was unusual someone went to look at the animal.

The amount of fluid drunk per day per species is given in several reference books (e.g. Inglis, 1980). For example, an adult rat weighing 250 g will drink 24–35 ml of water a day.

As much attention needs to be paid to cleaning the watering system as to hygiene in general. Water bottles and tops should be washed regularly in a suitable detergent cleaner although it is essential that all traces of this are removed before presenting water to the animal. It is possible to buy mechanical cleaning devices for this purpose. To minimize cross contamination the same bottle should be returned to the same cage if it is just being 'topped up'. To make this task easier a commercially made portable water dispenser is available.

FEEDING

Diets fed to laboratory animals are usually compounded to well-established formulas by commercial animal feedstuffs manufacturers. Usually the diet takes the form of pellets or cubes. Powdered diets (unless required for a specific purpose) are readily wasted and fouled by the animals and could result in unfair distribution to the animals low down the pecking order. Wet mashes are more readily accepted but are labour-intensive and sour quickly. Pelleted or cubed diet is therefore the most acceptable form to animals and workers alike, and goes some way to avoid the deliberate selection of a favourite ingredient by animals.

They are prepared by steam being injected into the diet, which is then forced, under pressure, through metal dies of a size appropriate to the species for which the diet is intended. The consistency of the pellets is important (Ford, 1977). Too friable, and much of it will be broken in the bag as well as enabling the animals to pull whole cubes or large bits through the bars. Mice, rats and hamsters are very adept at making nests entirely from diet! This is also extremely costly and wasteful and can make a nonsense of feeding trials.

The diet is usually supplied to the animal in a hopper or basket suspended on the outside of the cage or in a special compartment incorporated in the cage lid. This minimizes wastage if the hoppers are not overfilled and ensures a regular supply of good quality diet for the animals. Sometimes, rabbits will develop the annoying and wasteful habit of continually scratching their food out of their hoppers, a vice all too readily copied by other rabbits.

Although diets are manufactured to a nutritionally balanced formula compiled for each species, this can only be relied upon provided the diets are stored correctly and used before the 'expiry date' — in temperate countries usually six months from the date of manufacture. They should be kept in a cool, dry, sunlight-free area to ensure minimum destruction of the nutritional value of the diet before it reaches the animal.

In formulating the diet allowances must be made for the loss of nutrients due to such factors as storage, sterilization and the interaction between individual components. Diets formulated from natural ingredients may be contaminated with organisms potentially pathogenic to laboratory animals. Although one of the processes involved in manufacture is pasteurization, this may not be sufficient for SPF and germfree animals, where even minor contamination could have profound effects. Exposing the diet to gamma radiation at the rate of 2.5–4.0 megarad has been proven to produce sterility without significantly affecting the nutritional quality by, for example, destroying proteins or vitamins. Arrangements for the diet to be irradiated are now part of the normal service available from diet manufacturers. However, there is an additional charge for this resulting in a significant increase in the overall cost of the diet. It is expensive and futile to feed conventional animals irradiated diet.

The nutritional requirements of the most commonly used laboratory species are well documented and discussed in Chapter 12. Perhaps it is worth mentioning one or two idiosyncrasies. Guinea pigs, along with man and other non-human primates, cannot synthesize vitamin C (ascorbic acid) in their own tissues and therefore require a daily intake. This can either be fed in the form of supplements such as fruit and vegetables, added to the drinking water or fed in the diet, which has been fortified by the addition of extra ascorbic acid at the time of manufacture. This need is particularly important in pregnant and lactating females.

SPF and germfree animals do not have a normal gut flora and can lack the bacteria responsible for the synthesis of some vitamins, notably K and the B complex, and care must be taken to ensure that sufficient levels are supplied in the diet.

Some animals such as rats, mice, rabbits and guinea pigs practice coprophagy. This is the selection of specially formed faecal pellets which are removed directly from the anus by the animal and reingested. Coprophagy is often wrongly regarded as casual eating of faecal pellets but in fact it is an essential part of the animal's physiology which ensures the conservation of water and the B vitamins, and improved utilization of protein. Interference with coprophagy by deliberate confinement or by alteration of the normal gut flora as with Caesarian derivation can have profound effects. Some antibiotics will interfere with normal gut flora and treatment with these compounds should not be undertaken lightly.

It is thought that an imbalance in the ratio of some minerals such as that of calcium to phosphorus can lead to kidney complications in the rat (Clapp, Wade and Samuels, 1982; Harwood, 1982; Meyer, Blom and Søndergaard, 1982).

Animals are usually fed *ad libitum*, i.e. food is available at all times to the animal. This is convenient and gives the reassurance of knowing that food is always present. However, it runs the risk of producing obese animals which eat in excess of their bodily requirements, particularly when there is no opportunity to exercise. Boredom may also be a significant factor. It has been shown that maintaining rats and mice on a restricted food intake can reduce the incidence of tumours (Roe, 1981; Coneybeare, 1980).

Any feeding regime that changes the incidence of natural tumours could be of significance in toxicology studies and the assessment of carcinogenic effects of compounds.

Feeding supplements will probably be necessary only when dealing with the larger species, particularly monkeys. Although pelleted diet is fed this is usually accompanied by a selection of fruit and vegetables — partly to alleviate boredom and to reinforce some dietary requirements. In addition vitamin preparations may also need to be fed, for example New World monkey's requirements for vitamin D3.

Occasionally animals simply decide to stop eating even though there are no clinical signs of why this should happen. Rabbits often do this and need tempting. Introducing hay (autoclaved if necessary) can provide the stimulus to start eating. Nutritionally it can have little value, particularly if it has been rendered sterile, but perhaps it provides sufficient bulk for the gut to work again.

Postoperative animals will sometimes need encouragement to regain their normal appetite. Cats, for example, will often find food with a strong smell, such as pilchards, an incentive to eat. It may be considered an advantage to feed dogs bones to discourage the formation of plaque on their teeth and eliminate the time-consuming problem of descaling.

Feeding supplements must always carry with them the risk of possible contamination. To treat them beforehand may interfere with the nutritional benefits and the very reason why they are being fed, or may render them unacceptable to the animal. Obviously the answer is to avoid supplements where possible, but sometimes the risk has to be taken.

ENVIRONMENT

Much effort has gone into producing animals that are genetically and microbiologically uniform. This standardization produces animals that are less variable in their response. The number of animals required in an experiment is thus reduced because of the reliable and reproducible results obtained. Of equal importance is the environment within which these animals are kept.

The most commonly used laboratory animals are mammals and because they are homoiothermic they are capable of reacting and adapting to changes in their immediate surroundings, such as the everyday variations in temperature, light, humidity, air movement and noise. This can manifest itself in physiological and behavioural traits, for example change in activity patterns, metabolic rate, body temperature regulation, breeding performance. These responses may affect experimental findings so that it is desirable to stabilize the environment to minimize these unwanted variations.

The animals' environment is twofold: the macro environment — the conditions within the room housing the animals — and the micro environment — conditions within the animals' most immediate environment, i.e. the individual cage. Although it is comparatively easy to satisfy oneself about the former, the latter is not always taken into account and can reveal a very different set of variables. The cage size, material and flooring, provision and nature of nesting material, stocking densities and position of the cage on the rack or shelf can all create very different environments within the same room (Raynor, Steinhagnen and Hamm, 1983).

The more closely controlled the environment, the more sophisticated the equipment and the more elaborate and costly the monitoring, maintenance and emergency back-up systems have to become. There are no half measures — you either have good environmental control or you do not. Installation of equipment unsuitable for the demands that will be made on it will only result in frequent breakdowns in the system and the possible consequence of this on experimental work.

The factors which constitute the animals' environment will be discussed briefly below. A more detailed account can be found in Chapter 5.

Heating and ventilation

These are two of the most expensive systems to install and operate. The temperature range recommended for a species is usually within its thermoneutral zone, that range of temperature over which heat is neither lost nor gained by the animal. Experimental situations may cause the animal temporarily to lose its ability to regulate its body temperature (e.g. the administration of certain drugs or compounds or a postoperative state) and it is therefore very important that it is kept at a suitable and constant ambient temperature. Within reason, the more finely tuned the temperature control the better (± 1 or $2°C$). Incoming air can be heated to the desired level electrically or by hot water heat exchangers or steam coils. The choice of method will depend upon individual circumstances.

The volume of air changed per hour in the room will be decided on the size of the room and the total number of animals, in other words its maximum holding capacity. This normally falls between twelve and 20 air changes per hour but will vary according to the usage of the room and may be increased to around 140 air changes per hour in a linear air flow room. To reduce costs it is possible to run a system where a certain percentage of the air is recirculated, although this can pose certain risks and is not usual.

The positioning of the inlet and outlet grills must be so that air is evenly distributed around the room and to each cage irrespective of position. Those on the bottom rows require particular attention.

Light

This is probably the easiest and cheapest aspect of the environment to control. The light and dark period can easily be regulated automatically for each room.

The most common light cycle is twelve hours light and twelve hours dark. This seems acceptable for most species whether experimental or breeding. The time clocks used for switching on and off can be set to vary the light and dark periods according to need, even to reverse night and day. Some laboratories have a low 'daylight' level of lighting which is increased only when staff are working in the room. Sometimes a low intensity of light, usually red, is provided, which will permit some work by staff without it apparently being noticed by the animals.

Light intensity is important especially to albino animals, where if too high it can result in retinal damage. Uneven exposure can also occur between the animals closer to the light source (usually the top rows) and those lower down. Lighting level must obviously be sufficient for staff to work comfortably and safely but may have to be a compromise to ensure that it does not adversely affect the animals who are subjected to these conditions every day, maybe over several years.

It is important to control lighting for breeding animals to keep them in breeding condition, for example regular oestrous cycles in the females. Most laboratory animals will breed throughout the year given the correct conditions and freedom from external seasonal variations in temperature and daylight hours.

Humidity

On the whole this is not as critical as the other environmental factors. Not very much is known about the effects apart from one or two species. In rats, low humidity levels are held to be responsible for a condition known as 'ringtail' — not to be confused with ringworm, which is parasitic in origin. Young rats are the most susceptible, with constricted 'rings' appearing down the skin of the tail which can sometimes result in the complete tail sloughing off. It is thought to be associated with the inability to control heat loss in environments with 40 per cent or below relative humidity.

Before the days of much attention to the animals' environment, you could always tell when an animal house had low humidity problems by their 'manx' rats. However, this is not a simple relationship as ringtail nowadays is hardly ever seen whatever the humidity control. For monkeys from the New World who are used to living in equatorial forests, high humidity levels can be important but this decreases as they become more conditioned to and breed in laboratories. Even so it is advisable to keep the levels at around 50–70 per cent.

Air conditioning

In the UK air conditioning has been spoken of as a luxury because as a temperate country we rarely achieve climatic extremes. However, this view is less common than it was and today it is thought that air conditioning may have a more significant role in animal maintenance than had been supposed. Certainly it can ensure more comfortable and arguably more alert staff. In some other countries air conditioning is an accepted feature of environmental control.

Smell

Smell plays an important role in the animals' relationship to one another, not only as a means of knowing other animals are present but as part of their breeding behaviour. Pheromones are odours produced by both sexes which can affect the development, reproduction and behaviour of other members of the same species and are vital to running a successful breeding colony. They serve to identify for the male those females that are in oestrus, i.e. ready for mating. Male rodents can modify or regulate the oestrous cycle. Female mice housed together, isolated from the male, will soon display irregular and extended oestrous cycles. This is easily rectified by the introduction of males to the near vicinity. Parkes and Bruce (1961) and Bruce (1962) demonstrated in mice the phenomenon of pregnancy blocking: placing a female that had been successfully mated with a different male caused the pregnancy to terminate.

Noise

Constant background noise is thought to be less traumatic for the animals than sudden intermittent ones, especially if high frequency. For this reason it has been suggested that playing music in the animal rooms deadens the sometimes unavoidable bursts of noise. In some cases where this was implemented it caused controversy amongst the staff, with disputes on the type of music that should be played and who should be the DJ. Sudden noises are caused by people and can be reduced by them. Rushing into an animal room and clattering around with cages, especially if they are metal, can be avoided with a little care and discipline. Thought should be given to apparatus used in or close to animal rooms, which can often emit ultrasonic sound. Although that is by definition beyond our hearing range, it might well be very distressing to the animal and could explain unusual and unexpected behaviour.

Establishing and maintaining a successful animal facility is not just a question of luck or fluke. It requires thought, care, hard work and expert knowledge from the animal unit personnel. Often, because a unit is running smoothly, it is taken for granted by the users, who do not always appreciate or indeed give a passing thought to the delicate balance that makes the difference between a successful and a ruined service. Perhaps that is to be expected, but users must keep in mind how easily thoughtless behaviour on their part could upset it. To maintain this balance requires cooperation from all. A good working relationship between animal staff and research workers of all levels is essential to this end.

REFERENCES

Baker, B. A. and Morris, G. F. (1980). A new crush-cage with rotating squeeze-back for restraining large primates, *Laboratory Animals*, **14**, 113–15.
Bruce, H. M. (1962). The importance of the environment on the establishment of pregnancy in the mouse, *Animal Behaviour*, **10**, 3–4.
CCAC (1980, 1984). *Guide to the Care and Use of Experimental Animals*, Vols 1 and 2, Canadian Council on Animal Care, Ontario.
Clapp, M. J. L., Wade, J. D. and Samuels, D. M. (1982). Control of nephrocalcinosis by manipulating the calcium: phosphorus ratio in commercial rodent diets, *Laboratory Animals*, **16**, 130–2.
Coneybeare, G. (1980). Effects of quality and quantity of diet on survival and tumour incidence in outbred Swiss mice, *Food & Cosmetic Toxicology*, **18**, 65–75.
Danscher, G. (1972). An instrument for immobilization of small experimental animals, *Zeitschrift für Vesuchstierkunde*, **14**, 69–71.

Festing, M. F. W. (1980). *International Index of Laboratory Animals*, 4th edn, Medical Research Centre, London.
Ford, D. J. (1977). Influence of diet pellet hardness and particle size on food utilization by mice, rats and hamsters, *Laboratory Animals*, **11**, 241–6.
Gärtner, K., Büttner, D., Döhler, K., Friedel, R., Lindena, J. and Trautschold, I. (1980). Stress response of rats to handling and experimental procedures, *Laboratory Animals*, **14**, 267–74.
Harwood, J. E. (1982). The influence of dietary magnesium on reduction of nephrocalcinosis in rats fed purified diet, *Laboratory Animals*, **16**, 314–18.
Hearn, J. P. (1977). Restraining device for small monkeys, *Laboratory Animals*, **11**, 261–2.
ILAR (1985). *Guide for the Care and Use of Laboratory Animals*, revised edn, prepared by the Institute of Laboratory Animal Resources Commission on Life Science, National Institutes of Health, Bethesda.
Inglis, J.K. (1980). *Introduction to Laboratory Animal Science and Technology*, Pergamon Press, Oxford.
Lantbruksstyrelsens författningssamling (1982). Veterinära bestämmelser: Vb 12, LiberTryke, Stockholm.
Lennox, M. S. and Taylor, R. G. (1983). A restraint chair for primates, *Laboratory Animals*, **17**, 225–6.
Meyer, O., Blom, L. and Søndergaard, D. (1982). The influence of minerals and protein on the nephrocalcinosis potential for rats of semisynthetic diets, *Laboratory Animals*, **16**, 271–3.
Owen, J. A., Tasker, R. A. R. and Nakatsu, K. (1984). A simple, less stressful rat restrainer, *Experientia*, **40**, 306–8.
Owens, W. E. and Berg, R. D. (1981). Derivation of a breeding colony of germ-free athymic mice by caesarean section and foster nursing, *Journal of Immunological Methods*, **42**, 115–19.
Parkes, A. S. and Bruce, H. M. (1961). Olfactory stimuli in mammalian reproduction, *Science*, N.Y., **134**, 1049–54.
Raynor, T. H., Steinhagen, H. and Hamm Jr., T. E. (1983). Differences in the microenvironment of a polycarbonate caging system: bedding vs raised wire floors, *Laboratory Animals*, **17**, 85–98.
Reigle, R. D. and Bukva, N. F. (1984). A method of restraining rats for intravenous injection using a flexible plastic bag, *Laboratory Animal Science*, **34**, 497.
Roe, F. J. C. (1981). Are nutritionists worried about the epidemic of tumours in laboratory animals? *Proceedings of the National Society*, **40**, 57–65.
Rucklidge, G. J. and McKenzie, J. D. (1980). A new metabolism cage suitable for the study of mice, *Laboratory Animals*, **14**, 213–16.
Sadjak, A., Klingenberg, H.G., Egger, G. and Supanz, S. (1983). Evaluation of the effects of blood smelling, handling, and anaesthesia on plasma catecholamines in rats, *Zeitschrift für Versuchstierkunde*, **25**, 245–50.
Schade, R., Petzoldt, R. and Friedrich, A. (1983). Der Einfluβ sozialbiologischer phänomene auf die Konzentration ausgewählter Serumproteine der Ratte, *Zeitschrift für Versuchstierkunde*, **25**, 251–62.

Laboratory Animals: An Introduction for New Experimenters
Edited by A. A. Tuffery
© 1987 John Wiley & Sons Ltd

CHAPTER 7

Hazards and Safety Aspects of Animal Work

M. W. SMITH
Central Animal Services, University of Cambridge

INTRODUCTION

Variations in the construction, environmental control, staffing, types of work and species of animals used usually make it difficult to lay down mandatory safety instructions applicable to all animal units. It is, however, useful to provide guidelines giving an indication of the goal towards which animal users should work in the maintenance of safe working conditions.

The general hazards inherent in work in an animal unit can be grouped together under four headings.

Physical damage

This includes bites, scratches, bruises and abrasions (from both animals and equipment). All such accidents and injuries should be considered as potential sources of infection and treated as necessary, while safe animal handling methods should be practised to avoid injury.

Accidental Infection by Micro-organisms

Animals, animal tissues and secretions are reservoirs or potential reservoirs of organisms which may be harmful to man. These may gain access to the human body through the skin (especially where abraded) or via the mouth, eyes or respiratory system. Some animal parasites are also capable of living on or in man (e.g. fleas, roundworm larvae and *Toxoplasma*) and all laboratory animals and their environment should therefore be treated with the respect and care due to a potentially infectious agent.

Special metabolites

Bites or stings from insects or venomous animals may introduce anticoagulants, venoms or other foreign metabolites into the body and produce extreme physiological or toxic reactions. Contact with animal urine, skin debris and fur may also produce asthmatic attacks or other allergic reactions in some individuals. Aerosols disseminating potentially toxic or carcinogenic materials may be produced by routine cleaning procedures, dilution of chemicals and general animal work.

Non-specific hazards

Under this heading are considered those hazards commonly found in, but not inherently exclusive to, animal units. These include radioactive materials, dangerous drugs, surgical and X-ray equipment and many other items frequently used in experimental work.

PREVENTIVE MEASURES

In considering the specific hazard groupings much can be done to prevent accidental injury or infection. Systems of work can be designed to allow for the separation of clean and dirty or dangerous procedures and for the provision of methods of adequate protection, sterilization and disinfection. Procedures for the administration of substances, withdrawal of body fluids and minor surgery should be clearly defined and adequately practised in advance.

Cleaning methods and experimental methods should be arranged to prevent excessive dust or aerosol particles which could carry infectious agents or allergens. Proper personal hygiene is essential, whilst care should be taken to ensure effective separation and disposal of potentially infectious waste materials, for example by incineration or autoclaving. Where such methods of disposal are not thought necessary the use of sealed leakproof heavy-duty plastic bags may be considered as an alternative or adjunct to existing facilities.

Where it is intended to work in defined areas of biological hazard, for example with radioactive materials or X-ray equipment, advice should be sought where necessary from radiation protection staff, the departmental safety officer and those skilled in the use of such equipment before embarking on a research procedure.

In the handling and use of animals themselves no one should begin such work unless competent to do so except under supervision. Without such safeguards, animal users are exposed to risks of accident or injury due to improper handling. Where possible a trained laboratory animal technician should assist in experimental procedures using animals.

Protective clothing

Appropriate protective clothing should be worn at all times depending on the type of work being carried out. This might consist of a simple protective overall but could also include some form of footwear or overshoes not normally worn elsewhere. Under certain circumstances it may be necessary to consider the use of safety boots. Protective gloves might be necessary and also (depending on the unit, the species of animals being used and the type of work) masks, goggles, visors or caps, etc.

Protective clothing used in an animal house should be removed on leaving the unit and should be disinfected where appropriate or laundered regularly and (if disposable) bagged and sent to an incineration centre.

Animal health

Sickness in the laboratory animals themselves also has potential safety implications and the member of staff in charge of the unit should always be informed if any form of illness unrelated to the research procedure occurs.

The prevention of infection in laboratory animals is also of importance to the research worker quite apart from the need to maintain meaningful experimental results. Thus no attempt should be made to introduce animals from sources other than those approved by the unit manager. All normal purchases should be made through accredited breeders or approved suppliers.

Farm animals

There is a tendency when considering laboratory animals to think only of small mammals. However, large farm animals are not infrequently brought into experimental use and these also can be dangerous and difficult to handle.

Certain special classes and sexes of stock, for example breeding males, females with young at foot, animals which are sick and/or separated from their fellows, animals newly arrived on strange premises, animals recovering from anaesthesia, thoroughbred horses, etc., are unpredictable in their behaviour and should be handled with special care.

Wherever farm animals are in use the usual animal attendants should be present to assist in the restraint, especially since some animals are apprehensive of strangers and are likely to react violently to sudden movement or noise.

No one should attempt to handle farm animals unless adequately trained to the necessary handling techniques. The handler should check before handling the animal that the restraint equipment is suitable for the work and is in good repair. Helpers should also be available and know what to do.

Bulls, stallions, rams and boars should only be handled when assistance is available. No person should enter a bull pen, stallion box or boar pen alone and when doing so a means of quick exit should have been decided on in advance.

Out of normal working hours, the handling of large animals (by those unfamiliar with them) when help is not at hand to deal with unexpected emergency and accident needs special arrangements.

Other important safety factors on farm-type premises include the prohibition of smoking (especially near food stores and hay barns) and the proper storage in a safe and secure manner for drugs, veterinary instruments and handling equipment.

Good lighting is essential, especially for winter or evening work, while attempting to approach heavy animals in dark or ill-lit boxes or pens is an extremely hazardous procedure. Even a normal contented animal has sufficient weight to knock down a would-be handler who is unaware of the direction in which the animal is moving because of poor light.

Primate units

Simians offer special problems because they are one of the most dangerous animal sources of human infection. Those expecting to work with simians should receive a proper course of instruction on the manipulation of these animals before such work is undertaken. Animal users should be aware of the major zoonoses potentially present in wild-caught simians, including: B virus (herpes virus simiae), Marburg virus, rabies, tuberculosis, salmonellosis, shigellosis, amoebiasis.

Although it is not normally the responsibility of the animal user to do so, care should be taken to ensure the proper isolation and screening of newly introduced animals before they are brought into experimental use.

Even when an animal is considered 'safe' to use, experimental workers should be aware of the possibility of undiagnosed or opportunist infection and treat all simians with the respect due to potential carriers of zoonoses. Whenever necessary simians should be handled under sedation using ketamine hydrochloride as the drug of choice.

In the event of being bitten or scratched by a monkey or injured by glassware containing monkey tissue, first clean the wound and make it bleed, before applying a dressing. Notify the person's general practitioner and ask him/her to keep the patient under observation for three weeks. Arrange for the animal concerned to be anaesthetized and examined for mouth or other lesions.

Infective animals or animal tissue

These should always be dealt with in some form of isolation unit or flexible film isolator appropriate for the work concerned and in cages or containers properly labelled.

There should be adequate provision to allow staff to follow a high standard of personal hygiene.

It may be necessary to screen staff for suitability before being allowed to do the work. Only authorized staff should use the facilities and all such persons should be able to recognize the signs and symptoms of any agent in use which is infectious to man. Where preventive immunization is possible this should be carried out in advance and all those in regular contact with animals should be covered by at least an antitetanus immunization.

Special provisions will be necessary to allow for essential repairs and if the working unit is barriered it will be necessary to arrange some form of easy communication system.

Procedures concerning general use, entry and exit, medical aid and protective clothing should be clearly laid down and understood.

Special arrangements will be necessary for the disposal of waste and all rules should be obeyed regardless of the status of the person concerned.

Anaesthesia and euthanasia

The use of general and local anaesthetics should be restricted to competent personnel as these substances can be dangerous unless used correctly.

Where possible, antidotes should be available (with instructions for use) in case of accidental administration to those using the material.

Postmortem procedure

Postmortem examinations of animals which die unexpectedly should always be performed in a room designated for this purpose and not in an experimental area.

The work surfaces used should be easy to decontaminate and wooden tables should not be used. Where potentially dangerous material is involved use should be made of adequately ventilated safety cabinets.

Disposal of carcases and animal tissues

Carcases of animals should be placed in a leakproof plastic bag. After sealing this should be placed in a second bag, then sealed in a leakproof manner.

If the carcase is to be incinerated in the same building or area no further precautions are necessary. If it is necessary to transport the carcase for some distance to the incinerator, it should be placed in its two covering bags in some impervious container or bin before leaving the original site.

If the carcases are from otherwise normal adults the use of a macerator may be acceptable.

Tissues and specimens for laboratory examination must be placed in a suitable labelled container and if not killed and fixed at the time should be appropriately labelled and designated as such.

It should be noted that there are certain organisms which are not killed by formalin fixation. Where bagged carcases cannot be disposed of immediately they should be placed in a cold room at 4°C in covered bins.

Special care is needed in storing carcases of animals killed with ether. Considerable quantities of ether may be absorbed by the animal tissues — enough to form an explosive mixture with the air inside a closed refrigerator which can be ignited by an electrical spark, and cause a serious explosive or fire hazard. Even putting such carcases into the incinerator can be dangerous, and for these reasons the routine use of ether for this purpose is not to be recommended.

GENERAL HEALTH AND SAFETY REQUIREMENTS

Medical examination may be considered necessary prior to employment for those employed full time in an animal unit and thereafter as required. An agreement to accept prophylactic measures may also be needed depending on the work being done.

It may be considered prudent for those working with animals to carry a card identifying their work. The cards could also carry a contact telephone number and details of any prophylactic medication carried out.

Concurrently with this it is advisable to consider the necessity for a range of routine procedures including the antitetanus injections already mentioned and also BCG, and chest X-rays.

Finally, apart from normal routine good hygiene, no one should eat, drink or smoke in an animal unit except in the special rest area provided.

First aid

In the event of accident, injury or ill-health animal users should seek the advice and help of a trained first aider who has access to a first aid box, the contents of which have been kept up to date.

Instructions for artificial respiration are usually prominently displayed and animal users should make themselves familiar with such procedures.

All accidents and injuries should be reported and entered on the appropriate registers and report forms.

Those working in an animal unit should be aware of the position of the nearest telephone, which should have nearby the names and telephone numbers of those officers designated to carry out specific safety functions. These include the unit safety officer, radiation protection officer, health officer and fire officer.

If necessary there should also be the name and telephone number of a specialist medical adviser familiar with any specific infectious hazards in that particular unit.

Fire precautions

Animal users should make themselves familiar with the fire precautions in the animal unit. This will include the means of emergency exit and the arrangements for sounding the fire alarms. The position and use of fire appliances should also be noted and any special instructions to be carried out in the event of fire.

Fire authorities will normally attempt to isolate a fire and will aim to save human life first. No attempt should be made to move animals unless the person in charge who is familiar with the animals concerned is present.

Firemen will not enter isolation units where work on dangerous pathogens is being carried out except for emergency rescue of personnel. They will also not move or open refrigerated containers containing dangerous pathogens. Such containers should be identified with fireproof materials.

General

Regulations controlling general activity within an animal unit are useless without a high degree of personal commitment on the part of the animal user.

Rules are of little use without a sense of responsibility on the part of each individual, whose actions may affect not only his own safety but that of others.

Safety policy statements

Recent legislation in the United Kingdom has made it mandatory for certain procedures to be followed with the aim of ensuring safe working conditions for everyone at his place of work. These requirements apply to laboratory animal facilities as much as to any other workplace and part of them can be fulfilled by setting up safety policies. Workers in other countries should familiarize themselves with the appropriate state or national legislation and ensure that any procedures required are set up and followed. Where there are no such legal requirements, the following account of British practice may suggest some approaches for local application.

Apart from a general Code of Safety Practice and under Section 2(3) of the UK Health and Safety at Work Act 1974, the person in charge of a working unit has a duty to ensure that his department (insofar as is reasonably practicable) is a safe and healthy place to work in. There is a similar duty to those not employed by the department but who may from time to time have an accepted reason for being present on the premises, for example research workers, students, other technicians and other visitors.

Such duties are acknowledged in the form of a policy statement which outlines how the organization of the department will be arranged to ensure such safe working conditions.

Notwithstanding the responsibilities of the head of department, it is also incumbent upon employees and others using the facilities to carry out any procedures required to maintain healthy and safe working conditions and to instruct any others working under their supervision in safe working practices.

Staff are expected to cooperate with the head of department and to act in a personally responsible manner in accepting that such cooperation is a continuing obligation. The obligation implies a willingness to seek competent advice when necessary and to take care when initiating a piece of work.

It also indicates a necessity for the head of department to impose restrictions on the time, place and method by which specific procedures will be carried out without creating too many obstacles to the efficient conduct of research.

The animal user is therefore legally required to cooperate in the arrangements approved by the head of department and his safety committee. These arrangements, which may include restrictions on entry to the building or to specifically defined areas of the building designated by recognized hazard warnings, should not be construed as an interference with academic freedom, and ignored.

It should be understood that in any animal unit there has to be a high degree of cooperation and courtesy between those employed there and those who carry out research, who are almost certainly not the employees of the animal unit curator but who are nevertheless required to comply with the rules and in return receive safe working conditions and facilities.

LEGAL CONTROLS AFFECTING THE USE OF LABORATORY ANIMALS

The major legal control affecting the use of experimental animals in Britain is the Animals (Scientific Procedures) Act 1986 (described in Chapter 2), and the major legislation concerned with safety matters is the Health and Safety at Work etc. Act 1974. Readers outside the United Kingdom should make themselves familiar with their own equivalent legislation where appropriate. Apart from the two acts mentioned, a number of other acts and orders have some implications with respect to safety. These are outlined briefly below.

The Veterinary Surgeons Act 1966

This act replaces most of the previous legislation. Provision is made for temporary registration of persons holding Commonwealth or foreign degrees. Provision is also made for exemption from restrictions under certain circumstances, for example:

Treatment given to an animal by its owner.

Rendering emergency first aid for the purpose of saving life or relieving pain.

The performance of certain minor operations commonly carried out on very young animals, i.e. castration of pigs or calves.

The carrying out of any experiment duly authorized under the Animals (Scientific Procedures) Act 1986.

The performance by a registered medical practitioner of an operation on an animal for the purpose of removing an organ or tissue for use in the treatment of human beings.

The carrying out or performance of any treatment, test or operation by a registered medical practitioner or a registered dentist at the request of a veterinary surgeon.

In 1967 under a separate regulation enacted under Section 19 of the 1966 Act, special provision was made for veterinary students to carry out clinical procedures under the supervision of a qualified person.

The term veterinary surgery means:

1. The diagnosis of diseases in and injuries to animals including tests performed on animals for diagnostic purposes.
2. The giving of advice based on such diagnosis.
3. The medical or surgical treatment or animals.
4. The performance of surgical operations on animals.

The Diseases of Animals Act 1950 and 1975 (Consolidated in the Animal Health Act 1981)

The Act defines the powers to the Minister of Agriculture, Fisheries and Food in relation to animal diseases and also the powers of local authorities.

It enables the Minister to make orders dealing with specified diseases in relation to particular species of animals.

Regulations are also laid down in connection with the importation of animals and the manufacture of therapeutic substances.

Amendments are made from time to time in various agricultural acts. Certain diseases are listed as 'notifiable' and of these a number are specially dealt with in specific orders, for example rabies.

Thus, any work to be carried out on notifiable diseases, on animal diseases not normally found in the United Kingdom and on conditions requiring importation of animals would require prior consultation with the divisional veterinary officer.

Agriculture (Miscellaneous Provisions) Act 1968

Only Part 1 of this act has any bearing on experimental animals and is intended to apply to food-producing animals normally kept under farm conditions.

The Act makes provision for the prevention of unnecessary pain and distress to livestock and includes regulations requiring adequate feeding, housing and husbandry of such animals.

The Medicines Act 1968

The Medicines Act 1968 and consequent regulations control the sale and use of medicinal products — that is any material given to a person or animal which it is expected will be of benefit to them. SI No. 2167, the Medicines (Exemptions from Restrictions on the Retail Sale of Supply of Veterinary Drugs) Order 1977, allows the sale of certain veterinary preparations containing materials otherwise only available on prescription to persons 'known to have animals in their care' without prescription. This allows the person in charge of an animal house to order such products without a countersignature.

The Misuse of Drugs Act 1971

The Act lists and classifies controlled drugs and lays down restrictions on the importation, production, supply and possession of such drugs.

The Act also gives the Secretary of State powers to make regulations preventing misuse and requiring special precautions for safe custody of controlled drugs.

The Misuse of Drugs (Safe Custody) Regulations 1973

Relevant portions are set out in detail below:

> Where any controlled drug (other than a drug specified in schedule 1 to these Regulations) is kept otherwise than in a locked safe, cabinet or room which is so constructed and maintained as to prevent unauthorised access to the drug, any person to whom this Regulation applies having possession of the drug shall ensure that so far as circumstances permit, it is kept in a locked receptacle which can be opened only by him or by a person authorised by him.

Although the Act does not cover sedative drugs such as barbiturates and codeine, these and any similar drugs should also be kept in safe custody to prevent their misuse.

The Misuse of Drugs Regulations 1973, Part III specifies the requirements for documentation and record-keeping and requires departments having controlled drugs to comply with certain requirements which are listed below.

A register must be kept. Entries must be made in chronological sequence as set out in the attached schedules. Each quantity of controlled drugs either obtained or supplied by a department must be entered in the registers.

A separate register or part of a register must be kept for entries of each class of drug and each of the individual drugs within that class, together with its salts. Any preparation or other product containing it or any of its salts must be treated as a separate class. Any stereoisomeric forms of a drug or its salts should be classed with that drug.

Requirements as to registers

Departments required to keep a register under Regulation 19 must comply with the following:

1. The class of drug to which entries on each page of a register relate must be specified at the head of that page.
2. Every entry in the register must be made on the day on which the drug is obtained or supplied.
3. No amendment or deletion of an entry must be made. Corrections must be by a marginal note or footnote and specify the date on which the correction was made.
4. Every entry and correction must be made in ink or other indelible material.
5. The register must be used only for entries concerning drugs.
6. Anyone required to keep a register must, on demand from the Secretary of State's inspectors, furnish particulars of drugs specified in Schedule 2 and obtained or supplied by him or in his possession. He must also be able to produce proper documentation.
7. Every register in which entries are currently being made should be kept at the premises to which it relates.

The regulations also require that all registers and books shall be kept for two years from the date on which the last entry is made.

The Rabies (Importation of Dogs, Cats and Other Mammals) Order 1974 and The Rabies (Control) Order 1974

These two orders control the importation and quarantine of *all mammals* which might introduce rabies.

Powers are also given to control the import and use of rabies virus. In most laboratories this legislation will mainly affect those working with imported wild caught monkeys.

The Diseases of Animals (Zoonoses) Order 1975

The order makes special provisions for the control of organisms of the genus Salmonella and genus Brucella, both of which constitute a human health hazard.

Both organisms are placed on the designated list, but no notification to the Ministry is required where such an organism has been deliberately introduced into an animal or bird for research purposes.

The Diseases of Animals (Approved Disinfectants) (Amendment) Order 1978 and Amendment Orders 1981, 1982, 1983

The Approved Disinfectants Order 1976 lists the disinfectants approved by the Ministry of Agriculture in the control of specific diseases and also for general disinfection. In each case the dilution rate is included.

Further orders are made periodically to update the lists.

REFERENCES

No references have been cited in the text, but the following list contains a selection of the more important sources of information relevant to the topics discussed.

Association of Veterinary Teachers and Research Workers (1977). Precautions in the use of large farm animals for experimental purposes, *Veterinary Record*, **100**, 351.

Clough, G. and Gamble, M. R. (1976). *Laboratory Animal Houses, a Guide to the Design and Planning of Animal Facilities*, Manual Series No. 4, MRC, London.

Collins, C. H., Harley, E. G. and Pilsworth, R. (1974). *The Prevention of Laboratory Acquired Infection*, PHLS Monograph Series, HMSO, London.

Department of Health and Social Security (1980). *Code of Practice for the Prevention of Infection in Clinical Laboratories and Post Mortem Rooms*, HMSO, London.

Health and Safety Executive (1981). *A Guide to the Health and Safety (Dangerous Pathogens) Regulations 1981* (Health and Safety Services Booklet HS(R) 12), HMSO, London.

Laboratory Animal Breeders Association (1982). *The LABA Accreditation Scheme Manual*, LABA, Margate, Kent.

McSheehy, T. (1976). *Control of the Animal House Environment Laboratory*, Laboratory Animals Handbook No. 7, Laboratory Animals Ltd, Theydon Bois, Essex.

Perkins, F. T. and O'Donoghue, P. M. (1969). *Hazards of Handling Simians*, Laboratory Animals Handbook No. 4, Laboratory Animals Ltd, Theydon Bois, Essex.

Seamer, J. H. and Wood, M. (1981). *Safety in the Animal House*, Laboratory Animals Handbook No. 5, Laboratory Animals Ltd, Theydon Bois, Essex.

University of Cambridge (1971). *Precautions to be Observed in the Use of Radioactive Materials and Machines which Provide Ionizing Radiations in the Laboratories of the University*, University Printing House, Cambridge.

Van der Hoeden, J. (1964). *Zoonoses*, Elsevier, Amsterdam, London and New York.

WHO (1971). *Health Aspects of the Supply and Use of Non-Human Primates for Biomedical Purposes*, Technical Reprint Series No. 470, WHO, Geneva.

CHAPTER 8

Animal Behaviour

D. E. BLACKMAN
Department of Psychology, University College, Cardiff

INTRODUCTION

The major purpose of this chapter is to ensure that its readers are fully aware of the importance of the behaviour of laboratory animals. Animals have observable behavioural dispositions and their behaviour can be thought of as *part* of them, in the way that say anatomical structures or physiological processes are part of them. It would be strange to try to conceptualize anatomy or physiology independently of living organisms as something which is somehow different from them. But, it seems to be the case that there is a tendency for us to do this when we think about behaviour, as if it were something separate, a kind of appendage to the organism defined in terms of anatomy, physiology, etc. However, behaviour is a similarly orderly and predictable part of the very definition of an animal. It is impossible to think of living animals without active physiological processes: it is equally inappropriate to think of them without active behavioural processes. Behaviour should be conceptualized therefore as a biological phenomenon in its own right, an important and ubiquitous part of intact living organisms.

It is impossible for us to gain a *direct* insight into the consciousness of animals in order to see how aware they are of their surroundings, to identify their fears and wishes, or to understand their feelings. It is of course equally logically impossible for us to see such things directly in other people, because awareness, aspirations and emotions are essentially private phenomena. Few of us of course would doubt their existence in other people, however. We base our view that they enjoy inner experience on three factors: (1) our intimate knowledge of our own internal mental lives; (2) the self-reports given to us about such things by other people, who use a shared language to communicate with us about their inner lives; (3) observations of how other people behave. The question of animal consciousness is rather more difficult from a philosophical point of view, however, and the nature and existence of it is still widely discussed (e.g. Griffin, 1976). First, we cannot see into a dog's inner life of consciousness by looking into our own in the way that we can do for the species of which we are members. Indeed, we know that animals have different visual, olfactory and other sensory systems, and we can only guess about the impact that these systems alone have on the nature of animal awareness. Secondly, we cannot talk to animals to gain access through the medium of language to their accounts of their consciousness. However, we can see what animals *do*, how they behave. The ways in which their behaviour adjusts to different conditions or is sensitive to different aspects of the environment can form a focus of an objective science. Though some would say that the behaviour of animals can be used to infer the existence of a private and inner world of experience, many scientists are rather more cautious. Few would positively deny the possibility that animals' behaviour reflects consciousness, but because of the problems raised above most scientists prefer to stick firmly to observable data and not make untestable inferences. The nature of animal consciousness is a philosophically difficult concept to address, but there is no special problem in investigating what animals do or what they can do. So scientific studies of behaviour as a naturally occurring biological phenomenon have developed without much emphasis on inner life and experience.

Scientific studies which explicitly set out to investigate the behaviour of animals are carried our principally by psychologists and zoologists. Although no sharp distinction should be made between these two approaches, it is perhaps fair to say that psychological studies tend to emphasize experimental investigations of animal behaviour in laboratory conditions while zoologists tend to stress observational studies of the behaviour of a wider range of species as it occurs in the animals' natural habitats. As a result, psychological studies are often about environmental influences on behaviour while zoological studies (often described as ethology) are more addressed to evolutionary influences on behaviour, though again no absolute distinction should be drawn between the two disciplines in this respect.

Behaviour is also studied scientifically in laboratories as part of several other biological sciences. For example, pharmacologists often set out to assess the effects of drugs on the behaviour of animals or use behaviour as a dependent variable in general pharmacological studies. Similarly, physiologists sometimes use the behaviour of animals to investigate the functioning of the central or peripheral nervous systems. Geneticists sometimes investigate genetic influences on behaviour, endocrinologists sometimes use behaviour as an indication of the activities of the systems in which they are interested, and toxicologists sometimes investigate the cumulative effects of substances such as pollutants on behaviour. The list can be extended throughout biology, emphasizing the fact that behaviour is indeed an intrinsic part of the natural domain of biology.

However, it is worth emphasizing here that the behaviour does not form 'part' of a laboratory animal only when it is explicitly investigated. For example, the routine techniques of animal husbandry such as breeding, dietary provision, cageing and so on may all interact with or disturb the general behavioural dispositions of animals. This may result in effects on the general well-being of animals, and this can in turn be relevant to any scientific study of other 'parts' of animals. Similarly, techniques used routinely in sciences which investigate phenomena other than behaviour can often have an influence on the behaviour of animals, such as the techniques for taking blood for assay, for recording non-behavioural biological data from conscious animals, etc. Influences on behaviour such as those arising from animal husbandry and non-behavioural scientific procedures can lead to animals becoming, by reason of their behaviour, unrepresentative of their species, and this may decrease the chance that results obtained in one laboratory will be replicable in another. Furthermore, uncontrolled capricious variability in such influences can result in large individual differences in the behaviour of animals in a laboratory, leading perhaps to larger variations in the non-behavioural data which can contaminate or obscure main experimental effects. In short, wherever influences on behaviour can act they do, and it is therefore part of good laboratory animal science to be fully aware of behaviour whether it is a principal focus of a particular study or not.

In this chapter some of the variables which affect behaviour will be briefly outlined. The intention here is not to present psychology and ethology as separate biological disciplines, but to review relevant examples of the implications of psychological and ethological knowledge for general laboratory practice. The chapter then turns to a short review of explicit studies of animal behaviour carried out in laboratory conditions, in the hope of providing a general introduction to the principal experimental techniques which are used in such studies. It also considers briefly the extent to which behavioural studies are carried out in the United Kingdom and their standing in the context of legislation which controls all experiments with animals in the UK.

ANIMAL BEHAVIOUR AND LABORATORY ANIMAL SCIENCE

As mentioned above, the behaviour of animals can be a relevant consideration in any laboratory in which biological phenomena and processes are investigated. Furthermore, behaviour can be influenced in a variety of ways, planned and unplanned. It may therefore be useful to consider briefly some of the influences on behaviour of which all workers in laboratories should be aware.

Phylogenetic influences

First, it is important to emphasize that animals are suited to their evolutionary niche by their behaviour as well as by, for example, their anatomical form. Thus animal species can often be characterized by general behavioural dispositions or even by quite specific patterns of behaviour which have been developed as a result of their evolutionary history. An understanding of these general dispositions and these species-specific behaviours is therefore relevant to good animal husbandry and good experimental practice.

One evolutionarily determined factor which has to be considered when choosing an animal species for a scientific purpose is of course its anatomical form. Are the animals sufficiently small to be accommodated within the facilities available, and are they sufficiently large to be handled satisfactorily for the scientific purposes for which they are being obtained? Questions such as these are so obvious as to hardly warrant discussion, but of course they *are* relevant in any laboratory. In general, biological research is facilitated by choosing an animal species which is appropriate in a number of ways, by considering features which that species brings to a laboratory as a result of its evolutionary history. The general behavioural dispositions of species or stereotyped species-specific patterns of behaviour which characterize some species are just as relevant. For example, behaviour shows diurnal rhythms, and in some species nocturnal behaviour has been evolutionarily adaptive. In many species there are other marked cyclic variations in behaviour, especially of course in females. It is therefore prudent to consider the possible impact of these variations which have evolutionarily adaptive significance on the experimental programme which is being contemplated. Of course, many of these rhythms are recognized as being shown not just in behavioural terms but in terms of other biological processes too. Other forms of behavioural variability in a species, however, can sometimes be even more important although they cannot be related to physiological or other processes so easily. For example, allowance should be made for some 'natural' behaviours such as sand-digging, food-hoarding or nest-building in some species, for to ignore what are sometimes described as the ethological needs of an animal can be as unfortunate in terms of its well-being as ignoring physical characteristics. It should be added here that our opinion of what *ought* to be congenial and comfortable for an animal is not always a good guide, and judgements must be based on a knowledge of the ecological niche of the behavioural repertoire of a species. To keep cages for example aesthetically clean may be good from the point of view of humans but can be severely disruptive of organized patterns of behaviour and behavioural well-being if an animal is territorial and marks out its domain with olfactory cues.

Thorpe (1963) has suggested that patterns of behaviour which have been evolutionarily adaptive fall into six main areas, namely nutrition, fighting, reproduction, social relations, sleep and care of the body surface. Species of animals are likely to 'have' characteristic patterns of behaviour associated with each of these ethological needs, and these must be considered when choosing an appropriate animal for an experimental programme. So an awareness of the characteristic behavioural repertoire of laboratory animals is undoubtedly important. In his delightful book, Barnett (1963) shows that there is a richness in the characteristic patterns of behaviour of even the ubiquitous laboratory rat which is both engaging and provocative in the present context. In some species quite specific patterns of behaviour have evolved, particularly in the context of reproduction. Indeed, one of the main components of Darwinian sexual selection is surely behaviour. For example, there would be little point in the evolution of elaborate taxonomic forms within the context of sexual selection if those features were not to be displayed by similarly ritualized patterns of behaviour. Ethologists have shown that many species-specific behaviours are released in an orderly manner by quite specific events which are often to be found within the *behaviour* of another individual in the species. If these events are not presented, the behaviour can be released as so-called vacuum activities, and it can also be displaced to other patterns of behaviour in some conditions. Species-specific behaviours such as these are perhaps most characteristic of birds, and the dynamics of their organization should therefore be borne in mind by those who work with them.

In general, mammals seem not to show such stereotyped species-specific behaviours which are released by quite specific stimuli. However, as noted above, they do show general behavioural dispositions which can be understood in terms of their adaptive significance. One such general disposition in mammals which might bear mention here is their tendency towards marked individual differences in behaviour. In the case of mammals, a successful evolutionary strategem has been the development of behavioural plasticity and flexibility. Thus the behaviour of individuals reflects to a large extent their ontogenetic as well as their phylogenetic history, their past experience as well as their evolutionary history. This adaptive strategy has of course been most obvious with the so-called higher primates. Considerable investment is made by them in their individual offspring, allowing for behaviour to develop a rich variety which shows itself in the very considerable differences in the behavioural repertoires of adult individuals. Although the influences on such variety will be considered later in this chapter, the variety itself is at least worth noting here, because it is this very variety which represents one of the evolutionary strategies of higher mammals and which these species therefore 'have'. The more such variety there is in the behaviour of individuals of a species, the more laboratory conditions should allow for the behaviour of individuals to develop through exposure to rich and varied environments if the animals' well-being is to be maintained.

Genetic influences

Within species, different strains may show different behavioural dispositions as a result of their genetic inheritance. It is well known, for example, that strains of mice differ very considerably in terms of their motor activity. Strains of rat have been shown to differ in terms of their emotional reactivity (see Broadhurst, 1963); indeed, this disposition has been developed in differing degrees as stable characteristics of different strains by selective breeding within an originally undifferentiated strain. To return to common knowledge, different breeds of dog show characteristic patterns of behaviour as well as different physical forms. Strain differences such as these are also relevant to planning good laboratory practice and appropriate experimental procedures in a wide range of sciences, and care should therefore be taken to consider them so that a good choice of animal can be made for experimental purposes.

Influence of physical form on behaviour

Although it has been emphasized above that behaviour should not be conceptualized too readily as a mere appendage of the animal's physical form, it is also true that the physical characteristics of an animal are relevant to its behavioural repertoire. An obvious example of this is to be found in the existence of specialized senses. The degree to which the specialized senses of sight, smell, taste, hearing and touch have developed within a given species is of course to be understood in terms of its evolutionary and ecological niche. Just as is the case with diverse physical characteristics and ritualized behaviour in sexual selection, there is of course a symbiotic relationship between physical form and behaviour in the case of the special senses. For example, well-developed eyes are associated with subtlety and predominance of control over behaviour by that form of stimulation. Naturally enough, the variety of specializations here is again relevant to laboratory animal practice. The point has more force perhaps when the sensory modalities of animals are better developed than they are in the humans who are working with them. It is natural to tend to overlook those things we cannot see, and it is particularly important to note that the chemical senses of taste and smell may often be far better developed in laboratory animals than they are in humans. Once more dogs come readily in mind, but researchers are also of course aware of the fact that quite specific pheromones may exist for sexual behaviour in mammals. Furthermore, the study of ultrasonic communication in animals (e.g. Sales and Pyke, 1974) reveals that many laboratory animals live in a world of rich variety with respect to sound which we cannot hear. The general point to be made here is that laboratory animal scientists should try to increase their awareness of the potential impact on the behaviour of their animals of sensory stimulation to which we are relatively poorly attuned. To take one practical example, the overassiduous use of strongly scented disinfectants or even ammonia by technicians can without doubt have very marked and disturbing effects on the behaviour of rats, and these disturbances may in turn affect the data taken in scientific investigations in a variety of ways, as discussed above.

Environmental influences

Bearing in mind the differential susceptibilities to stimulation in different sensory modalities discussed above, it should also be emphasized that the effects of environmental stimuli on behaviour can be profound. Sudden or intense stimulation in any modality can of course have very disruptive effects, and should therefore be minimized. However, environmental influences affect behaviour within their *context*. For example, a bunch of keys can be more disruptive when dropped on the floor of a laboratory that is marked by subdued and controlled conditions than when dropped in a laboratory with constant unpredictable noises. It is a matter of judgement therefore whether to strive in the organization of a laboratory for breathless peace or lively bustle; the optimal conditions are doubtless to be found somewhere between these two extremes, but the precise arrangements which are best may depend upon the nature of the experimental programme which is being conducted, and this factor should therefore also be deliberately considered.

A further aspect of environmental influences on behaviour is to be found in *social* interactions between animals. Some species have evolved in such a way that interactions with other members of the species are of prime importance in maintaining well-being. Not surprisingly, those species which have adopted strategies of investing in individuals and behavioural plasticity are particularly sensitive to disruptions in social interactions, as zookeepers know very well. Even with rats considered within standard laboratory conditions, this variable can have a considerable impact on behavioural organization. The behaviour of rats caged individually can be markedly different from that of rats housed in groups, as can their general physiological characteristics, and rats brought from different social conditions of housing will behave differently in new situations such as those provided in routine biological procedures. Again this variable can be relevant to the homogeneity and replicability of biological data.

The effects of experience

The behavioural repertoires of individual animals are cumulatively affected by their ontogenetic experiences. As emphasized above, this is of course especially true of higher mammals in which behavioural plasticity is of crucial evolutionary importance. The behaviour of animals adapts to changing conditions, so that for example initial disruptions produced by an environmental change may not continue if the changed circumstances continue. It is therefore wise to allow for such adaptations to become settled before introducing animals to experimental procedures. Experiments carried out on animals which for example have only recently arrived in a laboratory may be performed therefore in a context of general and progressive changes in behaviour which again may make data unreliable or variable.

Adaptiveness in behaviour is shown through processes of learning which have been extensively investigated by experimental psychologists. Some of the procedures used by them will be discussed later in this chapter. Learning can develop in two ways, usually described as Pavlovian or classical conditioning and instrumental or operant conditioning. With Pavlovian conditioning, changes in behaviour occur as a result of the way in which environmental events are related to each other. For example, an event which reliably precedes the giving of food may come to produce salivation in dogs, and this salivation is described as a conditioned response. In general, any event which is reliably related to a biologically significant event may come to elicit a pattern of behaviour similar to that originally produced only by the biologically significant event. It should be noted that such conditioning occurs not only when experimenters set out to encourage it: a chance relationship between events in the environment which satisfies the requirements of Pavlovian conditioning will lead to progressive and sustained changes in behaviour. In operant conditioning, behaviour is affected by its consequences. For example, if a pattern of behaviour is reliably followed by food it may become more frequent, but if it is followed by some noxious event it may become less frequent. These processes which strengthen or weaken behaviour are described as reinforcement and punishment respectively. If some other event signals the occasions when the behaviour will be reinforced or punished, that event may come in turn to exert an explicit control over the behaviour. For example, a rat may press a lever only when a noise is present if the noise signals a period when the behaviour may be followed by food reinforcement. Again, operant conditioning does not occur only when experimenters set out to investigate it, but can arise whenever signals cue the possibility that behaviour will be followed by a significant consequence.

The behavioural plasticity shown in studies of conditioning and learning is of course to be found in its most developed form in primates, but it should not be overlooked that rats and pigeons have provided subjects for many experimental studies of learning in animals, and it is therefore clear that the behaviour of rats and of other commonly used laboratory animals can be very significantly affected by the experiences to which they are exposed.

The influences on behaviour which have been listed above cannot of course be considered in isolation from each other. The behaviour of any animal at a given time will reflect *interactions* between the various forms of influence, though with different species the relative weighting of the influences will differ. Considered in interaction with each other, the various forms of influence make it possible for us to conceptualize behaviour as a naturally occurring biological phenomenon. Naturally the behavioural characteristics of species, strains and special morphological features must be considered in terms of their relevance before any experiment is conducted. To take a very simple example, guinea pigs may be ideal for much biological research but their characteristic neophobia requires that they be adapted very carefully to changed circumstances and probably makes them inappropriate subjects for intrusive procedures which are carried out repeatedly. It cannot be emphasized too strongly that influences acting together do not affect behaviour only when a technician or experimenter recognizes or wants to investigate them. Behavioural differences can emerge, for example, from different animal husbandry and laboratory routines. Perhaps most important is the fact that environmental changes or a lack of environmental control can lead to cumulative changes in behaviour as animals learn from the experiences with which they are perhaps inadvertently presented over time.

The behavioural characteristics of animals and of the perhaps unplanned effects of the environmental conditions in which they are kept should be constantly considered therefore even though there may be no specific scientific interests in behaviour in a laboratory. To do this is no more than humane from the point of view of the animal technician, since it ensures that an experimental animal is as well suited to its laboratory conditions as possible. It is also prudent from a scientific point of view, because behavioural variability may well affect the reliability and replicability of any scientific procedures. In particular the effects of unplanned learning should constantly be considered. For example, an event that reliably precedes a noxious stimulus will come to elicit a 'fearful' reaction. If animals are treated in a stereotyped way before a mildly painful experience (as, for example, by moving their cage to an experimental room a short but predictable time before an injection is given or blood is taken) then the animal may well become agitated before the experimental procedure. If care is taken to break the simple predictive relation of moving the cage, however, this behavioural disturbance may be reduced. The result is a calmer and more tractable animal, which may in turn lead to more reliable and consistent data. To take another example, this time from operant rather than classical conditioning, if care is taken to construct a consistent context in which reinforcement or reward follows a desirable pattern of behaviour, then that behaviour may occur more consistently. If a dog is petted and fondled with extra vigour and devotion when it has been tractable and cooperative in an experimental situation, the 'rules of the game' may lead to docility when there might otherwise have been a struggle when the dog is exposed to the procedure once more. A person unaware of the rules of the game might of course tend to pet a dog with particular emphasis when it had *not* been tractable, seeking perhaps to calm it down but in fact potentially increasing the probability that the problem will recur on a future occasion. These comments should not be taken in any sense to imply that one should be pleasant to animals only when they serve our purposes. The point is simply that maximizing the potential power of reinforcement within a humane context can often be extremely effective in producing good laboratory practice. The examples given here may seem almost trite, but they are intended to emphasize that there is a great deal of scope in laboratories to plan and exploit the ways in which the behaviour of animals can become affected by environmental events rather than allowing the behaviour to be trapped in unplanned or counterproductive environmental circumstances.

All these conditions apply with added force to experimental situations in which it is in fact *behaviour* which provides the focus of study, as in studies of pharmacological, physiological or other influences on behaviour. The reliability and replicability of experimental data on behaviour are certainly likely to be adversely affected if potential influences on behaviour extraneous to the experiment itself are overlooked.

This brief review can be concluded by two assertions. First, behaviour is a biological phenomenon which is affected in an orderly and understandable manner by a variety of influences. Secondly, an awareness of the behavioural characteristics of animals and of the influences upon their behaviour leads both to sensitive laboratory procedures and to more reliable and replicable data, to good laboratory science and to good biological science.

EXPERIMENTAL STUDIES OF ANIMAL BEHAVIOUR

This section reviews some of the principal methods used in experimental studies of animal behaviour in the laboratory. There are of course many different experimental techniques which can be used to address a wide range of scientific questions about behaviour, and this review must therefore be cursory. Silverman (1978) provides a very clear and well-organized account of experimental methods as used in the study of animal behaviour in the laboratory, and readers are referred to this source for amplification of the basic points to be made here. Some familiarity with standard texts on the science of animal behaviour, such as those by Bermant (1973) and Dewsbury and Rethlingshafer (1973), can also be useful, since such texts place their primary emphasis on research findings rather than on experimental techniques; they also set laboratory studies in the broader context of the science of behaviour by incorporating accounts of ethological studies in the field as well.

Laboratory studies of animal behaviour can be approached from two perspectives. The first is provided by an interest in *differences* between species. Such an approach might be used to amplify findings which emerge from more naturalistic studies or to compare behavioural processes such as learning ability in a systematic way. The second perspective is provided by a search for *similarities* between the behaviour of species, looking for laws of behaviour which may be extrapolated from the laboratory and the species studied there to less constrained situations and to other species including man. Here researchers are interested in identifying basic mechanisms which underlie behaviour, as for example in investigating experimentally the effects of social deprivation. When the latter perspective is adopted with a view to understanding behaviour in general, researchers should be aware of the so-called comparative dilemma, which weighs the moral cost of using animals against the ease of extrapolating principles identified from their study. This dilemma is not unique to experimental psychology. Put crudely, it is often easier and less morally objectionable to study the behaviour of conventional laboratory animals such as rats, but it is difficult to be confident that behavioural principles identified by such study will be relevant, for example, to human behaviour. A knowledge of the differences in the evolutionary and ecological niche of the behaviour of different species, such as the differences between rodents and primates for example, suggests that extrapolation of behavioural principles should only be made with very considerable caution. It may sometimes be safer to generalize to human behaviour those principles which have been identified with some other primates (although this is not necessarily the case), but these experiments bring with them high costs both in practical and in moral terms. Researchers seeking general behavioural principles therefore have to adopt a strategy which balances these two features against each other.

The first requirement of any scientific study of behaviour is to develop objective measures. This is not as easy as it may at first appear. Techniques are needed to produce reliable measures from the observations which may be made of behaviour. Ethologists studying behaviour in its natural habitat have developed very sophisticated systems for recording and counting unambiguously defined categories of behaviour. The definition of these categories must be couched very strictly in terms of what is observable, i.e. what the animal *does*. The categories have to be very carefully defined so that they can be reliably recognized and scored when they occur. In order to obtain a sample of behaviour from which to take measures a number of techniques are possible. Especially with the advent of video technology which makes it possible to consider a sequence of behaviour as often as is necessary, one can attempt to construct a complete account of a stream of behaviour as it is exhibited. Such an account is perhaps a first step in quantification, for it makes it possible to assess the probability that one category of behaviour will be followed by another. Indeed, it is in this way that stereotyped patterns of species-specific behaviour have been separated from more variable sequences of behaviour. However, most measures of behaviour are expressed not so much in terms of transitional probabilities between categories but rather in terms of the frequencies with which behaviour occurs. This can be done by sampling procedures. With event sampling, the number of times a particular behavioural event occurs is recorded. With time sampling procedures, the stream of behaviour is examined at regular discrete moments, and the category of behaviour observed at any given time is recorded and counted. An event sampling procedure might lead to the conclusion that an animal exhibits more examples of behavioural category X than of Y; a time sampling procedure might show that behavioural category X is observed at a greater frequency than Y. These two statements are different, and one might be true while the other is false. This emphasizes the arbitrary nature of any systematic attempts to quantify behaviour, but the trade-off here is between artificiality and reliability.

These basic procedures for observing and quantifying behaviour can be carried forward into laboratory studies which also collect data in the form of the frequency of occurrence of objectively defined categories of behaviour in different conditions. However, the greater control inherent in laboratory studies also sometimes makes it possible for data collection to be automated, as will be illustrated later. Also it is sometimes possible to measure in some way the strength or intensity of a pattern of behaviour rather than simply its frequency and probability of occurrence.

Once an objective measure of behaviour has been decided upon, the requirements for adequate experimentation on the behaviour of animals are in principle no different from those of other biological sciences. Naturally the experiments must be carefully designed to address a carefully formulated question or scientific hypothesis. If the effects of some procedure or variable on behaviour are to be investigated it is often necessary to incorporate statistical comparisons between the behaviour shown by animals in different experimental conditions. Thus if, for example, the effects of social deprivation on the behaviour of rats are to be considered, it will normally be necessary to have at least two different groups of rats, each exposed to a given experimental condition. Animals must be allocated to these experimental groups in a carefully regulated and unbiased manner in order to avoid systematic contaminating effects of individual differences in behaviour. A study such as this might use a time sampling procedure to record the frequency of occurrence of carefully specified behavioural categories such as scratching, eating, walking, etc., or the intensity of behaviour such as the duration of 'freezing' after a startling stimulus etc. Such a study might also of course wish to incorporate a measure of social behaviour resulting from the experimental procedures, such as the frequency of sexual activity when the rats which have been differentially housed are allowed access to a rat of the opposite sex. When the data have been collected, they are submitted to statistical procedures which make it possible to evaluate the significance of any differences in the scores obtained from the two groups. These statistical procedures make it possible to form conclusions about the effects of a procedure (independent variable) on the behaviour of animals (dependent variable).

In some behavioural experiments, it is possible to dispense with statistical designs by exposing individual animals to more than one carefully defined procedure and making comparisons between the behaviour obtained in each. However, such a strategy is unusual, and in general it has to be recognized that there is normally a degree of 'noise' in the data obtained in behavioural experiments, that is to say there is some degree of variability between, and also within, individuals. Statistical designs therefore predominate in experimental studies of the behaviour of animals.

It is now possible to consider briefly some characteristic methods for studying animal behaviour in the laboratory. As was stressed above, however, in all cases they should be used in such a way that (1) the scientific problem has been formulated in terms of observable behaviour so that objective measures can be taken, and (2) an appropriate experimental design allows for meaningful comparisons to be made between observations obtained from different conditions.

There are many simple methods available for studying the motor activity of animals. Photobeams can be placed across an experimental space (which might be the animal's home cage) and counts can be made of the number of beams broken by the animal as it moves about in that space. Other commercial recorders automatically record activity by picking up movements of a finely balanced cage in which the animal moves about ('jiggle platform'), or by sensing the animal's position at different times by means of simple telemetric devices. Gross patterns of feeding and drinking are also easily monitored; for example, the *amount* of food or liquid removed from its source can be measured, licks at a bottle can be detected by a simple electrical circuit, or an animal may be trained to operate a device to release a measured amount of food or water. It is possible to conduct preference tests here, by allowing animals to choose between two substances and measuring the amount taken of each, but in this case extreme care must be taken to avoid experimental artefacts such as animals' preferences for one of the *places* in which the substances are available and so on. Locomotor activity can be monitored by counting the number of arbitrary but carefully defined units of space which an animal enters, and if the experimental space is novel this provides a measure of exploratory behaviour.

The effects of environmental conditions on behaviour can be measured. For example, behaviour can be recorded, perhaps by one of the above methods, in different conditions of ambient stimulation (noise, light, etc.), and thus any differences in behaviour may be attributed to these environmental variations if the experiment has been carefully designed to avoid artefacts. It is also possible to record an animal's reaction to a sudden change in stimulation, perhaps by means of a jiggle platform used to detect 'startle', expressed as recorded movement. Sometimes the effects of aversive stimulation are assessed. For example, in the hotplate test, a rat's paw is placed on a warm or hot surface and a record is made of the *time* taken before the rat removes its paw from that stimulation. A test as simple as this has proved effective for assessing the effects of analgesic drugs, for these lead to longer periods of time before the paw is withdrawn. These longer reaction times after a drug must of course be interpreted cautiously in terms of other possible effects which the drug might have on behaviour. For example, if the drug reduces general activity as well as increasing the latency of paw withdrawal, the latter effect may be a reflection of the former rather than of any specific analgesic properties of the drug.

Emotional reactivity in rats is usually measured by means of a standardized open-field test. The rats are taken from their home cage and placed for a short time in a large circular arena which is brightly lit and subjected to fairly loud 'white noise'. The primary measures here are ambulation (number of floor segments entered) and defaecation (number of boluses excreted). Emotional rats ambulate less and defaecate more than less emotional rats in these relatively stressful conditions, and these simple measures appear to correlate with measures of emotional reactivity in other situations such as startle to novel stimulation, neophobia (i.e. time taken to approach a novel object) and exploration. Interestingly in the light of the discussions in the previous section regarding interactions between influences on behaviour, the open-field test is not a satisfactory measure of emotional reactivity in mice, probably because mice are more territorial creatures than rats.

Social behaviour is often difficult to measure and observations in this field are rarely automated. However, quite simple objective measures can be very revealing. For example, dynamic processes of interaction between primate mothers and their developing offspring can be detected simply by recording how often and for how long mother and infant are seen to be more than a given distance apart, who initiates the separations and who initiates the returns. The effects of social and isolated rearing or housing have been investigated with many species. Sexual, agonistic (i.e. aggressive) and territorial behaviour can all be investigated if careful and objective measures are taken of appropriate aspects of behaviour, and thus the effects of experimental variables on these aspects of behaviour can be detected.

As noted above, a major thrust in experimental psychology has been the study of conditioning and learning, the rudiments of which were outlined earlier. The term conditioning is often preferred by experimental psychologists because it emphasizes that behaviour is a function of the environmental condition to which it is exposed and does not presume that there are underlying learning processes within the animal which mediate these changes. Pavlovian or classical conditioning experiments were originally carried out with dogs, which were restrained in harnesses. Stimuli such as noises and lights were presented in a reliable relationship with biologically significant events such as food or shock. All stimuli were delivered independently of the animal's behaviour, that is to say the dog did not need to develop a conditioned response in order to obtain food. Such a contingency would be characteristic of operant rather than of classical conditioning. In spite of this independence of behaviour, however, the previously neutral event acquired control over behaviour, coming to elicit a response similar to that which was previously produced only by the biologically significant event.

Pavlov's early studies were very important in the development of psychology, because they helped to reorientate the developing discipline to observable behaviour rather than unobservable mental life. It is interesting, however, that Pavlov termed the conditioned salivation which was produced by a stimulus that preceded food 'psychic secretion'. It is also interesting to note that Pavlov thought of himself as a physiologist assessing changes in activity in the central nervous system through changes in behaviour, although none of his behavioural studies in fact included direct physiological measures. Pavlov's work was also important because it identified general laws of behaviour which transcend particular species of animals. Indeed, Pavlov extrapolated freely from his experiments in order to develop a theoretical account of behaviour in general.

Far more emphasis in contemporary psychology is placed on studies of operant or instrumental conditioning, where behaviour is influenced by its immediate environmental consequences. This process can be conceptualized in similar terms to those which are used to describe how taxonomic form is selected by the environmental consequences in an evolutionary perspective. Thus patterns of behaviour are selected by reinforcers and become more prominent characteristics of the behavioural repertoire of an individual. A leading exponent of operant conditioning has been B. F. Skinner, and the field has been extensively reviewed (e.g. Blackman, 1974). In general animals, usually but not always rats or pigeons, are placed in controlled environments in which extraneous stimulation can be minimized in order to study the effects on behaviour of specified environmental conditions. The animals are free to emit any pattern of behaviour within that environment, although in truth there is not a great deal of opportunity offered for an extensive variety of behaviour in such a constrained setting. However, one pattern of behaviour is identified by the experimenter as the 'response' whose frequency of occurrence is to be studied as a function of environmental consequences. For rats this is usually pressing a lever in the experimental chamber, and for pigeons pecking a key or disc which can be illuminated in different colours. These behaviours are chosen as models of behaviour in general: operant conditioners are not interested in lever-pressing in rats *per se*, but rats are able to emit this behaviour over long periods of time and with varying frequencies, and its occurrence can be easily detected automatically and reliably by means of a small microswitch device attached to the lever. Operant behaviour is followed by reinforcing consequences such as food or water, and the frequency of the behaviour increases because of the selective influence of these reinforcers. It is not necessary for the reinforcers to be delivered every time the behaviour is emitted. Indeed, extensive work in operant conditioning has revealed that reinforcers will exert their effects when they occur only occasionally. However, characteristic patterns of operant behaviour then emerge as a function of the schedule according to which reinforcers are related to responses. Some schedules generate very high sustained rates of operant behaviour while others maintain very low rates. Some schedules engender highly organized patterns of responses in time. As noted above, operant conditioning experiments also investigate the ways in which lights and noises can gain control over behaviour by setting the occasion for a response to be followed (perhaps only occasionally) by reinforcement.

The techniques of operant conditioning provide a very powerful technology for the experimental analysis of animal behaviour. They generate interesting behavioural baselines against which the effects of further variables can be investigated. This can be illustrated here by reference to their impact in behavioural pharmacology. It has been shown, for example, that the effects of drugs may depend not just on their pharmacological properties and their effects on physiological systems, but also on the nature of the behavioural baseline against which their effects are being assessed. Some schedules generate patterns of behaviour that are very resistant to change by drugs, while others generate more sensitive baselines. In other studies, the reinforcing effects of drugs have been assessed by investigating their selective effects on patterns of behaviour which lead to their administration. Such research makes obvious contact with interests in addictive behaviour and drug abuse. Further studies have investigated how classically conditioned emotional responses ('fear') generated by a stimulus preceding a slight unavoidable shock interact with different patterns of operant behaviour, demonstrating that schedule-dependent effects can be seen here as well as in behavioural pharmacology. Such experiments in turn lead to the possibility of using these behavioural disruptions as a baseline against which to assess the effects of anxiolytic drugs. The technology of behaviour offered by operant conditioning has also been used in very subtle ways to investigate the sensory and perceptual capacities of animals. For example, it has been possible to identify the minimum intensity of light which is necessary if a pigeon is to detect it, and it can also be shown that pigeons are subject to some of the visual illusions which humans see. Again, therefore, the experimental techniques can be used to assess the effects of drugs on sensation and perception using animals as experimental subjects.

Some operant conditioning experiments have investigated the way in which animals' behaviour can be selected out by consequential *reductions* in the frequency of aversive events. This is described as free-operant avoidance conditioning. The modulating effects on organized patterns of operant behaviour have also been studied when aversive events are made dependent upon that behaviour, i.e. punishment. Behavioural baselines generated by such procedures can also be used to assess the potential analgesic or anxiolytic effects of drugs.

Another very interesting possibility opened up by the techniques of operant conditioning is the study of the *internal* stimulus properties of drugs. If one pattern of behaviour is reinforced with food only when a drug is given and another (for example pressing a different lever) is reinforced only when no drug has been injected, an animal's behaviour will come under the control of the drug administration, just as it would come under the control of a light which signalled reinforcement contingencies in a similar way. Experimenters can then assess the effect of different substances to see to what extent they are perceived as being similar to the training drug. Such drug discrimination experiments can also investigate how two different drugs can be used to establish a discrimination, and also therefore the extent to which a novel substance appears to be more similar to one training drug rather than the other.

The study of operant conditioning is of very considerable interest to psychologists in its own right. It reveals the sensitivity of behaviour, even in animals, to environmental influences. The impact of the behavioural technology on other biological sciences has been illustrated here largely by reference to pharmacology, but it is not in any way confined to that science. The technology of operant conditioning is both powerful and complex, and has potential uses in other sciences too. The environmental stimuli that impinge upon an animal are very carefully controlled, and the automated sensing of responses and delivery of experimental stimuli and reinforcers can lead to very strong control being exerted over behaviour and therefore to very consistent behavioural effects. As a result, it is in this area that it is sometimes possible to dispense with the conventional statistical designs discussed earlier. Individual animals can be used in operant conditioning experiments as their own control, by exposing them to different specific experimental conditions sequentially, although of course the possible contamination of sequence effects must be carefully counterbalanced. However, operant conditioning experiments are very demanding of apparatus and expensive of time. They require specialized test chambers in order to assess the influences of environmental events, and automated control apparatus which is often now to be found in the form of dedicated microprocessors or digital computers. In order to allow the carefully controlled environments to exert reliable and consistent effects on the behaviour of individual subjects, these animals have to be exposed to the experimental procedures regularly and for quite prolonged periods of time, for example in daily sessions of one hour over a period of months. The control of the general laboratory procedures, as discussed in the preceding section, is therefore of paramount importance if reliable behavioural effects are to be obtained in these experiments. So although the technological impact of this technique for other sciences is considerable, it cannot be introduced lightly. For experimental psychologists, however, one of the main fascinations of operant conditioning is the subtlety and orderliness which it reveals in the relationships between environment and behaviour.

When the demands of operant conditioning experiments as discussed above are too great, a number of other experimental techniques are available for assessing the effects of reinforcers on behaviour. For example, animals can be trained to traverse simple mazes in order to study their learning abilities. As a matter of fact, recent research has shown that rats can perform surprisingly well in quite complex maze situations: for example, in a radial maze in which sixteen arms each lead from a common central point and each have a reward at the other end, rats can learn to traverse all the arms in no consistent order but nevertheless without much backtracking even when no olfactory cues are allowed to influence them. Their behaviour thus shows a very considerable sensitivity to spatial cues. Avoidance behaviour can also be studied outside the conventional operant conditioning procedures. Passive avoidance is when animals refrain from doing something which has previously led to an aversive consequence, like stepping from a platform onto a grid floor through which a slight electric current is passed. Active avoidance can be studied by means of a shuttlebox apparatus, in which an animal is required to cross from one side of the apparatus to the other in order to avoid a slight shock when a signal indicates that the shock is imminent. Both these types of avoidance behaviour are readily learned by rats. In fact, it is more difficult to train rats to press a lever in order to avoid shock than to engage in these activities. This is an interesting example of a phylogenetic intrusion into studies of environmental control over behaviour. Presumably the 'natural' (i.e. phylogenetically influenced) behaviour of rats is to run away from a threatening situation rather than to stand still, a tendency which is of course exploited in the shuttlebox apparatus in comparison with the lever-pressing situation discussed earlier.

A further example of avoidance behaviour in which phylogenetic influences can be important is provided by studies of conditioned taste aversion. Animals are remarkably adept at learning to avoid novel tastes if these are associated with nausea, even though the nausea may be artificially induced by means, for example, of injections of lithium chloride. The adaptive significance of this is obvious. Most animals are to some extent neophobic, and so if they encounter a novel edible substance they are likely to take only a small amount. If this substance is in fact toxic, this small amount may make them ill but may not kill them. It is therefore evolutionarily adaptive to associate taste and nausea, since it leads to subsequent avoidance of the dangerous substance. Such taste aversions can develop in quite 'degraded' conditions in comparison with other examples of conditioned behaviour, for example when there is a considerable delay between the taste and nausea or when taste and nausea are paired only once. The study of conditioned taste aversion is a wonderful example of the ways in which the behaviour of animals must be conceptualized in terms of an interaction between phylogenetic and ontogenetic influences on behaviour.

Studies of conditioned behaviour can be pursued, naturally enough, in a variety of ways and with different animals. For example, some researchers have compared the learning ability of different species in different forms of mazes. It should be noted here that it is sometimes difficult to be 'fair' to different species by finding equivalent conditions of reinforcement, by allowing for differential receptor sensitivity etc. Different topographical patterns of behaviour can be investigated as a function of environmental conditions. For example, some researchers have investigated swimming in mazes rather than running through them, for a variety of good reasons. However, perhaps the above discussion is sufficient to give some flavour of the techniques which are available and of the purposes for which they may be used. It should be emphasized again here that more extensive accounts are available in the references cited previously.

A final and less specific area of research to which psychologists have contributed by studying the behaviour of animals in laboratories is in relation to the concept of stress and its disruptive effects on animals. This is a difficult field to summarize briefly, but Archer (1979) reviews the contribution of behavioural studies to an area of research which has been marked by interdisciplinary efforts. In general, research using psychological, physiological and endocrinological measures has manipulated the extent to which animals are able to cope with strains on their well-being exerted by the pressures of environmental events. This makes it possible in turn to investigate the effects of different techniques for coping with such strains. The phenomenon of stress is important, and it has been extensively investigated. However, this area of research is but one small part of experimental psychology, although many commentators give it undue prominence. This is mainly because of the ethical implications of research which inevitably exposes animals to situations with which they find it difficult to cope and which in psychological terms disrupts their behavioural organization.

To conclude this section, three general statements can be offered. First, the experimental study of animal behaviour in laboratories is a well-defined and important part of contemporary psychology, and many techniques have been developed and used in standardized ways. Secondly, the experimental study of behaviour can be used in an attempt to find behavioural differences between species, or general principles of behaviour which can be extrapolated beyond the laboratory. Thirdly, the experimental techniques which have emerged can be used to further psychologists' analyses of those events which influence behaviour, but they can also be used as tools to explore the effects of other biological influences on behaviour.

RESEARCH ON ANIMAL BEHAVIOUR IN THE UNITED KINGDOM

It may be helpful to conclude this chapter with a brief indication of the scope and context of laboratory investigations of animal behaviour in the United Kingdom.

It should be realized at the outset that only a few psychologists in this country in fact have research interests in the field of animal behaviour. The majority of psychologists have no direct interest in or involvement with animal research, and psychology is therefore rather different in this respect from many other biological sciences in which comparative interests dominate. Indeed, the role of animal research within psychology has been a matter of debate and controversy within the psychological community. For example, the British Psychological Society, which is the national organization representing both the scientific and professional interests of psychologists in the United Kingdom, set up a working party to discover the scope of animal research in psychological laboratories, by identifying how many psychologists had active research interests in animal behaviour and what general procedures they used, and also to consider whether there are any special ethical issues involved in such research beyond those involved in animal experimentation in general.

The report of the Society's working party was published in 1979. It surveyed all departments of psychology in universities and polytechnics in the United Kingdom, together with research laboratories which had overt and specifiable interests in behaviour. The report shows that in 1977 some 215 psychologists held licences under the 1876 Cruelty to Animals Act, thus enabling them to conduct animal experiments which might cause an animal pain or suffering. Approximately 9000 animals were being used for psychological research in November 1977, of which 6000 were rats, 1000 were mice, 800 were birds and 300 were primates. Approximately 7000 animals per year received drugs in psychological experiments, 5000 were subjected to some form of surgery, 4000 were exposed to mild electric shock, 8000 were deprived of food or water as an incentive procedure in conditioning experiments and 6000 were exposed to other forms of deprivation. The detailed analysis of these figures by species can be seen in the report. Thus the scale of overtly psychological research with animals in the United Kingdom is quite limited. The annual statistics of experiments on living animals in Great Britain published by the Home Office indicate larger numbers of animals being used in behavioural experiments. These data presumably include experiments carried out largely within other disciplines such as pharmacology, toxicology, physiology and genetics, in which behaviour forms some part as a dependent variable. The Home Office statistics are not easily interpreted on this matter, however, because animals which are exposed to more than one procedure tend to be summarized together rather than being broken down so that one can see a pattern of how behavioural interests interweave other major biological sciences.

In reviewing the ethical considerations inherent in psychological research with animals, the report of the working party of the British Psychological Society concluded that there were no ethical issues which are unique to psychology in this respect. Of course there are differing views within the discipline of psychology, as in the community at large, about the scientific value of such experiments, and there are also differing views about the ethical acceptability of experimenting with animals at all. The working party felt that the dominant criterion to be used in evaluating psychological research with animals should be expressed in terms of a calculus of the possible scientific gain to our understanding of behaviour in comparison with the amount of stress or suffering to which animals might be exposed. Clearly procedures which promise good prospects of important scientific contributions to our attempts to understand behaviour but which carry very little cost in terms of any suffering on the part of animals are more acceptable than studies which promise little scientific gain but cause much suffering, or indeed than those which promise much but still might entail a good deal of suffering. Following its exploration of the problem, the British Psychological Society has set up an ethical review committee which is charged with the task of keeping these general matters under consideration within the Society.

The legal context of experiments with animals in the United Kingdom has recently changed as a result of the replacement of the 1876 Cruelty to Animals Act by the 1986 Animals (Scientific Procedures) Act. Like scientists in other disciplines, psychologists who wish to carry out experiments on living animals must now hold appropriate licences issued by the Home Office under the 1986 Act. A Personal Licence is required to establish that a psychologist is a suitable person and is suitably qualified to carry out experiments. A Project Licence is also required, for which a psychologist, like other researchers, must submit an account of the scientific background of the intended work, specifying the objectives of the research and indicating its potential benefit. The number and species of animals to be used must also be specified. A summary of the procedures to be employed is also required, with an indication of their likely severity judged on a three-point scale of mild, moderate or substantial. Appropriate veterinary and other arrangements must also be specified.

This is not the place to review the details of the new legislation controlling experiments and other scientific procedures with living animals in this country. However, it should be emphasised here that psychological and behavioural research is clearly included within the scope of the new legislation. This is a marked improvement on the situation which applied previously, for the position of psychological and behavioural experiments was slightly ambiguous in relation to the 1876 Cruelty to Animals Act. For example, that Act related explicitly to experiments which caused pain, and made no reference to experiments which, though not causing pain as such, might nevertheless be expected to result in suffering, as for example with studies of the effects of maternal deprivation on the psychological development of primates. Another source of ambiguity was to be found in the fact that psychologists, like other scientists, were required to say how their research was designed 'to advance by new discovery physiological knowledge or knowledge useful for prolonging life or alleviating suffering' if they were to be granted a licence to conduct experiments under the 1876 Act. This requirement to justify psychological research in terms of its contribution to another discipline (physiology) emphasizes the fact that the domain of experimental biology did not include studies of behaviour when the previous legislation was drawn up over a century ago. This historical fact led to a further anomaly in the legal control of psychological experiments under the 1876 Act: in order to obtain Home Office licences, psychologists were required to obtain the support of a professor of physiology, medicine, surgery or anatomy, and were therefore required to have their research intentions evaluated by scientists from other disciplines. Although it is possible to see some merit in such an arrangement, it did not apply equally to those scientists who were themselves physiologists or anatomists, for example. In spite of these anomalies, psychological experiments with animals were controlled by custom and precedent under the terms of the 1876 Act: experimental studies of the behaviour of animals were scrutinised and controlled and psychologists were granted licences in the same way as obtained in other fields of animal experimentation. The current legislative framework is much more satisfactory, however, in that psychological research is explicitly included within it in exactly the same manner as are other biological sciences. For example, the 1986 Act specifies that a project licence may be granted for work which is undertaken for 'the advancement of knowledge in biological *or behavioural* science' (5 (3) (d), emphasis added).

Psychologists tend to be aware of the fact that many people are particularly disturbed about the ethics of psychological and behavioural experiments with animals. Their experiments seem often to be bracketed with cosmetics research rather than with medical or biological science, and the implication is that they are scientifically and morally less worthy than experiments carried out in other biological sciences. Psychologists have certainly found themselves in the forefront of the current highly charged debate about animal experimentation, sometimes to their surprise and frustration. For example, many critics imply that psychological experiments are synonymous with experiments on the effects of stress, though in fact few psychological experiments investigate stress and some non-psychological experiments do investigate stress. To some extent, hostility to behavioural experiments may result from misunderstandings about the nature of psychology as a scientific discipline. In the context of this heated and often ill-informed debate, it is greatly to be welcomed that the scrutiny and control of psychological experiments have now been explicitly incorporated within the remit of current legislation.

CONCLUSION

This chapter has sought to do three things: (1) to explain why all laboratory scientists should be aware of the dynamic influences on the behaviour of animals; (2) to provide an introductory review of the principal experimental techniques used in the study of animal behaviour in the laboratory; (3) to provide a brief perspective on the prevalence and the moral and legal standing of experiments on animal behaviour.

A constructive way in which to end a contribution such as this may be found by making reference to an important and readable book by Dawkins (1980). In this the author addresses the question of animal suffering in the science of animal welfare. She provides an exemplary example of how an awareness of the biological context of animal behaviour and a familiarity with the techniques of experimental psychology can be used to make a sensitive contribution to the task of evaluating animals' comfort and welfare in laboratories, intensive farms, or indeed anywhere else. Dawkins shows how one can construct questions about the preferences, wishes or suffering of animals in behavioural terms, and how these questions can then be addressed by experimental research. Dawkins' book is permeated by the spirit and orientation which the present chapter, in a more modest way, has set out to try to foster.

REFERENCES

Archer, J. A. (1979). *Animals under Stress* (Studies in Biology No. 108). Arnold, London.
Barnett, S. A. (1963). *A Study in Behaviour*, Methuen, London.
Bermant, G. (ed.) (1973). *Perspectives on Animal Behavior*, Scott, Foreman & Co., Glenview, Illinois.
Blackman, D. E. (1974). *Operant Conditioning: An Experimental Analysis of Behaviour*, Methuen, London.
British Psychological Society (1979). Report of a working party on animal experimentation, *Bulletin of the British Psychological Society*, **32**, 44–52.
Broadhurst, P. L. (1963). *The Science of Animal Behaviour*, Penguin Books, Harmondsworth.
Dawkins, M. S. (1980). *Animal Suffering: The Science of Animal Welfare*, Chapman and Hall, London.
Dewsbury, D. A. and Rethlingshafer, D. A. (eds) (1973). *Comparative Psychology: A Modern Survey*, McGraw Hill, New York.
Griffin, D. R. (1976). *The Question of Animal Awareness*, Rockefeller University Press, New York.
Sales, G. and Pyke, D. (1974). *Ultrasonic Communication*, Chapman and Hall, London.
Silverman, P. (1978). *Animal Behaviour in the Laboratory*, Chapman and Hall, London.
Thorpe, W. H. (1963). *Learning and Instinct in Animals*, Methuen, London.

Laboratory Animals: An Introduction for New Experimenters
Edited by A. A. Tuffery
© 1987 John Wiley & Sons Ltd

CHAPTER 9

Animal Handling and Manipulations

P. SCOBIE-TRUMPER

The Animal Unit, University of Surrey, Guildford

INTRODUCTION

The skill of handling laboratory animals will not be acquired easily. It may take many hours of practice on just one species to become professional. Observation of the points listed below will help to achieve this.

1. Try to find some background information on handling the particular species from books, films, slides, etc. Watch experienced handlers at work and observe the pitfalls.
2. Try to overcome fear or anxiety for it is almost impossible to work under these conditions. If the handler is not relaxed then neither will the animal be.
3. Incorrect handling of animals may cause physical damage and will certainly lead to stress, all of which will render them unsuitable for experimentation.
4. Never subject an animal to any procedure until it has been successfully relaxed and restrained.

This chapter will carry on to explain in detail how to deal with various species, pointing out the individual differences. Care should be taken to note the points on animal welfare as well as those on 'how not to get bitten'!

RAT

Handling

It will be found that with one or two exceptions most strains of laboratory rat will possess a very placid temperament. Even when upset by various treatments

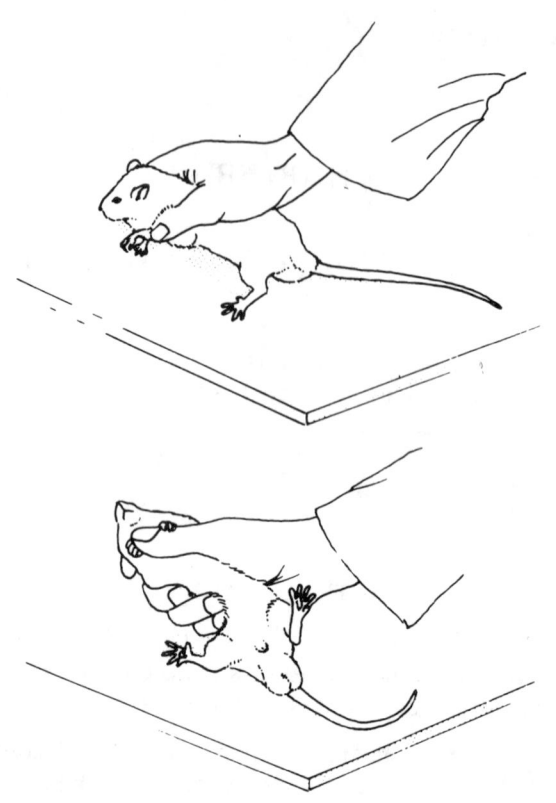

Figs 9.1 and 9.2   The rat — pick up by grasping around the shoulders and turn over onto its back

or not being regularly handled for long periods, rats can be calmed and relaxed within a short time by correct handling.

The adult rat should not be picked up by the tail as this can cause stress to the animal and is not at all necessary. Picking up by the tail can at times cause injury to the animal because in panic it will spin round in the air thus stripping the skin off the tail at the point of contact.

Rats should be grasped firmly round the shoulders, the fingers winding round the abdomen. The animal should then be turned over onto its back and the thumb placed under its chin. In this position most rats will relax (Figs 9.1 and 9.2). If the animal is large, i.e. over 250 g, then the other hand can be used to support the rump. Animals that appear to be struggling at this stage may be gently rocked backwards and forwards; this will have a calming effect.

The degree of success of this method will depend on the amount of pressure applied when initially grasping the rat: too much will impair the respiration, not enough and the rat will struggle and scratch because it feels unsafe. This method should be used when handling the rat for examination or transfer from cage to balance, etc. Rats should never be handed from one person to another as it is likely they will scratch during the transfer.

Great care should be taken when handling rats in the advanced state of pregnancy, giving plenty of support to the rump. When handling rats with a litter always wait for the mother to move off the nest if the young are suckling and remove the mother from the cage when handling the neonates. Hold rats over a cage or bench in case they should fall or be dropped. It is unlikely that a rat will suffer much injury as a result of a fall from normal bench height, but it may possibly become contaminated by contact with the floor.

Restraint

Before performing any procedures such as injections, oral dosing, ear-marking, etc., it is desirable to restrain the rat as this is the time it will bite and scratch, the latter being the most common injury to unskilled operators. An adult rat can get its jaws round the average human finger and the laceration caused is by the victim pulling away on feeling the bite. Experienced operators freeze when feeling a bite thus reducing the injury to two small punctures.

The best method of restraint both for the operator and the rat is that known as scruffing, i.e. making use of the loose folds of skin on the neck and back of the animal. The animal is after all used to being picked up in this fashion by its mother. The animal should be restrained in the left hand thus leaving the right hand free to manipulate the syringe, needle, etc.

The animal should first be picked up in the right hand, leaving the head and shoulders slightly protruding (Fig. 9.3), then with the left hand the scruff

Fig. 9.3   The rat — 'scruffing' — pick up with the right hand, leaving head and shoulders clear —

156   LABORATORY ANIMALS

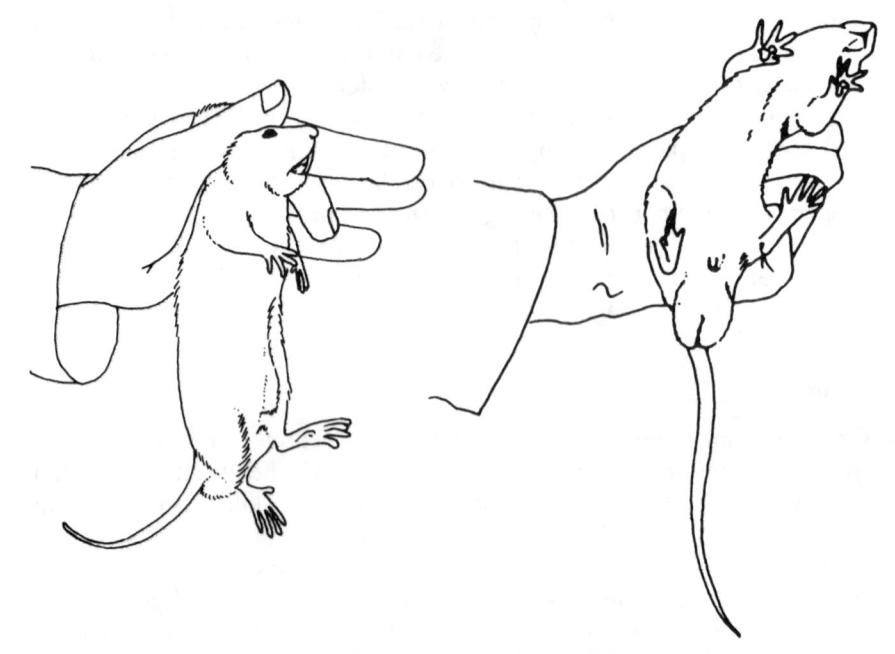

Fig. 9.4 — transfer to the left hand, taking up the scruff between the thumb and index finger —

Fig. 9.5 — now catch up the loose fur and skin down the back with the rest of the fingers, and turn the animal over for inspection or injection

should be taken between the thumb and index finger (Fig. 9.4). The rest of the fingers can then catch all the loose fur and skin down the back to the base of the tail (Fig. 9.5). The rat should then be turned over onto its back to present a perfect target area for injections. While in this position the rat is not only immobilized but also very relaxed. This method of restraint should be used for most injections, ear-marking, etc.

The method of restraint for oral dosing is slightly different in as much as the head and neck only are restrained, not the back. This is achieved by taking hold of the scruff with the thumb and middle finger leaving the index finger erect (Fig. 9.6). With the rat hanging relaxed in this position, the index finger is then placed on the animal's head to pull up the nose. This will achieve the desired position, i.e. a straight line between the mouth and the stomach (Fig. 9.7). This method needs practice and until perfected operators should not attempt to pass a dosing needle into the stomach.

An alternative hold for oral closing is described in Chapter 13 (p. 227), but I recommend this procedure as less stressful.

Fig. 9.6   The rat — hold for oral dosing — take hold of the scruff with the thumb and middle finger, allowing the animal to hang relaxed —

Fig. 9.7   — now pull up the head with the index finger, and the rat is in the correct straight line position for oral dosing with a needle. (The head should be raised further back than is shown in the drawing)

MOUSE

Handling

It must be stated that the one unfortunate problem with handling mice is they will almost certainly try to bite the operator. After many weeks of handling they will become reasonably tame but even then they cannot be trusted. Certain strains are more temperamental than others. Weanling mice of all strains are particularly difficult.

Care must be taken not to excite the mice before opening the cage; avoid sharp movements and high-pitched noises such as whistling. If possible slide the cage lid back a fraction or just enough to get a hand in. A mouse can jump several feet into the air, but rarely becomes injured as a result of a fall.

The mouse is too small to be held by the shoulders or waist and in any case it objects to being confined in this manner and will either bite or escape. It should be picked up by the tail about one inch from the rump (Fig. 9.8), and not suspended for any longer than necessary. In this way the mouse can be examined or transferred from cage to balance, etc. Keeping hold of the tail, the mouse can then be placed on a flat hand for further examination (Fig. 9.9). Several mice may be scooped up in cupped hand but this can present a risk of being bitten, especially if the strain is known to be lively.

Restraint

The mouse is one of the few laboratory animals that will not relax when being handled and restrained so great care must be taken to follow the method described below. Using the right hand, catch the mouse by the tail about one inch from the rump, place it on a surface where it can get a grip with all four feet, but keep hold of the tail. The best surface for this purpose is probably the wire lid of the cage. Placing the mouse across the mesh so that it can grip with all four feet will ensure that it remains still long enough to be taken by the scruff. If the mouse attempts to turn back then the tail must be pulled gently and it will go forward again (Fig. 9.10). At this stage, using the thumb and forefinger of the left hand, the mouse can be scruffed. This must be carried out quickly and deliberately in one movement (Fig. 9.11). Care must be taken to gather enough of the loose skin at the back of the animal's neck for if it is able to turn its head it will certainly bite. When the mouse is scruffed the tail can then be held against the base of the thumb by the rest of the fingers (Fig. 9.12).

This method should be used for i.p. injections, oral dosing, etc.; for intravenous injections or tail bleeding a restraining cage should be used.

When the mouse bites it usually ends up being thrown into the air by the operator pulling away on feeling the nip. This often results in severe injury to the mouse.

Fig. 9.8 The mouse — initial pick-up with a hold on the tail about 1 in. from the rump

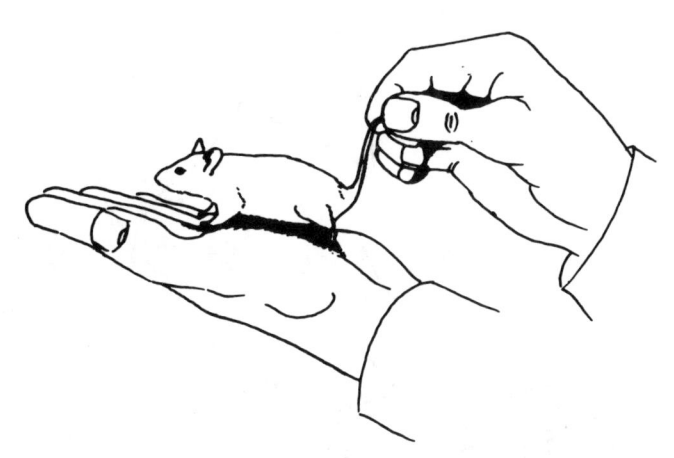

Fig. 9.9 — still holding by the tail, transfer to the flat hand for further examination

160　　　　　　　　　　LABORATORY ANIMALS

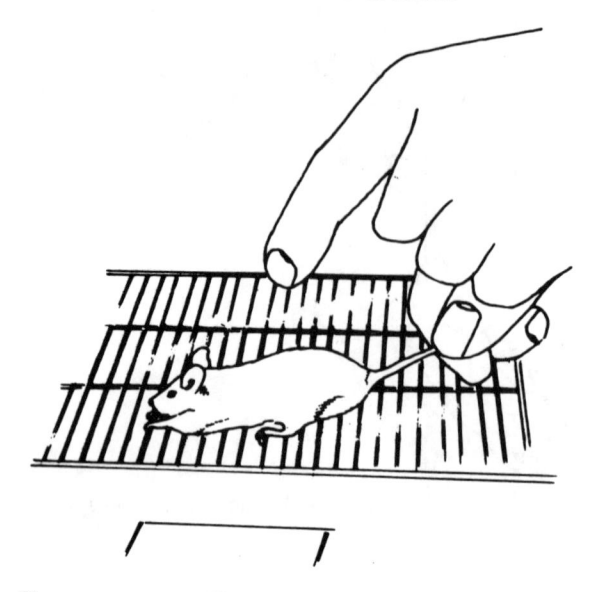

Fig. 9.10　The mouse — scruffing — the approach, and —

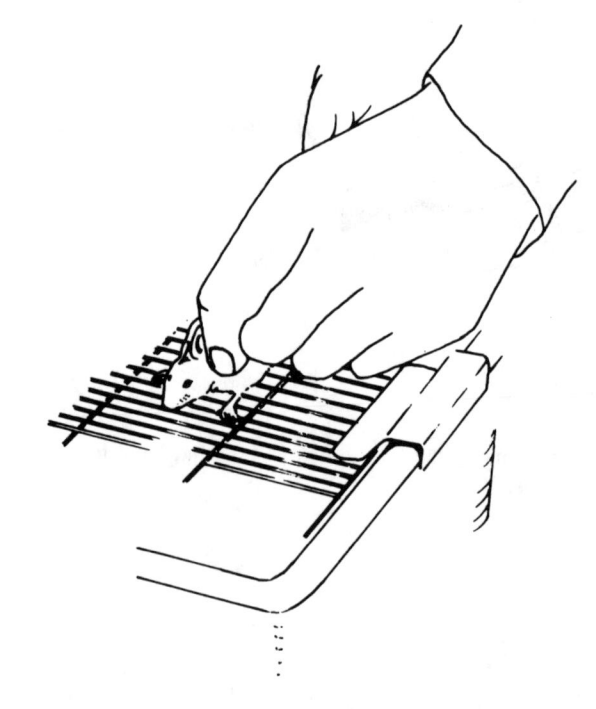

Fig. 9.11　— the hold — between thumb and index finger, high up behind the ears —

ANIMAL HANDLING AND MANIPULATIONS 161

Fig. 9.12 — tuck the tail in with the little finger and the animal can be turned over for inspection or inoculation

RABBIT

Handling

Unfortunately, inefficient handling of the rabbit is still all too frequent and this can result in fatal injury to the rabbit and quite serious injury to the operator. Most rabbits are very tame and do not bite, but occasionally a vicious one will be found. The signs of a nasty rabbit are: grunting, almost growling, backing into a corner of the cage bringing the front feet up to scratch, and finally biting.

There are several efficient methods of handling the rabbit, however, and these are some important points to remember before commencing:

1. The rabbit has very powerful back legs and long sharp claws which can inflict bad scratches on the operator.
2. Rabbits commonly suffer fatal back injuries from incorrect handling or falling.
3. Operators should always wear overalls with long sleeves to cover the arms and wrists, and remember to keep the face away from the rabbit even when it is restrained.

Taking the rabbit out of the cage can present problems at times because some of the cages will be at different levels. Short people may not be able to reach or see sufficiently to remove the animal safely without the aid of a small platform.

To remove the rabbit from the cage the operator must grasp the ears plus a

Fig. 9.13 The rabbit — removal from the cage

Fig. 9.14 The rabbit — carrying position, with head tucked under the arm and the hindquarters firmly supported

generous handful of loose skin behind the neck, place the other hand underneath the belly keeping the hind legs well away and then lift out the animal (Fig. 9.13). To carry the rabbit keep one hand holding the ears and scruff and place the head under the arm, the other hand being used to support the rump. The rabbit seems to panic less if it cannot see what is happening (Fig. 9.14). When returning the rabbit to the cage always put the hindquarters in first so that it does not see the cage and try to jump in. On no account should the rabbit be picked up or carried by the ears only.

Rabbits will panic when put down on a smooth surface but will be quite happy when placed on rough material where they feel more secure, for example towelling, sackcloth or wire mesh.

Restraint

For examination or i.p. injections the rabbit may be turned over onto its back as follows: grasp the ears and scruff with the right hand and lift, turn the animal over with the left hand and then move the left hand to the lower abdomen and thighs to secure the hind legs (Fig. 9.15). In this position most rabbits will remain very calm and relaxed.

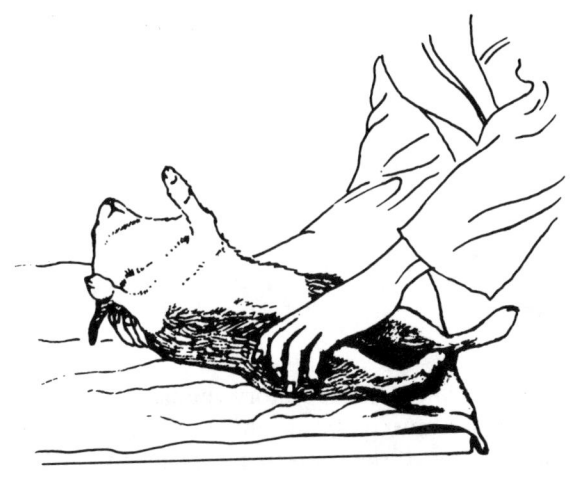

Fig. 9.15 The rabbit — supported on its back on a covered bench for i.p. injection

GUINEA PIG

Handling

Guinea pigs are very shy, nervous animals and do not take readily to being handled. When approached by the operator they will rush round the cage or pen in panic, thus making catching very difficult. Great care must be taken not to let a guinea pig drop or fall as it is unlikely that it will survive.

The animal should be approached from the front and back with both hands, one hand then grasping over and round the shoulders, the other hand supporting the rump (Fig. 9.16). Most pigs will struggle and squeak at this stage so care must be taken to maintain a firm hold. The animal may be more relaxed if the thumb and forefinger are placed either side of the jaw (Fig. 9.17). It must be stated that, although the guinea pig has quite large teeth, it seldom bites.

The methods illustrated should also be used for restraint.

Fig. 9.16 The guinea pig — picking up with a hold across the shoulders
Fig. 9.17 The guinea pig — hold, supporting hindquarters

CAT

Handling

The cat is the most difficult of all laboratory animals to handle. It is suspicious of strangers so some time must be spent getting to know the animals individually — they are all different. If the cat is unaccustomed to being handled it can present a dangerous risk to the inexperienced operator. Persons who do not like cats or who are afraid should not attempt to undertake this task. Cats have five danger points: four sets of claws and very sharp teeth; these they will certainly use unless adequately restrained.

A friendly cat can be held by tucking the animal under one arm, the hand gently restraining the front paws (Fig. 9.18). A cat should not be carried any distance as it is possible it will escape from the handler. Instead a transport basket should be used. When releasing the cat from the basket remember to lift the lid slowly so as not to frighten the animal, also make sure all doors and windows are closed. A cat can easily jump ten feet up or down without sustaining injury.

Restraint

It is possible for one person alone to restrain a docile cat as follows: grasp the scruff of the neck, the arm being parallel to the spine, with the cat's front feet

ANIMAL HANDLING AND MANIPULATIONS 165

Fig. 9.18 The cat — carry position for a friendly animal

Fig. 9.19 The cat — restraint on a bench for a docile animal

firmly on the table (Fig. 9.19). An intractable cat will need two persons to restrain it successfully, one to hold the scruff of the neck in one hand, the two front paws in the other. The other person can then restrain the two hind paws. The cat should be restrained always on a table or bench with a non-slippery surface.

Difficult cats

Unfortunately some cats supplied to laboratories may be wild and therefore most difficult to manipulate so the use of thick leather elbow-length gloves and a face vizor may be necessary for protection. It must be remembered that a wild cat will attempt to inflict injury to the eyes and face. In the case of wild cats, if the experiment permits then a tranquillizer can be administered. If all else fails then, as a last resort, the animal can be trapped or netted and placed in a crush cage as used for primates.

Kittens

Kittens can be restrained by wrapping them in a thick towel; this will trap the claws and help to keep the animal immobile. Kittens tend to present less of a problem than adult cats and are usually more subdued when restrained.

DOG

Handling

Most dogs respond to confident, gentle handling and with adequate reward being given their cooperation will increase. The operator should approach the dog from some distance away and crouch down to reduce his height to somewhere near that of the animal. As the dog approaches the operator should

Fig. 9.20 The dog — stages in the application of a bandage used as a muzzle

slowly let the dog see the palm of his hand. Never should the hand be brought down from above the animal's head. Talking to the dog all the time to give it confidence the operator should then bring his hand up from the neck slowly to the muzzle so the dog can sniff it. The hand can then be rubbed along the muzzle and the side of the face while the dog is talked to in a calm, steady tone; next the hand can be moved down the neck and back to the hindquarters. If the dog is happy and relaxed at this stage, then it should present no problems.

Unfortunately not all dogs are quite so amicable and particularly nervous dogs may take weeks of coaxing before submitting to any procedures. A strange dog should be approached slowly with great caution, making no sudden or unexpected movement. If a dog has attempted to bite then it is kinder to the dog and safer for the operator to muzzle it. This can be achieved by fitting a wire dog muzzle or using a 2 in. bandage which is placed round the jaw and nose several times and tied round the back of the head (Fig. 9.20).

These days most dogs used for research will have been bred in a commercial laboratory so are unlikely to be vicious, but should one meet a really intractable animal it may be necessary to use a dog noose. This consists of a length of rope running through a tubular pole, forming a loop at one end. The loop is slipped over the dog's head and pulled tight from the other end of the pole. A second operator can then muzzle the dog (Fig. 9.21).

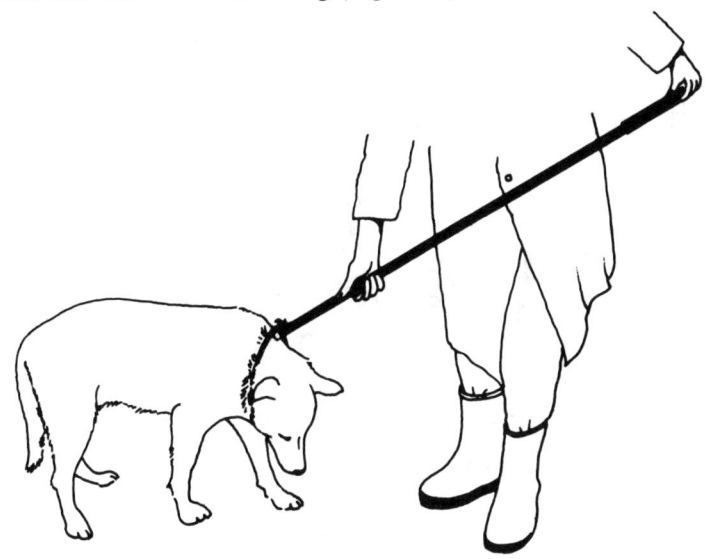

Fig. 9.21   The dog — use of a dog noose for an intractable animal

Restraint

Large dogs should be restrained on the floor as they will feel unsafe on a table, also they are not easy to lift. The operator should stand or kneel to one side of the dog then take hold of the nape of the neck or collar with the forearm in line with the spine, the other hand under the abdomen (Fig. 9.22).

Smaller breeds are best restrained on a table or bench, making sure the surface is steady and non-slippery. The operator should take hold of the nape of the neck or collar, the forearm in line with the spine, the other arm under the abdomen to give support (Fig. 9.23).

Fig. 9.22 The dog — restraint of a large animal on the floor

Fig. 9.23 The dog — restraint of a smaller animal on the bench

Some points to remember are:
1. Dogs are creatures of habit so soon become familiar with strange procedures.

2. Dogs become attached to particular people so operators should not be changed if possible.
3. Always give time for the dog to settle in and get to know the people and the routine before using it for experiments. Practise handling techniques with all new dogs before experimentation.
4. Do not let people handle dogs unsupervised if they are inexperienced or in any way frightened.
5. Remember it is quite possible that a sick dog can be an ill-tempered dog.

CHICKEN

Handling

Chickens will not present many handling problems with the exception of wing flapping which, if not controlled, will result in injury to the birds. An adult hen may be picked up by approaching it from the front, placing both hands either side and spreading the fingers over the wings to prevent flapping (Fig. 9.24). Alternatively one hand may be placed under the bird, the legs secured between the fingers, then the bird lifted and tucked under the arm to prevent flapping (Fig. 9.25).

Some large adult cocks may be quite vicious and will attempt to attack with the spurs so operators should wear long sleeves to protect their arms.

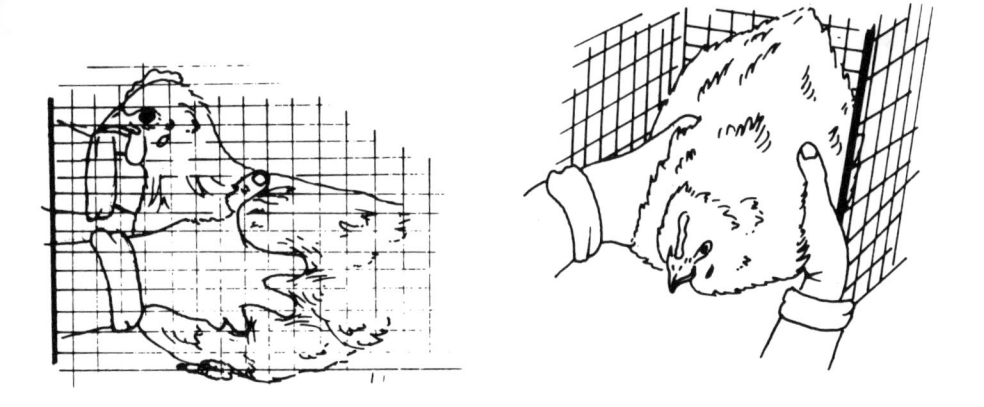

Fig. 9.24 The chicken — removing the bird from its cage — two views

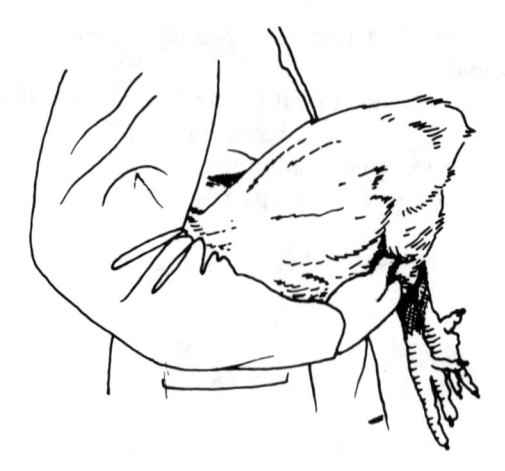

Fig. 9.25   The chicken — carrying a bird

Laboratory Animals: An Introduction for New Experimenters
Edited by A. A. Tuffery
© 1987 John Wiley & Sons Ltd

CHAPTER 10

Euthanasia

C. J. GREEN
Division of Comparative Medicine, Clinical Research Centre, Harrow

INTRODUCTION

Euthanasia involves the killing of an animal with a minimum of physical and mental suffering. It may be required for several reasons: it may be necessary to terminate an 'acute' experiment in which the animal has been anaesthetized throughout perhaps eight hours and different tissues are required for biochemical or radioisotope assay; groups of animals may have been kept for years and are to be autopsied to examine, say, the incidence of tumours; or it may even be necessary because the animal is ill or is considered likely to suffer unduly if kept alive. There are several options to be considered but these fall into two main groups: (1) the physical and (2) the chemical methods. The animal may be killed instantly or it may pass quietly into an unconscious state and then die without regaining consciousness. In selecting a method, the experimentalist must first aim to avoid frightening the animal and avoid transmission of apprehension amongst any group of animals to be killed. Secondly, he should consider the safety of personnel and avoid using aesthetically unpleasant methods which add to the strong emotional distaste felt by those involved. Finally, he must avoid damage which could interfere with postmortem investigation and analysis.

Whatever method is used, therefore, the animals should be handled gently, taking care not to frighten or antagonize them unnecessarily. They should always be removed from the holding room and never killed in the presence of another live animal. Any blood which escapes from one animal during euthanasia should be cleaned up and the area washed with disinfectant before another animal is brought in.

PHYSICAL METHODS

These mostly involve breaking the continuity of the spinal cord in the neck or rendering the animal unconscious by stunning or penetrating blows to the head. Physical methods, if carried out correctly by trained personnel, are often the least distressing to the animal since it is rendered instantly unconscious but they are often too aesthetically distasteful to recommend without reservation.

Dislocation of the neck is a quick and painless technique which is useful for mice, young rats, guinea pigs, rabbits, young kittens, young puppies and birds. However, excessive bruising and local haemorrhage of the neck and upper respiratory tract caused by this method may interfere with subsequent fixation of the lungs and interpretation of histological sections.

Rapid freezing is used when it is important to minimize enzyme activity for subsequent biochemical estimations of tissues. The animal is plunged head first into a beaker of liquid air or nitrogen, held for a few seconds until frozen solid and then decapitated. The method is only suitable for small animals such as mice which are rendered instantly insensible by the very low temperatures. This again cannot be recommended in any but special situations as it is aesthetically unpleasant.

Stunning followed by exsanguination is the method of choice where collection of blood uncontaminated by anaesthetic agents is required. A blow on the head should only be used to stun animals which have relatively thin skulls, for example rabbits or young kittens. Blood is then collected by severing the neck (jugular veins and carotid arteries). Larger animals may be stunned by electrical methods or by commercially available captive bolt pistols, compressed air guns, humane stunning cartridges or 'knocker-stunners'. The animals are rendered unconscious for long enough to bleed them out.

Electrocution is effective for dogs if performed properly but, perhaps because of the high conductivity of their coats, is unsuitable for cats. It is essential to anaesthetize the animal first with a low voltage a.c. current between two head electrodes before applying a much higher voltage shock between one head and a hind-leg electrode.

CHEMICAL METHODS

Euthanasia using chemical methods is simply an extension of anaesthesia to irreversible cardiac arrest. It follows that inhalational or injectable agents may be used.

Inhalational agents

Ether is commonly used to kill numbers of rodents but it has several disadvantages. It is doubtful whether this can be regarded as a humane method since the irritant nature of this agent to skin and mucous membranes at high concentrations is stressful to the animals. The explosive risk presents a particular laboratory hazard since bags full of dead animals may build up a high concentration of ether which can easily be fired by sparking connections in refrigerator or deep-freeze cabinets and during incineration. The respiratory tract is severely affected (bronchial secretions and pulmonary oedema) and this will inevitably interfere with interpretation of histological sections. Finally, it cannot be used if lipoid estimations of tissues are to be made.

Carbon monoxide inhalation causes rapid death by combining with erythrocyte haemoglobin to produce fatal anoxia. The onset is so rapid that loss of consciousness occurs before the animal is stressed. The animals are killed either in a chamber to which household gas is supplied or in which carbon monoxide is formed from the chemical interaction between crystals of sodium formate and sulphuric acid. Carbon monoxide concentrations of 0.5–14 per cent have been used. Animals collapse and are unconscious within 40 sec, failure of the respiratory centre occurs in about 2 min, and cardiac arrest follows in 5–7 min. If normal haemoglobin is necessary for the study, then clearly this method cannot be used.

Carbon dioxide at concentrations above 60 per cent in O_2 causes loss of consciousness, paralysis of the respiratory centre and cardiac arrest. It must be emphasized that CO_2 is an anaesthetic agent and is not simply displacing O_2 so that the animals die of asphyxia. As it is non-irritant, it is undoubtedly the inhalational agent of choice for euthanasia of small animals, including all rodent species, rabbits, cats and small pigs. It can be piped into a killing chamber, clear polyethylene bags or into the animals' holding cages enveloped in suitable bags, where the gas will rapidly fall by gravity to build up suitable concentrations around the animals. Unconsciousness is induced quickly, peacefully and painlessly within 30 sec of exposure, but it is wise to leave the animal in near 100 per cent CO_2 gas concentrations for a further 20 min to ensure that it is dead. Neonates are particularly tolerant of carbon dioxide and can survive up to 30 min exposure to the gas (90–100 per cent concentrations). They should therefore be placed in disposal bags into which carbon dioxide has been piped before they are sealed.

It has been assumed in the past that carbon dioxide euthanasia would produce minimal changes to organs. However, it is now known that oedema of perivascular connective tissue and extravasation of blood into interstitial spaces form artefacts in lung sections after carbon dioxide euthanasia. Similar lesions have been noted in lungs from mice and rats anaesthetized with $CO_2:O_2$ (1:1) for periods longer than 2 min and this suggests a loss of endothelial integrity resulting from a rapid fall in plasma pH.

Chloroform should not be used in laboratories since it is dangerously hepatotoxic to personnel and animals. Even trace concentrations carried to breeding rooms in the ventilation system will interfere with the breeding programme in rodent colonies.

Halothane at high concentrations (>4 per cent) within a chamber or bag will rapidly anaesthetize the animals and cardiac arrest will occur within 90 sec. The bags should be sealed to prevent environmental pollution but it is safe to incinerate the animals in these bags afterwards.

Enflurane at high concentrations induces anaesthesia even more quickly than halothane and cardiac arrest occurs in 60–90 sec. The same precautions should be taken to prevent environmental pollution but the animals can be safely incinerated.

Injectable agents

Animals are most humanely killed by intravenous, intracardiac, intraperitoneal or intrathoracic injections of barbiturates.

Pentobarbitone (18 per cent concentration), preferably by rapid intravenous or intracardiac injection at 60 mg/kg, is the most satisfactory agent for most animals. The intraperitoneal or intrathoracic routes are reasonable alternatives for intractable subjects or where intravascular injection is difficult, but higher doses (80–150 mg/kg depending on species) should be used, and the animals must be carefully observed afterwards to ensure that they are dead. It is important to auscultate the heart to ensure irreversible arrest, since some animals, particularly cats, may appear to be dead but, after a short time or after a stimulus such as being dropped into a disposal bag, their heart may start beating normally again.

RECOMMENDED PROCEDURES

Dogs

The simplest and most humane way of killing dogs is by rapid i.v. injection of a concentrated solution (180 mg/ml) of pentobarbitone at 40 mg/kg. Respiratory paralysis and cardiac arrest are produced within seconds and without vocalization or struggling. Dangerous animals must be adequately restrained either physically or chemically with, for example, ketamine (40 mg/kg i.m.) before attempting euthanasia. The intracardiac route is often then the most convenient for injecting lethal doses of pentobarbitone (40 mg/kg).

Cats

Cats of all ages can be killed with carbon dioxide or carbon monoxide in a suitable container and suffer minimal discomfort. Alternatively, pentobarbitone (180 mg/ml) can be given by the i.v., i.p. or intrathoracic routes at a dose rate of 50 mg/kg. If the animals are considered to be dangerous to handle, they can be covered in thick towelling and injected with ketamine (40 mg/kg i.m.) prior to removal from their box for injection with pentobarbitone (30 mg/kg by the intracardiac route).

Rabbits

These may be killed by a sharp blow at the back of the neck with a short stout length of wood if other more aesthetically satisfactory methods are contraindicated. Alternatively, the neck may be dislocated if the head is bent sharply backwards at the same time as the legs and body are jerked sharply downwards with the other hand — but this must *never* be attempted on live animals until practice has been gained on dead rabbits. Both methods provide instant and painless death if carried out properly but can be barbaric in the hands of the novice.

A far more satisfactory method is to inject pentobarbitone (60 mg/kg) into an ear vein, after ensuring proper physical restraint. Rabbits struggle if exposed to volatile agents in euthanasia chambers, and even appear apprehensive if exposed to carbon dioxide or carbon monoxide, so these too should be avoided if possible.

Guinea pigs

Guinea pigs can be killed with carbon dioxide or carbon monoxide in a chamber, or with pentobarbitone injected at 90 mg/kg i.p. If it is not possible to use these techniques, then the neck can be dislocated and the animal killed humanely in this way. The right hand is placed over the front of the animal's head and the neck is gripped between finger and thumb. The arm is swung downwards vertically to one side and the weight of the animal dislocates the neck at the bottom of the drop.

Hamsters

These are best killed with carbon dioxide or carbon monoxide in a killing chamber or with pentobarbitone injected at 150 mg/kg i.p.

Rats

Young rats may be stunned on the edge of a bench or be killed by neck disarticulation. Adults are best killed either in a killing chamber with carbon dioxide or carbon monoxide, or by i.p. injection of pentobarbitone (100 mg/kg).

Mice

Mice may be killed humanely by dislocation of the neck. The mouse is placed on a flat surface, a pencil or pair of forceps is placed across the back of the neck and dislocation is effected by a sharp backward jerk on the base of the tail.

Alternatively, mice can be killed in chambers or bags with carbon dioxide or carbon monoxide, taking care to ensure that the pups too are dead, or by i.p. injection of pentobarbitone (150 mg/kg).

AUTOPSY AND CARCASE DISPOSAL

In all cases where no hazard is presented it is best to examine the carcase immediately after an animal has been killed. Autolytic changes occur during freezing down to $-20°C$ and ice crystal formation disrupts soft tissues so that histological artefacts may be created if the animals are simply placed in a deep-freeze cabinet. Further damage will occur when they are removed and allowed to thaw out slowly.

Some animals, however, present, a risk either to personnel or to other animals in the area. It is important to emphasize yet again the danger of flammability posed by animals which have been killed with ether and then placed in sealed bags in refrigerators or deep-freeze cabinets. The only way to avoid this risk is to ban the use of ether completely in an animal house and ensure that there are better and safer agents available.

All non-human primate tissues should be handled with extreme care. To avoid the risk of dangerous aerosols being released, they should only be autopsied by experienced staff wearing proper protective clothing and working within a protective hood under aseptic conditions. Any tissues including blood should be carefully labelled as dangerous and transported in sealed containers triple covered in plastic bags. The carcases should then be triple bagged and the outer plastic cover should be dunked in strong antiseptic. They should then be incinerated immediately.

Similar precautions should be taken with animals which have been exposed to dangerous pathogens and either naturally or deliberately infected. The hazard presented here of course may be to other animals, to human handlers or to both. Under no circumstances must these be stored in deep-freeze cabinets under the mistaken impression that the low temperatures will kill infectious agents. Many bacteria can survive long periods at $-20°C$ while viruses are very resistant to slow freezing. Again, the animals should be placed in strong plastic bags, dunked in strong antiseptic and then incinerated immediately.

Finally, the disposal of carcases contaminated with radioactive materials presents a special hazard. It is often necessary to store these carcases for several days in a clearly marked refrigerator to allow the radioactivity to decay before the bodies can either be incinerated or disposed of by macerator and washing into the sewage system. The carcases must again be triple bagged, the bags must be decontaminated to avoid spillage and they must then be clearly marked with dates, the nature of the isotope and its half-life, and the total radioactivity measured before the carcases are placed in the cold cabinet.

CHAPTER 11

General Aspects of the Administration of Drugs and other Substances

H. B. WAYNFORTH
Charing Cross & Westminster Medical School, London

INTRODUCTION

Chemical substances that are commonly administered to laboratory animals are often grouped under familiar categories. Thus drugs are generally considered to be compounds which have pharmacological activity and are used therefore in medicine; carcinogens are substances which produce cancer; teratogens induce congenital malformation. Others can be grouped under pesticides, cosmetics and food additives and still others have biological applications; these may be hormones, enzymes, blood products such as heparin, vaccines, antitoxins and similar compounds. While these are familiar to most of us, there are other chemicals both organic and inorganic in nature which are not conveniently grouped as having any application in particular but are nevertheless used for medical or scientific purposes. As far as the body is concerned, it does not distinguish between groups and treats a chemical substance as just a chemical substance though it does not necessarily treat each one in an identical manner. For convenience, all of these substances will now be referred to as drugs. Drugs are administered to animals for a variety of reasons. These include:

(a) to study their activity
(b) to evaluate their safety
(c) to prepare animal models
(d) to act as a marker for biological activity
(e) to treat and prevent disease.

It is probably true to say that in most branches of the biomedical sciences, drugs are administered by scientists empirically, perhaps as a result of a recommendation, without any real appreciation of the basis of that recommendation. Yet stemming from such a lack of appreciation there could be a serious loss of information or even the complete absence of the desired result. Circumstances are known, for example, where, when a substance apparently had little or no effect, this was due not to an intrinsic inability but to factors in the body modifying its passage to the site of action. Simply increasing the dose adequately may elicit a good reaction. Understanding those factors which govern whether a chemical administered to an animal will have an observable effect or not, and if it does then to what likely extent, involves the somewhat complex study of pharmacodynamics. However, an understanding in simple terms of the basic principles of this can help a scientist achieve a schedule of drug administration which has a greater chance of producing a desired effect.

THE BEHAVIOUR OF DRUGS IN THE BODY

A desired effect from a drug will be more likely to be obtained by producing and maintaining an effective concentration of the drug at its site of action. How a drug is moved from outside the body to its site of action and the factors that interact with it to allow it to express its action can be described by considering three interrelated processes.

Entry of drugs into the body

Translocation

In order to arrive first at the general blood circulation and then at its site of action, a drug has to cross a series of cell membranes. How well it does this determines its effectiveness. These biological membranes are, with minor exceptions, fairly consistent in structure, being composed of a bimolecular layer of phospholipid molecules bounded on both sides by a thin layer of globular protein. The membrane is punctured in an irregular manner by water-filled pores which vary in size. Simple diffusion is the major way that drugs cross the membrane and this can be shown in particular for lipid-soluble compounds and for water-soluble compounds which have adequate lipid solubility. The rate of movement in these cases is proportional to the concentration gradient of the drug across the membrane. Substances which are small and water soluble only generally gain access to the cell interior by passage through the open pores in the membrane, and here passage depends on hydrostatic or osmotic differences on either side of the membrane. It is important to grasp the concept of lipid solubility of a drug since this is by far the most important determinant for its translocation and movement through cell membranes.

Another important concept is that of ionization. Many drugs are either weak organic acids or weak bases and exist in solution as ionized and/or unionized forms. The non-ionized fraction is generally lipid soluble and therefore readily diffuses across cell membranes, while the ionized molecules are not significantly soluble in lipid and are virtually excluded from membrane diffusion. Each acidic or basic group in a drug has a dissociation constant or pKa value. When the pH of the local environment in which the drug finds itself and the drug's pKa value are equal, 50 per cent of the drug will be ionized and 50 per cent unionized. Importantly, in the case of an acid drug, raising the pH of the solution increases the degree of ionization, while lowering the pH decreases the degree of ionization, resulting in more of the unionized, lipid-soluble fraction being present. The converse holds for basic drugs. Also, each one unit pH change to the acid side of the pKa, for an acidic drug, produces a tenfold increase in the unionized fraction, while a similar but opposite change occurs when shifting the pH one unit to the alkaline side. Again, the converse is true for a basic drug. This shows therefore that large changes in lipid solubility are produced by small changes in pH. The majority of pharmacological drugs have pKa values between three and eleven and their degree of ionization therefore will depend on what sites in the body they become located in. For example, a weak acidic drug such as aspirin (solium salicylate) has a pKa of 3.4 and will be almost completely unionized in the strongly acid conditions of the lumen of the stomach of animals such as the rat and dog. Therefore it will diffuse easily and quickly through the mucosal cells into the surrounding tissue. There it will encounter a pH of about seven and consequently will become changed into the lipid-insoluble ionized form. It follows that for both weak acids and weak bases, the total concentration of the drug will be greater on the side of the cell membrane where it is more highly ionized, since now, being relatively lipid insoluble, it will tend to accumulate rather than move rapidly from that location.

Other ways in which substances cross the membrane barrier are by active transport and facilitated diffusion. In both cases a specific carrier is thought to combine with the substance and carry it across the membrane, but only in the case of active transport is this done against a concentration or electrical gradient. On occasions, particularly with large drug molecules, translocation is facilitated by pinocytosis, a mechanism which has much in common with active transport.

Rate of absorption

In most instances, the initial aim of administering a drug is for it to be absorbed and to enter the blood circulation in an effective amount, because it is probable that the effect of a drug is closely correlated with its plasma concentration. How well this is done is affected by several factors, notably the physical form of the drug, the route of administration and those physicochemical characteristics already discussed (e.g. lipid solubility).

Form of the drug. Drugs completely dissolved and given in aqueous solution are more rapidly absorbed than those dissolved in an oily solution or given as a suspension or in a solid form (e.g. tablet or powder). Chemicals injected in a solution of high concentration are absorbed more rapidly than those in solutions of low concentration.

Routes of administration. The rate of absorption is also influenced in direct proportion to the area of the absorbing surface and the blood supply to the site of absorption. The common routes and their characteristics are:

(a) Intravenous. Chemicals injected directly into the blood circulation have a predictable concentration, are completely available for distribution to the target and other tissues and produce an extremely rapid effect. Moreover, the rate of introduction of the chemical into the circulation can be controlled precisely. This can be important, for example, in the administration of an anaesthetic such as sodium pentobarbitone, whose effective dose range is not far removed from its toxic level; and in the injection of drugs into sick animals, which may handle drugs differently to healthy ones.

(b) Intraperitoneal. Passage of drugs into the circulation from this route is relatively fast due to the very large surface area for absorption and the abundant blood supply to the organs and tissues of the peritoneal cavity. Absorption is about four times slower when compared to an intravenous injection. It should be noted that drugs placed intraperitoneally will, for the most part, be absorbed into the portal circulation. As a consequence they will be subjected to the metabolizing activity of the liver before gaining access to the general circulation. Therefore, some modification of the drug dose could be expected.

(c) Intramuscular. Muscles are abundantly supplied with blood vessels and absorption is rapid but slower than from an intraperitoneal injection. The muscle site is sometimes important. Injection into the deltoid muscles may result in a 20 per cent increase in the rate of absorption compared to injection into the gluteal muscles. Absorbed material will pass directly into the general circulation.

(d) Subcutaneous. Absorption from this route is usually slower than after intramuscular injection. Solid drug forms can also be effectively administered by subcutaneous implantation. The peak concentration in plasma of a water-soluble drug can be expected within 30–60 min after it has been injected by the subcutaneous or intramuscular route.

(e) Percutaneous. This route presents special problems because of the relative impermeability of the epidermal layer (*Stratum corneum*). The degree of permeability varies considerably with the species, that in the pig, for example, being 30 times less than for the rabbit or rat. The efficiency with which chemicals penetrate the epidermis is mostly dependent on the vehicle of administration. Absorption is fastest when the vehicle is an oil or an oil in water emulsion, particularly when the oil is of plant or animal origin (e.g. peanut and olive oils). Absorption is also rapid when organic solvents such as acetone are used. Special sorption promotors such as dimethylsulphoxide and dimethylformamide can aid absorption markedly.

(f) Intragastric. Absorption takes place along the whole length of the gastrointestinal tract but it is in the small intestine, with its large and abundantly vascular absorptive surface, that most drugs are absorbed after intragastric and oral administration. Some initial absorption may occur in the stomach if the chemical remains there in a mostly unionized form and is also lipid soluble. When the drug reaches the intestine, and if it has a pKa value above three and below about eight, it is very well absorbed. Most compounds cross the intestinal barrier by passive diffusion but some, such as water-soluble nutrient molecules and structurally related chemicals, cross by specific carrier systems. A functional property of the stomach, viz. the rate of emptying into the intestine, can markedly affect the rate of absorption. A delay occurs when the stomach is full and the pylorus closed, which prevents material from passing into the intestine. Animals with gastrointestinal disease or fever also show delays in gastric emptying times and therefore a delay in absorption of drugs. Since many compounds are solids or suspensions when administered intragastrically, they can only be absorbed after the solid has disintegrated and the resultant small particles (or suspensions if initially in this form) have undergone dissolution in the surrounding fluids. These steps, which also occur with solid formulations placed in parenteral sites (e.g. subcutaneous), are frequently rate-limiting for the overall absorption process and indeed in some instances absorption can be so protracted that effective plasma levels are never achieved. Dissolution can be aided by using the salt form of the drug or by decreasing particle size. Certain types of drugs placed in the stomach also run the risk of inactivation by the highly acid conditions or by the enzymes and bacteria such as are found in the stomach of ruminants. Further, enzyme inactivation can also be expected in the intestinal mucosa, and bacterial inactivation in the lumen of the intestine. Most drugs are absorbed to a greater or lesser extent from the gastrointestinal tract, but because they must pass through the liver via the portal system they may be modified to some degree by the liver's metabolic activity before they reach the general circulation.

(g) Other routes. These have specialized applications and include intrathoracic, intracardiac, epidural, intraarticular, intrathecal and pulmonary. Absorption of volatile agents by the lungs is extremely rapid because of the enormous surface area and the abundance of the blood supply. It has limited application but is important for inhalation anaesthesia and the administration of pharmacological drugs in vapour or aerosol form.

Extent of absorption

The factors which determine how much of the drug finally reaches the circulation are the pH at the absorbing surface, the pKa of the chemical compound, its lipid solubility, its concentration, its rate of dissolution and the blood supply to the absorbing surface and its surface area. The absolute absorption, in terms of plasma levels, can be determined in practice by comparing the area under the plasma drug concentration versus time curve, obtained after intragastric administration or administration by one of the other routes, with that obtained after intravenous administration. Absorption of all of the drug can only be assumed after an intravenous injection. For other routes it can be very variable. For example, only some 50 per cent of an aqueous dose of tetracycline given orally reaches the plasma compared to an intravenous injection. A similar picture can be shown for the absorption from the stomach into the general circulation of the active ingredient of aspirin tablets. In contrast, almost all of a solution of the antibiotic gentamycin reaches the circulation after an intramuscular injection. However, it should be noted that even if all of a drug reaches the circulation a very slow rate of absorption from the site of administration may result in a particular dose having an ineffective or too low a plasma concentration, giving the impression that the drug is inactive. In reality this may be erroneous, since suitably adjusting the dose or modifying the drug form may be all that is needed to make it effective. Estimating the time at which the peak concentration of the drug in plasma is reached gives a close approximation of the rate of absorption. As might be expected after intragastric administration of many compounds, the maximum concentration in the blood may be reached only after two or three hours compared to 30–45 min after an intramuscular injection. However, both the rate and extent of absorption from the gastrointestinal tract are highly variable among species and even within species. For example, twice as much chloramphenical is found in the plasma of the cat than in the pig at peak concentration, after oral administration of the same dose, while it cannot be detected at all in the goat.

The distribution of drugs

When a chemical compound has reached the blood circulation, the concentration at the site of action will depend on the rate of distribution and on the rate of elimination of the drug from the circulation. However, before a compound arrives at its site of action, it has to cross the capillary wall. Compared with other cell membranes, chemical substances of small to intermediate molecular size cross the capillary endothelium very rapidly and almost instantaneously if they are highly lipid soluble. Water-soluble compounds with molecular weights up to about 6000 leave the blood through pores in the capillary wall and pass into the extracellular tissue fluid. An exception is seen in the capillaries of the brain where the pores are sealed, forming the well-known blood–brain barrier, and as a consequence the entrance of highly ionized water-soluble molecules is virtually excluded from the brain. The converse is found for the permeability characteristics of the kidney and liver, which relate to their unique functions, and the pores may be considerably larger allowing many large molecules through with ease. A factor which tends to decrease the active concentration of the compound in the blood is protein-binding. A variable but often significant portion of many drugs become physically bound to plasma albumin and occasionally to other blood constituents. It is an important concept that only unbound 'free' drug can pass through the capillary wall. However, protein-binding is a freely reversible process, and when free drug leaves the circulation an equivalent amount of bound drug becomes free to restore the balance. The major consequence of protein-binding is that a drug's effectiveness may be markedly reduced. Moreover, increasing the dose to compensate for the binding can be hazardous since the initial dose may partially saturate the binding sites so that the higher dose results in less of the compound becoming bound and consequently a large amount remaining free and active. Doubling the therapeutic dose of phenylbutazone, for example, results in a relative tenfold increase in free drug in the plasma which leads to a proportional increase in its toxicity rather than in its therapeutic action. It also follows that an animal whose plasma protein composition is altered by, for example, malnutrition, fever or disease or other major stress conditions will show binding characteristics and drug reactions different from normal, possibly with adverse effects.

Once a chemical has passed from the blood into the extracellular fluid, its further passage into the intracellular fluid of the cells which constitute its site of action, and where it may have to combine with specific receptors before it initiates a response, depends on the physicochemical characteristics already discussed (e.g. lipid solubility). The amount of drug reaching the site of action is closely associated with the mass of the target organ or tissues relative to the total body mass and the blood flow to these. Organs such as the brain, heart, kidney and liver are perfused at high rates whereas bones and the subcutaneous fat receive blood at very low rates. The rapidly perfused organs quickly acquire a drug concentration equal to that in the arterial blood, particularly if the drug is freely lipid soluble. However, one consequence of high lipid solubility here is that the drug may quickly cross back into the blood and be redistributed into non-target tissues, in particular into fat tissue which acts as a depot. The action on the rapidly perfused organ tends therefore to be quick and short as the drug becomes redistributed and its concentration at the target organ falls. Thus the rapid but short-lasting action of anaesthetics such as sodium thiopentone and halothane, for example, is due to their very high degree of lipid solubility.

Metabolism and excretion of drugs

The termination of a biological effect of a chemical compound is brought about by its metabolism and eventual excretion. The rate at which these occur will be influenced by whether the drug is extensively bound to blood proteins (80 per cent), the degree of perfusion by blood of the eliminating organ, the activity of the drug-metabolizing enzymes and the efficiency of renal excretion.

Metabolism

As perhaps might be expected, those physicochemical characteristics that are useful for the absorption and distribution of a chemical compound are the opposite of what are required for its elimination from the body. Metabolism changes substances generally into inactive metabolites which are more polar (i.e. water soluble), more ionized and less lipid soluble than the parent compound. In some cases, however, the initial processes of metabolism may actually create a biologically active substance from inactive parent compounds, or they may produce a metabolite with activity different from that of the parent compound. However, even these compounds will eventually be metabolized and in most cases inactivated and processed for excretion. The liver, with its contingent of microsomal enzymes, is the major organ engaged in metabolism, though kidneys, lungs, blood, intestine and the gut microbial flora all contribute to a lesser or greater extent. Different species can differ widely in their metabolic processes and this is due to differences in their drug-metabolizing enzymes. For example, cats lack liver glucuronide-forming enzymes while some strains of rabbits and rats possess a plasma atropinesterase which is absent in man and in mice. It should be noted that some degree of lipid solubility is required for a substance to be metabolized by the liver. Highly ionized, water-soluble compounds are not metabolized and usually are excreted unchanged. Very few drugs undergo complete metabolism within the body, and unchanged drug levels can be measured, a fact made use of in forensic science! The rate of metabolism is a determinant in the duration of action of a drug and for many drugs the rate, within normal limits, is directly proportional to the concentration of free drug plasma. However, when drugs saturate the capacity of the metabolizing enzymes inactivation becomes constant and maximal, in spite of any increase in the drug concentration that might occur because of further administration of more drug. Drug effects therefore may be different from that expected for the dose given.

In animals, as in man, there are several hundred drugs which are known to stimulate the synthesis of microsomal enzymes. These 'enzyme inducers' can be loosely grouped into those that structurally resemble phenobarbitone and those that are similar to the carcinogenic polycyclic hydrocarbons. Chronic (i.e. longer term) administration of a chemical compound may thus stimulate and increase its own metabolism and indeed that of other compounds, and interactions may occur between concurrently administered compounds. Similarly, inhibition of microsomal enzyme activity can also be achieved by drug interactions. It is also worth noting that patterns of disease such as liver disease, fever, some infectious diseases and altered haemodynamics produced by, for example, shock-induced reduction of blood flow can markedly alter drug liver metabolism with consequential alterations in drug effect.

Excretion

The kidney, via the urine it produces, is the major route of excretion of drugs. Hydrophilic drugs are often excreted in the urine unchanged whereas lipid-soluble drugs usually undergo metabolism to a more water-soluble form before they can escape in the urine. Three main factors affect excretion in the urine — filtration, tubular secretion and tubular reabsorption. Ultrafiltration of the blood plasma at the kidney glomerulus is by passive diffusion. Because the membrane pores of the glomerulus are very large, they allow the passage of molecules with a molecular weight of up to 68,000. However, albumin is excluded to a large extent and substances bound to it are not filtered. At the proximal tubule surface, active secretory mechanisms ensure that as free drug passes into the tubule lumen, dissociation occurs, liberating more drug or materials bound to albumin with further active transmembrane passage. Passive concentration-gradient associated secretion as well as other special transport mechanisms are also found in the proximal tubules. As water is reabsorbed during its passage through the renal tubules, the concentration of drugs in the glomerular filtrate increases, establishing a concentration gradient which favours the reabsorption of drugs back into the renal circulation. However, as reabsorption is proportional to lipid solubility and the degree of ionization and is determined by the pH of the tubular fluid and the pKa of the drug or metabolite, the more polar and highly ionized drugs are excluded from reabsorption and are thus excreted. The normal urinary pH of carnivores such as the dog and cat and of the common laboratory animals, which are fed a diet high in protein and grain, is acidic, while the urinary pH of herbivorous animals is generally alkaline. The degree of tubular reabsorption under these conditions can therefore be evaluated. For example, in alkaline urine, acidic compounds ionize and are less well reabsorbed than would be the case if the urine were acid. The urinary (tubular fluid) pH can be manipulated by various means and this is used to advantage, for example in clinical medicine, to enhance or retard excretion of pharmaceutical drugs.

The presence of renal disease can make an important difference to the body's handling of drugs. In some instances, the rates of metabolism, the elimination of drugs which are normally excreted in the unchanged form and the extent of binding to plasma albumin may all be reduced. This could result in abnormally high and possibly toxic levels of active drug in the body. It is well to remember that renal disease is prevalent in middle-aged and old laboratory animals, probably because of the excessively nutritious diets they are fed, and their use when administering drugs must be carefully considered if accurate drug effects are to be achieved.

Although renal excretion is the main route for the elimination of waste compounds, a large number of drugs are excreted partly or even mainly via the bile. Polar water-soluble compounds with molecular weights above 300 are likely to be excreted via this route in significant amounts, particularly in rats and dogs. Less is excreted by this route in guinea pigs, rabbits and humans, which are classed as poor biliary excreters. Since compounds excreted in the bile end up in the lumen of the duodenum, glucuronide conjugates may be hydrolysed by the enzymes produced by the gastrointestinal bacterial flora, resulting in release of the parent compound. If sufficiently soluble, this can now be reabsorbed by the intestinal mucosa, reenter the general circulation, via the portal system, and exhibit pharmacological or other activity. Thus, this enteropepatic circulation of certain chemical compounds may contribute to an untoward drug effect from the 'increased' drug dosage that ensues (e.g. a prolonged effect of the drug).

In conclusion, dose is a quantitative term which describes the amount of a chemical compound which is administered and which gives a particular biological response. Factors which influence the effective concentration of a free drug at its site of action, and therefore its biological response, include the size of the dose administered, the route of administration, the release and absorption of the drug from its dosage formulation, the extent of plasma distribution, the penetration through cell membranes to the specific receptors and the rate of metabolism and excretion of the drug. Admittedly, in many instances, it will not be possible to gather all this information, which is designed to provide the investigator with a very good idea of how his drug will behave in the animal after it has been administered. But at least by being aware of and actively incorporating as many of the basic principles of pharmacodynamics as possible, he will ensure a more successful conclusion to any investigation than by merely using the purely empirical approach.

SOME PRACTICAL CONSIDERATIONS FOR THE ADMINISTRATION OF DRUGS

Suitable vehicles for injection

Water-soluble compounds should be dissolved in distilled water or preferably physiological saline (0.85 per cent w/v sodium chloride) for injection via all routes of administration. Distilled water may cause some haemolysis when given intravenously and physiological saline is much to be preferred. Balanced salt solutions of various kinds, such as tissue culture media, can also be used without adverse effect. Some less soluble compounds may require a more complex solvent to prepare them for injection and one of the following combined with distilled water or physiological saline can be tried: 50 per cent (v/v) dimethylsulphoxide or dimethylformamide; 10 per cent (v/v) ethyl alcohol; 60 per cent (v/v) propane-1:2-diol (propylene glycol); 10 per cent (v/v) Tween 80 (polyoxyethylene (20) sorbitan mono-oleate); 60 per cent (v/v) N-hydroxy-ethyl lactamide; 10 per cent (v/v) polyoxyethylene (23) lauryl ether; 15 per cent (v/v) glycerine and polyethylene glycols of molecular weights 200–600, providing not more than about 15 mg/kg of neat solvent is administered in total. The percentages given are generally maximum concentrations of these solvents and the use of lower concentrations is desirable. Other solubilizing agents such as cremophor E L (polyethoxylated castor oil) are available but have been prepared for special applications, often to overcome very difficult solubility problems, and their use in particular circumstances would have to be judged empirically. It should be borne in mind that some of these solubilizing agents will have pharmacological activity of their own. In reality, the problem of solubilizing drugs can be extremely complex and may require considerable expertise. For insoluble compounds, a homogenous suspension will need to be prepared. Mixing the compound with 10 per cent (w/v) aqueous acacia (gum arabic), a pharmacologically inactive agent, is often employed, particularly in the pharmaceutical industry. This should be prepared by adding the acacia to distilled water and allowing an overnight period for it to dissolve. The compound to be administered should be mixed with the warmed (37°C) acacia solution, preferably by using a pestle and mortar, though it can be done less efficiently by whirly mixing, and injected as soon as possible. Gum tragacanth (0.5 per cent w/v) can be used similarly. Another agent which might be useful is 0.5 per cent (w/v) aqueous carboxymethyl-cellulose. If it is required to inject a suspension intravenously it must be done with care and the particle size should be very small and finely dispersed. Injections of suspensions can never be absolutely accurate because of the tendency of particles to sediment.

Administration of lipid-soluble compounds can be done using various types of oils, but vegetable oils such as peanut, olive and corn oils are much preferred to mineral oils (e.g. paraffin oil). Oils retard the absorption of drugs from the site of administration and also cannot be safely used to inject via the intravenous route. If it is desired to inject fat-soluble compounds intravenously then this can be done as a suitable emulsion. An example of this is a 15 per cent (v/v) emulsion prepared with any vegetable oil mixed with purified soybean lecithin and a non-ionic water-soluble polyoxethylene–oxypropylene polymer (e.g. Pluronic F-68) and emulsified with distilled water. The emulsification process is fairly critical and relatively complex. Fat-soluble compounds can of course also be administered as a fine suspension in water.

Volumes for injection

The actual amount that can be injected will depend mainly on the route of administration and on the size of the animal. The peritoneal cavity will accept the largest amounts (e.g. 5 ml in a 200 g rat and 50 ml in a 20 kg dog), closely followed by the stomach. The subcutaneous and intravenous routes will accept about the same amount as each other, but it should be noted that too much fluid injected intravenously (e.g. over 1 ml in the adult rat) can sometimes cause a fatal pulmonary oedema. With subcutaneous injections, too much fluid will result in a leakback through the needle hole in the skin and thus significant loss of the computed dose. Only relatively small amounts can be placed successfully into intramuscular sites but many sites can be utilized simultaneously to increase the total dosage. In general, irrespective of the route of administration, the smallest volume should be used that is compatible with solubility of the compound and accuracy of the dose (see Table 13.1, p. 00). Detailed recommendations for injectable volumes in species of different sizes are given in Chapter 13.

The pH of solutions for injection

In all laboratory animals a working range is generally in the region of pH 4.5–8.0. The stomach can tolerate more acid conditions providing the acid concentration is similar to that found in the stomach, but more alkaline solutions are often not well tolerated in this location. The blood has buffering capacity and therefore tolerates the widest pH range, followed by the intramuscular and the subcutaneous routes respectively.

Nature of the injection solution

Drug solutions that cause pain are best injected by the intramuscular route since muscles are poorly innervated with pain fibres. However, distension of muscles with large volumes is painful. Some solutions of drugs can be highly irritating, often because of adverse pH (e.g. sodium thiopentone solutions have a pH of 11) and usually these can only be safely injected intravenously. If they inadvertently get administered subcutaneously they will often produce necrosis and sloughing of tissue at that site. Sterile abscesses may occur with some drug forms.

In theory, solutions for injection should be sterile otherwise an infection or fever can result. In practice, particularly with rodents, rabbits, birds and other small laboratory animals, more often than not little or no attempt is made to keep the solution to be injected sterile although gross contamination should be avoided. Moreover, whereas in humans and often in the larger species the surface of the area to be injected is swabbed with an antiseptic solution, this is rarely done with the small animals. In spite of this there rarely seems to be any unfavourable outcome to injecting non-sterile solutions in these small species. At the present time therefore, although it would be wrong to recommend the injection of non-sterile solutions into small laboratory animals, it can nevertheless be a matter for discretion and informed judgement.

The rate of injection

All injections should be made slowly, particularly with intravenous injections, where the animal should be watched closely for any adverse reactions. However, with certain parenteral anaesthetics, for example sodium thiopentone, it is usually recommended that half the computed dose is administered quickly. This is because it is highly lipid soluble and needs to arrive in concentrated form at its site of action for its initial anaesthetic effect.

Size of needle for injection

The needle of the smallest diameter that is compatible with the viscosity and nature of the final form of the compound to be administered and the size of the animal must always be used. Too large a needle will leave a relatively large needle hole and needle track and may result in loss of fluid by leakback and inaccuracy of the dose. For aqueous solutions, a 25 gauge needle or smaller is recommended but a 21 or 19 gauge may be used in dogs and larger animals. For viscous solutions and for suspensions it is usually necessary to use 21 down to 16 gauge needles in all animals. Luer-Lok needles and syringes are recommended for difficult to inject fluids, to prevent inadvertent and sometimes explosive separation of the needle from the hub of the syringe. Again, detailed recommendations are given in Chapter 13 and Table 13.1.

SOME PRACTICAL CONSIDERATIONS FOR CHOOSING THE RIGHT EXPERIMENTAL ANIMAL

Laboratory animals are used for experimentation and for the toxicity testing of chemical substances because society has determined that man must not be so used. Although it is the basis of much discussion, there seems little doubt that results from laboratory animals can be extrapolated successfully to man. This has been particularly emphasized by retrospective clinical studies in man in which the toxic side-effects of drugs observed in man are compared to those found initially in animal studies. It must be remembered, however, that there is no animal species which shows exactly the same response as man and therefore extrapolation of results must be done only after careful consideration of and allowing for the known idiosyncracies of the test animal. For example, the mouse, rat and rabbit have inactive steroid-17-hydroxylase in the cortex of their adrenal glands and therefore cannot synthesize cortisol efficiently. Consequently, they produce corticosterone as their main adrenal glucocorticoid. Such animals are quite different in several features from man, as well as from the guinea pig, dog and monkey, who produce mainly cortisol. In a further example it is known that the inhibition of NaK-ATPase by ouabain needs drug concentrations in the rat some 100–1000 times that required for man, cattle and dogs. Thus, transfer of results to man can only be done after careful quantitative and qualitative corrections.

In many cases in which the question arises of a suitable choice of animal for a particular study, the answer would appear to be simple. The animal species and strain to choose is the one which has appeared frequently in the literature and used for a procedure similar to the one contemplated by the investigator. The rationale for this choice is that much will already be known and published about that animal species and strain and its use will have been proven. However, there may be a lie to this argument in that the original choice of the best animal may have been based on inadequate and inappropriate initial considerations, even perhaps on such a flimsy one as 'that species was the only one available in the animal house at the time'! Although the subsequent results obtained with this species may fortuitously be adequate, more consideration given to the choice of animal might have resulted in a strain being chosen which showed a considerably better sensitivity and response. In spite of this, it must be said that this method of choice has become traditional and will no doubt continue to be considered as a reasonable compromise procedure.

The use of a particular species or strain for one purpose does not necessarily make it suitable for another and the best way to proceed to make a choice, especially if the proposed procedure is a new one, can be described by a few basic principles. Paradoxically, the first consideration, in practice, when making a choice is not whether the species or strain will give a suitable response but whether it is easy to keep! The questions to ask are whether it: (1) is easily available, in adequate numbers; (2) can be maintained in captivity without too much trouble; (3) has reasonable space requirements for which adequate facilities are available; (4) is easy to handle; (5) is easy to feed; (6) can be kept in reasonably good health under the conditions available; and (7) can be produced reasonably cheaply with respect to a definitive budget. Any or all of these may limit initial choice. Following this, the investigator should enquire what is known about the species or strain. He may want to know how suitable is its anatomy, physiology, biochemistry, behaviour and genetics for instance, and how closely these resemble similar parameters for human beings. A list of characteristics of mammalian species commonly used in biomedical work is available either in individual scientific papers where they arise empirically as a result of the experiments made, or from major reference sources. The rat and mouse are by far the most popular choice of laboratory species, while the guinea pig, rabbit, dog and primate have been given less consideration because of their more complicated maintenance requirements. The ubiquitous use of rats and mice has resulted in many strains being highly standardized with respect to their characteristics and use. Inevitably this has simplified the procedure of choice, a move that can only be welcomed by experimental scientists and those involved in the testing of pharmaceutical and other products.

An important consideration for choice, particularly in relation to clinical medicine, is whether there exists an animal which naturally mimics a disease condition in man, or whether there is an animal model in which the animal can be treated with a drug in a way which causes it to simulate a disease condition in man. Examples of the first category are the Brattleboro rat (di/di), which naturally develops diabetes insipidus; the nude mouse (nu/nu), which is athymic and immunodeficient; the kyphoscoliotic mouse (ky/ky), which spontaneously develops a skeletal abnormality resembling scoliosis in humans; several rat strains which develop various forms of diabetes mellitus; a strain of chicken and one of the mouse which display hereditary muscular dystrophy; and strains of dogs which are naturally deficient in various factors needed for normal blood clotting. Examples in the second category are arthritis induced in rats by injections of the bacterium *Mycoplasma arthritides*; diabetes mellitus induced in rats with alloxan and other chemicals; amyloidosis induced in several species by administration of casein; vitamin E deficiency induced in rhesus and other monkeys by dietary manipulations; and ulcerative colitis induced in guinea pigs with carrageenin. Very many more examples can be found and it is evident that if an appropriate animal model of an abnormal condition exists, whether naturally occurring or induced, use of that model will facilitate a study of that condition.

Some animals have natural attributes or have undergone extreme adaptations such that their study will facilitate fundamental research. The giant nerve cells discovered in the squids *Loligo* and *Sepia* have considerably helped research into nerve conductance. The finding that active secretion was a major function of the kidney tubule was initially suggested from work done in the dog, but proof was only obtained by a study of the goose fish, *Lophius piscatorius*, which peculiarly had neither a kidney arterial supply nor glomeruli and yet was capable of active secretion. As an example of extreme adaptation, fishes in the Arctic and Antarctic live in sea water the temperature of which is sometimes below that of the blood of these animals, suggesting that they should in reality be frozen! That they are not can be shown to be due to large molecular weight peptides in their body fluids which moderate the effect of such low temperatures. Study of such animals makes it possible to evaluate basic physiological processes better. These examples bring into focus that it may be worthwhile, in particular circumstances, investigating the experimental use of an uncommon species rather than a common one. Such uncommon animals as the chinese hamster, the pica, the multimammate mouse, the short-tailed opossum, the armadillo and various hibernating animals (e.g. bats, hedgehogs and squirrels) have all been advocated as having a possible use in numerous investigations, and conditions for their successful maintenance in the laboratory have been described.

Many other factors should be considered before finalizing one's choice of species and strain. For example, health status may be an important consideration, as may longevity of the species, the inbred or outbred nature, the sensitivity and reactivity to various chemicals, and even the ease of injection by the intravenous route if this route is important to the investigation. Ideally, the choice of laboratory animal and the factors affecting that choice should be closely allied to the purpose for which the animal is required. However, it is rarely possible to incorporate all the factors on which a choice should be made. In many cases, economic and other pressures do not allow for the best choice at all and a compromise is the best that can be done. But whatever choice is finally made, it should have been as a result of a conscious effort to understand the factors affecting choice and their limitations. The ultimate proof of the right choice, made on the basis of such conscious effort, can only be found in its practical application. In most cases a preliminary short-term practical trial should be carried out and sometimes this can be done with as few as just two or three animals. The results from this could validate the reasons for one's choice and may prevent a very costly failure as a result of making an inappropriate choice of experimental animal.

THE PREPARATION OF A PROTOCOL

In toxicological evaluation and testing of chemical compounds, a strict protocol is invariably prepared so that the compound and the test animal can be identified precisely if required, both for publishing the results and for carrying out subsequent similar tests. Such an identification procedure should also be the goal of all scientists in any biomedical discipline and an example will be given of typical requirements for such a protocol.

Identification of the drug

A record sheet should be prepared so that the test article or drug can be identified by its chemical name, its common name (if different), its structural formula, its purity and/or concentration, its impurities, its molecular weight and its method of synthesis. Physical characteristics such as appearance, colour, melting point, boiling point, specific gravity and vapour pressure at room temperature should be noted. The method by which the definitive solution of the compound is to be prepared for administration should be described. Information should be available on conditions of storage of the native compound and/or the chemical either in solution or as otherwise prepared for administration. For example, what is its stability relative to the effect of light, heat, air and water? How should it be stored in between use? Information should also be available on the safety of the product — has it any known toxicity? are there any special handling precautions needed? is there an antidote to personal contamination, either chemical or behavioural (e.g. immediate showering)? Finally, the method of disposal of the compound after use should be clearly indicated. Depending on the type of study, it may not be necessary or even possible to obtain a record of all these parameters but whatever points of identification can be made should be collected together in one simple document where the information is easy to retrieve.

Identification of the animal

A procedure similar to that for identification of a drug should be carried out for the test animal. The species, strain, origin of supply, age, body weight, sex and health status should be noted. The environment in which the animals are maintained during test should be described. This will include the average and range of temperature, humidity, ventilation and lighting. The type of diet and water and how these are administered should be noted. In some studies it may be essential to know the component analysis of the food and water and this must be obtained from the supplier if they are not to be determined by other means. It may be important to acclimatize and condition the test animal for a period before use, or perhaps to immunize or treat it against certain diseases, and if so these must be recorded. How the animals can be physically identified (e.g. ear marks or named neck collars) should be described. It is also useful to note how the animals have been allocated to the cages (e.g. randomly), how many animals are in each cage of a known dimension and also how the cages have been allocated to the racks or to their position within a room (e.g. again randomly). A record should also be made of what to do if the animal becomes ill or is found dead, if only for the instruction of other staff involved.

The justification for recording such parameters for the complete identification of both and drugs to be administered and the test animals can be obtained from studying the following section on the effect of variables. The need for such detailed specification can be summed up in one word — reproducibility. The ultimate aim of the experimental or toxicological scientist is to publish the results in one form or other. True confirmation of the results of a single experiment by a repeat of the work, either by the author himself or by workers in other laboratories (usually considered as a logical necessary step in the method of science), can only be obtained if sufficient attention has been given to recording and eventually publishing as fully as possible those various parameters of identification which have been described above. Thus it is logical to assume that results can only be reproduced precisely if the conditions under which those results were initially obtained are themselves reproduced faithfully.

THE MODIFICATION OF DRUG ACTION BY BIOLOGICAL AND ENVIRONMENTAL VARIABLES

It is insufficiently appreciated that the behaviour and functioning of an animal is dependent on the quality of both the external environment and the internal biological factors which act on it. The well-being of a laboratory animal can be defined as that steady state in which the animal is in physiological equilibrium and in which it is not experiencing any significant stress. Stress itself can be defined as one or more factors which upsets the steady state of an animal, inducing compensatory mechanisms which, if inadequate or inappropriate to return the animal to 'normal', may lead to pathological conditions. All environmental and biological variables are examples of factors which can stress the animal. Changes in these factors bring about changes in the animal's behaviour and function which, if slight, may be considered as inconsequential, that is the animal remains substantially within our definition of normal. On the other hand, these changes may matter a great deal, producing an animal which is abnormal and responds thus. The administration of a drug to an animal for which we are able to define these environmental and biological factors will cause it to respond in a specific way relative to the action of the drug. If the drug is readministered at some future time, the same response can only be expected if the variables influencing the animal are the same as before. The response, therefore, is a product of the interaction of the drug with an animal which has a defined set of environmental and biological factors. Change these factors and the response to the drug will change. Environmental and biological factors which influence the animal's response are unfortunately inherently variable. Under normal circumstances this is compensated for by the scientic method, which is to use control animals and statistical analysis. However, this in itself only assures the investigator that the response observed has a high chance of being truly representative only in the conditions under which the drug was administered. Therefore it is important to try to define the set of conditions which can be acceptable as 'normal' and to understand the effect on the animal when these conditions or variables are allowed to alter.

In many cases, investigators make no attempt to define their animals, often taking on trust an animal handed to them, and possibly maintained, by the animal house or facility with which they are associated. This is in stark contrast to their work with chemicals, where all the facets of a particular technique, say a quantitative biochemical measurement of the activity of an enzyme, are carefully studied in order to get the conditions such as temperature and substrate correct and defined precisely. This implies that scientists generally ignore or are ignorant of the impact on their experimental results of numerous variables that can affect their animals, whereas they recognize the impact on their data of small alterations in biochemical and other biomedical test systems. The seriousness of this omission cannot be overstated and examples can be found in the literature where such an omission has led to confusion and discrepancies, particularly between different laboratories. In extreme cases, it has led to a complete invalidation of the results!

Environmental variables

Clough in Chapter 5 discusses in some detail the role played by a number of important environmental variables in the response of laboratory animals to test procedures. Temperature and humidity have obvious influences upon such factors as activity and metabolism generally, but more subtle effects such as changes in the cardiotoxicity of isoprenaline and absorption of topically applied preparations can be very marked under the right conditions.

Lighting, sound, smell and diet (in Chapter 12) have all been shown to have important effects which might influence an animal's response to the planned test procedure, drug or toxic material. Retinal damage in albino rats, audiogenic seizures in mice, pheromone effects in female mice and tumour incidence or kidney disease in rodents are examples of other confounding effects.

Behaviour and social factors

The role of behaviour as an important component of animal's make-up is discussed by Blackman in Chapter 8, and behaviour and social factors comprise an important part of the influences affecting an animal's response during an experimental study. One can identify animal/cage interactions (type, size and shape of cage), animal/animal interactions (size of animal groups; length of time a group has been established; isolation from a group) and animal/man interactions (handling procedures and the frequency and skill of such procedures). Many studies have demonstrated the influence of each of these situations upon an animal's response. (See, for example, the following references in addition to those given in Chapter 5: Chance (1947); Symposium (1971); Symposium (1976); Fox *et al.* (1979); Gartner *et al.* (1980); Frohlich, Walma and Souverign (1981); Yoshida, Hagihara and Ebashi (1982).) It should also be noted that the transporting of animals demonstrably disturbs them, and for all these problems acclimatization periods are needed to restore the animal to 'normal'.

Disease

Clough (Chapter 5) discusses this factor in general terms, and describes current approaches to control of animal health. This is perhaps the most important variable which can influence the use of laboratory animals in biomedical work, and it can be illustrated by considering the effects of some common disease organisms or conditions.

Haemobartonella and *Eperythrozoon* are rickettsial parasites which are often found in animal tumours. Rats harbouring *Haemobartonella* organisms are resistant to implants of tumours already infected with this organism. Infection with these bacterial species variously causes hyperphagocytosis, increased plasma 1gG and 1gM levels and an almost complete inhibition of the response to interferon. Procedures which cause immunodepression such as the administration of cortisone, splenectomy and irradiation activate the organisms, which will then produce disease. *Eperythrozoon* markedly potentiates any infection with mouse hepatitis virus, lactic dehydrogenase virus and the virus of lymphocytic choriomeningitis, all of which have a common pathogenic pathway involving replication in Kuppfer cells prior to invasion of the liver parenchyma. It is also well established that both organisms complicate studies of rodent models of malarial disease.

Tyzzer's disease is probably caused by *Bacillus piliformis* (Fries, 1981). The disease can confound immunological experiments involving graft v. host reactions since liver necrosis is a feature of both. It has been suggested that Tyzzer's disease has been responsible for more ruined cancer research than any other mouse disease. Irradiation or other immunodepressant measures can unmask clinically silent infections with lethal consequences. Challenging rodents with immunodepressive doses of cortisone is a well-known way of testing for the subclinical presence of this and other disease organisms. Since Tyzzer's disease affects the liver, any resulting alterations in the activity of the liver drug-metabolizing enzymes can change the pharmokinetics of chemical compounds administered to these animals. For example, significantly increased blood levels of Warfarin have been found after its injection in mice in which Tyzzer's disease was causing liver damage. Immunological research is particularly vulnerable to disease effects in animals. Some bacteria and viruses, for example Group A streptococci and ectromelia and lymphocytic choriomeningitis viruses, produce immunodepression while some, such as *Corynebacterium parvum*, produce immunosuppression. Thus an infection with these or any other organism which modulates the immune response is likely to affect experimentation and toxicological studies. The disease organisms discussed above are only the 'tip of the iceberg', and interfering effects of diseases caused by such other well-known organisms as, for example, *Pseudomonas aerugenosa, Proteus* spp., *Staphylococcus* spp., *Escherischia coli*, Sendai virus, lactic dehydrogenase virus and the rat urinary tract parasite *Trichosomoides crassicauda* have been well documented. In spite of this, our knowledge is rudimentary and probably the full implications of the complicating role of disease in experimental and other studies are yet to be elucidated. It should now be readily apparent that serious scientific research and testing can only be carried out with confidence if animals of the highest health status are used.

Other variables

The list of variables mentioned above, and discussed by Clough (Chapter 5), is by no means complete, but represents probably some of the most important. Age, sex, pregnancy, circadian and ultradian rhythms, drug tolerance, medication, anaesthesia and the effects of parameters concerned with pharmacodynamics, such as the state and form of a drug, the type of vehicle, the route of administration and the rate of injection, are all examples of variables which are well known under various circumstances to adversely influence scientific investigations.

The enormity of the consequences that the effects of variation of biological and environmental parameters can have on results is no better illustrated than in a study by Chance (1947). He showed that the toxicity of certain sympathomimetic amines could have a tenfold level of variation unless the right combination and conditions of the following variables were obtained simultaneously; the strain of mouse, the state of hydration of the animals, the room temperature, the degree of confinement, the number of mice in the cage and the sex of the animals. More recently, it has been found that an adequate development of adjuvant arthritis in rats in response to an injection of *Mycoplasma arthriditis* was obtained only in the absence, though not necessarily in combination, of dietary mineral deficiency, overcrowding, group housing of high male:female ratios and infection.

Reducing the effect of variables

The situation for animal experimentalists can perhaps be adequately explained by analogy with that for users of organic and inorganic chemical compounds. Very many of these compounds are now produced in remarkably pure form so that interaction between such chemicals will produce results which will be due only to the chemicals themselves. This of course was not always the case and the situation when purification techniques were not as advanced was that most chemicals were produced with variable numbers and amounts of impurities. Interactions between these chemicals therefore could be influenced by the impurities. It would not be illogical to consider an experimental animal as a chemical, contaminated by a variable number of chemical, biological and environmental impurities. Laboratory animal science has now progressed to a point where many of these 'impurities' have been reduced. Thus, specified pathogen free animals and germfree animals are freely available and the environmental conditions under which laboratory animals apparently show a relatively homogeneous functioning of physiological, biochemical, immunological and other parameters has been defined with some degree of precision. In reality, therefore, the experimentalist requires a normal animal, which by definition is one which is not experiencing the action of stress factors and is behaving according to an accepted pattern of behaviour (that is, what the majority of people consider as normal behaviour). It is of course impossible in most instances to eliminate the 'impurities' and indeed the analogy must stop before this point since a perfectly normal animal is in fact constituted by a complex mixture of 'chemicals' and other factors all interacting in a definite pattern.

It is attempting to achieve this normal pattern that should concern the animal experimentalist. The obvious way to do this is to use animals which are already available with some of the 'impurities' reduced, that is healthy SPF animals kept in standardized conditions accepted as normal for the species and strain, and then to standardize as far as possible other known variable factors according to any information available to the investigator or as a result of logical or even intuitive reasoning. Providing a detailed record is kept of the standardized conditions under which the animal is maintained, the credibility and reproducibility of results will inevitably be strengthened. It should perhaps be emphasized that ultimate credibility will depend on results being reproducible by other workers and this will only be achieved satisfactorily if the environmental and biological conditions under which the experimental animals were standardized are published fully. The documented failure of this to happen at the present time is undoubtedly a prime reason for the apparent scepticism among investigators concerning the ability to reproduce in a straightforward manner, and in some cases even not at all, animal methods and studies published in the scientific literature.

One of the greatest problems confronting the investigator is to determine which variable factors should be considered for standardization relative to his experiment. Clearly, not all such factors are important in every case. Even if a choice had been made, the extra effort which may be required to standardize the animals and conditions could be considerable in certain circumstances and possibly could be achieved only after what might be considered by the investigator as much 'non-productive' preliminary testing. It is perhaps understandable therefore that the scientist is under much pressure to continue to disregard taking any but the simplest measures for controlling the common variables. However, whereas this might have been acceptable a decade or two ago, the amount of factual scientific evidence that has now accrued on the effect, sometimes devasting, of uncontrolled variables on the outcome of experiments makes dilatory and evasive action on the part of the investigator no longer acceptable. It is now certain that in many cases finer control of experimental variables will markedly improve results, which will then become more meaningful and trustworthy.

REFERENCES

Included in this list are a number of sources not specifically cited in the text but which deal usefully with related topics.

Animal Models of Human Disease. Fascicle 1–14 (1972, ff.). The Registry of Comparative Pathology, Armed Forces Institute of Pathology, Washington DC.

Baggot, J. D. (1977) *Principles of Drug Disposition in Domestic Animals*, W. B. Saunders, Philadelphia.

Barnes, C. D. and Eltherington, L. G. (1973). *Drug Dosage in Laboratory Animals*, 2nd edn, University of California Press, Berkeley.

Brander, G. C. and Pugh, D. M. (1971). *Veterinary Applied Pharmacology and Therapeutics*, 2nd edn, Bailliere Tindall, London.

Brown, A. M. (1967). The future of the defined laboratory animal, Symposium, *Carworth Europe Collected Papers*, Vol. 1, Carworth Europe, Huntingdon, UK.

Brown, A. M. (1969). Uniformity, Symposium, *Carworth Europe Collected Papers*, Vol. 3, Carworth Europe, Huntingdon, UK.

Burgen, A. S. V. and Mitchell, J. F. (1978). *Gaddum's Pharmacology*, 8th edn, Oxford University Press, London.

Chance, M. R. A. (1947). Factors influencing the toxicity of sympathomimetic amines to solitary mice, *J. Pharmacology and Experimental Therapeutics*, **89**, 289–97.

Conney, A. H. (1967). Pharmacological implications of microsomal enzyme induction, *Pharmacological Reviews*, **19**, 317–66.

Deeny, A. A. (1983). The laboratory animal: an equation of quality and experiment, *Animal Technology*, **34**, 53–60.

Foster, H. L., Small, J. D. and Fox, J. G. (1983). *The Mouse in Biomedical Research*, Vol. 3, Ch. 13, Academic Press, New York.

Fox, J. G., Thibert, P., Arnold, D. L., Krewski, D. R. and Grice, H. C. (1979). Toxicology studies 11. The laboratory animal, *Food and Cosmetic Toxicology*, **17**, 661–75.

Fries, A. S. (1981). *Tyzzer's Disease and the Importance of Inapparent Infection in Biomedical Research*, Department of Pathology, Royal Veterinary and Agricultural University, Copenhagen.

Frolich, M., Walma, S. R. and Souverign, J. H. M. (1981). Probable influence of cage design on muscle metabolism of rats, *Laboratory Animal Science*, **31**, 510–12.

Gartner, K., Buttner, D., Dohler, R., Friedel, R., Lindena, J. and Trautschold, I.

(1980). Stress response of rats to handling and experimental procedures, *Laboratory Animals*, **14**, 267–74.
Gartner, K., Hackbarth, H. and Stolte, H. (1982). *Research Animals and Concepts of Applicability to Clinical Medicine*, Experimental Biology and Medicine Monograph 7, S. Karger, Basel.
Goodman, L. S. and Gilman, A. (1980). *The Pharmacological Basis of Therapeutics*, 6th edn, Macmillan, New York.
Hegreberg, F. and Leathers, C. *Bibliography of Induced Animal Models of Human Disease. Bibliography of Naturally Occurring Animal Models of Human Disease*, Department of Veterinary Microbiology and Pathology, College of Veterinary Medicine, Washington State University, Pullman.
Hughes, H. C., Jun. and Lang, C. M. (1978). Basic principles in selecting animal species for research projects, *Clinical Toxicology*, **13**, 611–22.
ICLA (1978). Recommendations for the specification of the animals, the husbandry and the techniques used in animal experimentation (1978), International Committee for Laboratory Animals, *Bulletin* No. 42.
Illman, O. (1965). A plea for more unusual laboratory animals, *Food and Cosmetic Toxicology*, **3**, 203–8.
Paton, W. D. M. and Payne, J. P. (1968). *Pharmacological Principles and Practice*, J. & A. Churchill, London.
Roe, F. J. C. (ed.) (1983). *Microbiological Standardisation of Laboratory Animals*, Ellis Horwood, Chichester.
Rogers, H. J., Spector, R. G. and Trounce, J. R. (1981). *A Textbook of Clinical Pharmacology*, Hodder & Stoughton, London.
Schurr, P. E. (1969). Composition and preparation of intravenous fat emulsions, *Cancer Res.*, **29**, 258–60.
Sollman, T. (1942). *A Manual of Pharmacology*, 6th edn, W. B. Saunders, Philadelphia.
Symposium (1971). Diseases of laboratory animals complicating biomedical research, *American J. of Pathology*, **64**, 624–769.
Symposium (1976). Environmental and genetic factors affecting laboratory animals. Impact on biomedical research, *Federation Proceedings*, **35**, 1123–65.
Symposium (1978a). Recommendations for the choice of suitable experimental animals, *Publications of the Gesellschaft fur Versuchstierkunde*, No. 7, Basle.
Symposium (1978b). Results in laboratory animals — are they valid for man? *J. Royal Society of Medicine*, **71**, 675–96.
Waynforth, H. B. (1980). *Experimental and Surgical Technique in the Rat*, Academic Press, London.
Working Committee for Genetics/GV-Solas (1979). Recommendations for the choice of suitable experimental animals, *Publications of the Gesellschaft fur Versuchstierkunde*, No. 7, Basle.
Working Committee for the Biological Characterization of Laboratory Animals/GV-Solas (1985). Guidelines for specification of animals and husbandry methods when reporting results of animal experiments, *Laboratory Animals*, **19**, 106–8.
Yoshida, H., Hagihara, Y. and Ebashi, S. (1982). Models and quality control of laboratory animals, *Advances in Pharmacology and Therapeutics* 11, **5**, 295–352.
Yoxall, A. T. and Hird, J. F. R. (eds) (1979). *Pharmacological Basis of Small Animal Medicine*, Blackwell Scientific Publications, Oxford.

Laboratory Animals: An Introduction for New Experimenters
Edited by A. A. Tuffery
© 1987 John Wiley & Sons Ltd

CHAPTER 12

Feeding and Watering

MARIE E. COATES
The Robens Institute, University of Surrey, Guildford

INTRODUCTION

The laboratory animal is entirely dependent on the curator or experimenter for its food supply. It has no opportunity to forage for itself so the diet must provide an adequate amount of every essential nutrient. The food must be palatable and be presented in a form that the animal will accept and be able to eat in sufficient quantity to satisfy its hunger.

Stock diets are commercially available for most species of laboratory animal and may be of variable or fixed formula. In the former the manufacturer may vary the ingredients according to current costs and availability in order to achieve a nutritionally adequate product at least cost. Such diets are satisfactory for the breeding or maintenance of a colony for which economy is the prime factor but for the research worker who requires consistency of performance the composition of the diet should not vary. Even diets made to a fixed formula may not be strictly consistent from one batch to another if the ingredients are natural materials which are subject to seasonal or varietal differences in their composition. To ensure complete consistency over long periods of time diets must be compounded of chemically pure ingredients. These so-called 'purified' diets are expensive, but the cost must be set against the risk of interference with the experimental programme by variation in the standard conditions. Furthermore, the objectives of the investigator may call for modification of one or more components of the diet, and this can be most readily and accurately achieved if the basal diet consists of purified ingredients.

In formulating diets for experimental animals certain basic nutritional principles must be adhered to as outlined below. Tables of the nutrient composition of typical ingredients of laboratory animal diets have been prepared by several official bodies (see, for instance, Harvey, 1970; Aitken and Rankin, 1972; National Research Council, 1982).

FORMULATION OF LABORATORY ANIMAL DIETS

Quantitative requirements for nutrients

Although the quantitative requirements for nutrients of laboratory animal species cannot be stated exactly, recommendations based on the available scientific evidence have been made and have proved satisfactory in practice (Clarke *et al.*, 1977; Rechigl, 1977; National Research Council, 1974, 1977, 1978a, b, c). Many of the definitive recommendations refer to the rat and requirements of other species have been less well documented. Nutrient requirements are influenced by factors such as age, strain, environment and physiological status. Growth, pregnancy and lactation all impose increased demands for nutrients, which are only partially satisfied by increasing food consumption. It may be necessary to provide breeding and growing animals with diets of higher nutrient density than those fed to non-breeding stock. Conversely, overfeeding can be as harmful as deficiency, particularly in long-term experiments. Formulas based on those for domestic animals tend to be overgenerous, particularly in energy content, since the circumstances in which laboratory animals spend their lives are very different from those of farm animals.

Specific nutrients

Protein

It is not strictly true to say that an animal requires protein. Proteins consist of amino acids, and different proteins are made up of different combinations of amino acids. An animal needs a source of amino acids to build its own body proteins. It cannot utilize proteins as presented in the diet until they have been broken down into their component amino acids or small peptides by digestive enzymes in the stomach or small intestine. These are then absorbed and used to synthesize tissues enzymes, antibodies, etc. However, protein synthesis cannot occur until all the necessary amino acids are present at the same time. Amino acids are not stored in the body, so if one is in short supply the others are wasted and the animal's requirement is not satisfied.

Thus the qualities required of a dietary protein are that (1) it should be readily digestible so that its component amino acids are released for absorption and (2) the amino acids so released should be in the right quantities and proportions to be utilized for protein synthesis. Animals are able to synthesize some amino acids, for example serine, from simpler compounds. These are termed 'non-essential' or 'dispensable' (i.e. it is not essential that they be present in the diet). The so-called 'essential' amino acids cannot be synthesized by animal tissues (e.g. lysine) or are not synthesized in sufficient quantity to satisfy requirements (e.g. methionine) and must be provided in the diet.

There are no really ideal food proteins with the possible exception of whole egg. Most are lacking in one or more of the essential amino acids and few are completely digestible. In diets of natural ingredients it is customary to use a mixture of proteins so that deficiencies of one may be compensated by excesses in others. Some amino acids are commercially available at relatively low costs, and can be added to diets to make good small deficiencies. Typical amino acid compositions of some proteins commonly used in diets for laboratory animals are given by Clarke *et al.* (1977) and the National Research Council (1982).

Digestibility of food proteins depends on their ready exposure to proteolytic enzymes. Heating can impair digestibility either by hardening the protein or by encouraging chemical reactions between amino acids and other components of the diet. For example, the Maillard reaction between reducing sugars and amino acids is responsible for the browning and impaired digestibility of milk proteins.

The 'crude protein' value is a chemical measurement and gives no information about nutritional quality, which can only be assessed by elaborate biological tests. As a rough guide, it may be assumed that the biological value of the proteins used in stock diets is about 60 per cent that of their crude protein value. For proteins such as casein or lactalbumin used in purified diets the figure is approximately 90 per cent.

Energy sources

Energy is released in the course of metabolism of fats, carbohydrates and amino acids. Metabolizable energy (ME) is a measure of the amount of dietary energy that is retained and utilized by the animal. The ME value of fats is about 38 kJ/g(9 kcal) and of starches and amino acids rather less than half that value (17 kJ or 4 kcal/g).

Fat is not itself an essential nutrient except as a source of essential fatty acids and a means of facilitating the absorption of fat-soluble vitamins. It improves palatability and can be used to increase the energy density of a diet by replacing an equivalent amount of carbohydrate. Amino acids in excess of requirement are utilized as an energy source but proteins are usually the most costly ingredients of a diet so it is uneconomic to provide them purely on the basis of their energy content.

An animal's energy requirement is dependent on its body size rather than its body weight, and is expressed in terms of 'metabolic body size', i.e. kg body weight$^{0.75}$. The energy requirement for maintenance is essentially similar over a range of animal species and has been estimated to be about 450 kJ/kg body weight$^{0.75}$. For growth, pregnancy and lactation the values are higher and have been given as 1200, 600 and 1300 kJ/kg body weight$^{0.75}$ respectively (Clark *et al.*, 1977). These amounts should be adequate for laboratory animals, most of which are confined to cages in a comfortable environment and have little need to expend energy on heat production or physical activity. Intake of energy in excess of requirements leads to obesity and may also be associated with an increased incidence of spontaneous tumours. Both conditions can have serious consequences for long-term studies. (Roe, 1981; see also p. 219).

Protein:energy ratio

Within broad limits, when animals are fed *ad libitum* they usually eat enough to satisfy their energy requirements. Thus the amount they eat of a high-energy diet is likely to be less than of a low-energy diet. In consequence their intake of the other ingredients will be proportionately less and may fall below requirements unless their concentration is increased. Conversely, if the dietary energy content is very low protein constituents will be used as a source of energy. On the basis of studies with purified diets, protein:energy ratios of the order of 1 g crude protein to 70 kJ have been tentatively suggested to support reproduction in the rat (Clarke *et al.*, 1977). A similar value was proposed for growing rabbits given a natural diet containing white fishmeal (Davidson and Spreadbury, 1975). However, in studies with growing rats and mice given diets of natural ingredients, variation in the energy content from 9.62 to 12.07 kJ/g did not appear to influence food intake and performance was best on the diets with higher energy content irrespective of the protein:energy ratio (Ford and Ward, 1983a, b). In experiments with growing rats given a series of diets with protein:energy ratios ranging from 1:1 to 2:1 (per cent:MJ/kg) the animals ate less of the high-energy diets but did not completely adjust their food intake to give constant intakes of energy. In consequence, parameters such as body weight, abdominal fat and relative organ weights were greater in the rats given the high-energy diets. No common protein:energy level for optimal performance could be established (Edwards *et al.*, 1985; Edwards and Dean, 1985).

Fibre

The term 'dietary fibre' refers to a variety of carbohydrates that are virtually indigestible and therefore unavailable to the animal consuming them. They include lignin, cellulose and hemicelluloses which constitute the fibrous parts of vegetable matter and also bulky substances like pectin and guar gum. Fibre is not an essential nutrient, but is important in maintaining the physical properties of the digesta. It prevents the luminal contents from compacting into a hard mass, and so facilitates contact with digestive enzymes. It improves palatability of the diet and helps to maintain a regular rate of transit through the alimentary tract. Ruminants obtain a contribution to their energy requirements from cellulose and hemicelluloses, which are degraded to volatile fatty acids by microbial action in the rumen. Rabbits and other herbivores may derive a small amount of energy from fibre subjected to microbial digestion in the caecum and large intestine (Parker, 1976).

The ingredients of diets for natural materials contribute adequate amounts of fibre and there is no need for further addition. It is usual to include about 3–5 per cent fibre in purified diets and larger amounts can be used if it is required to dilute the energy density of the diet.

Vitamins

The essential nature of vitamins in animal metabolism has been well documented, and only matters of particular importance in laboratory animal nutrition will be mentioned here. The effects of deprivation of a vitamin may not be manifested in the same way in all animal species, as indicated in Table 12.1. Animal tissues usually carry reserves of the fat-soluble vitamins A, D, E and K and water-soluble vitamin B_{12}, so that dietary deficiencies may not become apparent until those reserves are exhausted.

Vitamin A occurs naturally in fish oils, but is unstable unless the oil contains an antioxidant. Dry powders, in which the vitamin is dispersed in a protective vehicle, are the preferred form of supplementation. Carotenoid precursors of vitamin A occur in plant materials and practical diets are likely to contain carotenoids in ingredients such as maize and dried grass. The most active precursor is β-carotene, which theoretically has half the biological potency, weight for weight, of retinol. However, the rate of conversion achieved in practice is very poor and it is unwise to regard the contribution of dietary carotenoids as a significant source of vitamin A. This is especially true of diets for the cat, in which the mechanism for conversion of carotenoids to vitamin A is virtually non-existent. Excessive amounts of vitamin A can be harmful. Adverse signs including skeletal deformities, severe haemorrhages, retarded growth and foetal abnormalities have been recorded as a result of massive doses (Deuel, 1957a; Moore and Wang, 1945; Cohlan, 1953). Although these effects are unlikely to be encountered in normal feeding practice, care should be exercised to avoid oversupplementation.

The two forms of vitamin D, ergocalciferol (D_2) formed on irradiation of plant sterols and cholecalciferol (D_3) from animal sources, are equally active for most mammals. It has long been known that birds are unable to utilize vitamin D_2 efficiently (Remp and Marshall, 1938) and there is more recent evidence that for New World monkeys vitamin D_2 is similarly ineffective (Hunt et al., 1967). Both respond normally to vitamin D_3.

Vitamin D is formed in the superficial tissues on exposure to ultraviolet light. If Ca and P are provided in sufficient amounts and in the correct proportions (i.e. a Ca:P ratio of more than unity), most laboratory animals require little or no dietary vitamin D. It is usual to include it in the diet as a precautionary measures, but over supplementation must be avoided. Vitamin D is toxic in

Table 12.1 Effects of vitamin deficiencies in laboratory animals

Vitamin	Chemical name	Effects of deprivation	Comments
A	Retinol (1 i.u. = 0.30 µg retinol or 0.34 µg retinyl acetate)	Poor growth. Reproductive failure. Visual disturbances — impaired dark adaption. Keratinization of epithelia — particularly the cornea. Xerophthalmia (rat). Increased susceptibility to infection.	Some carotenoids have vitamin A activity. Theoretically 0.6 µg β-carotene = 0.3 µg retinol but this is not usually attained in practice.
B_1	Thiamin	Retarded growth. Polyneuritis, resulting in incoordinated movements (ataxia), neck retraction (opisthotonus) in birds, convulsions. Cardiac failure.	Thiamin is concerned in carbohydrate metabolism. Requirement is reduced if dietary energy is supplied as fat.
B_1	Riboflavin	Retarded growth, loss of hair, Opacity of the cornea. Neurological abnormalities. 'Curled-toe' paralysis (chick). Dermatitis (dog, rhesus monkey).	
—	Nicotinic acid	Loss of apetite, depressed growth rate. Black tongue (dog). Rough coat, anaemia (pig). Perosis (chick).	Nicotinic acid can be synthesized in the animal body from tryptophan. Severe deficiency is only encountered on diets low in tryptophan.
—	Pantothenic acid	Loss of hair pigment (achromotrichia), Adrenal necrosis, loss of appetite. Blood-stained whiskers, 'spectacle eye' (rat). Dermatitis (chick).	
B_6	Pyridoxine	Dermatitis. Microcytic anaemia. Hyperexcitability, convulsions, marked decrease in growth, decreased antibody production.	Pyridoxol, pyridoxal and pyridoxamine and their corresponding phosphates are all equally active.
—	Biotin	Dermatitis. Paralysis. 'Spectacle eye' (rat).	Biotin forms a complex with avidin, a factor in raw

Table 12.1 (contd.)

Vitamin	Chemical name	Effects of deprivation	Comments
		Reproductive failure. Alopecia (mice). Perosis (chick).	egg white. Severe deficiency is only observed if raw egg white is included in the diet.
—	Folic acid complex. Pteroylmono-glutamic acid (and allied compounds with more than one glutamic acid residue)	Poor growth. Macrocytic anaemia. Leucopaenia.	Many species appear to derive some folic acid from microbial synthesis in the gut. In such animals, severe deficiency only occurs if sulphonamides are included in the diet.
B_{12}	Cobalamin	Retarded growth. Reproductive failure.	Requirement for vitamin B_{12} is reduced on diets containing ample methionine or choline.
C	Ascorbic acid	Scurvy (primates, guinea pig).	Most other species can synthesize enough in tissues and do not need a dietary source.
D_2 D_3	Ergocalciferol Cholecalciferol (1 i.u. vitamin D = 0.025 µg cholecalciferol)	Rickets in young animals. Osteoporosis in adults. Black feather pigmentation (chick).	Vitamins D_2 and D_3 are equally effective for most mammals. Birds and New World monkeys utilize vitamin D_2 very inefficiently. Animals and birds synthesize vitamin D if exposed to u.v. radiation. Dietary requirement is therefore reduced if they are maintained in bright sunlight.
E	The tocopherols (1 i.u. = 1µg synthetic dl-α-tocopheryl acetate)	Muscular dystrophy. Reproductive failure (except in cattle and sheep). Encephalomalacia. Exudative diathesis (chick). Liver necrosis, yellow fat (rat, pig, mink).	Some signs of deficiency (e.g. muscular dystrophy, encephalomalacia) can be prevented by antioxidants and others (e.g. exudative diathesis, liver necrosis) by selenium.
K_1 K_2 K_3	Phylloquinone Menaquinone-7 Menadione	Haemorrhagic disease, prolonged blood-clotting time.	Vitamin K_2 is synthesized by micro-organisms in the alimentary tract.

excess, causing bone resorption and deposition of Ca in the soft tissue. Although these severe effects have only been reported with exceptionally high amounts (Deuel, 1957b), vitamin D accumulates in the tissues and even moderate overdosage may be harmful in long-term studies. It is also efficiently carried over from dam to offspring, so for experimental work requiring animals to be depleted of vitamin D the amount in the maternal diet should be restricted.

The animal's requirement for vitamin E is profoundly influenced by other components of the diet. Vitamin E occurs in natural materials in the form of tocopherols, of which α-tocopherol is the most important and the most active. It is involved in maintaining the integrity of cell membranes (Molanaar et al., 1972), although its precise metabolic role(s) remains unclear. The tocopherols have antioxidant properties and so protect other dietary components such as vitamin A that are susceptible to oxidation. The requirement for vitamin E is increased in the presence of polyunsaturated fatty acids (PUFA) and diets designed to deplete animals of vitamin E usually contain fats with a high proportion of PUFA. An interrelationship exists between vitamin E and selenium (Scott et al., 1974), and many though not all of the affects of vitamin E deprivation can be relieved by dietary supplements of selenites. For example, selenites counteract liver necrosis in the rat and exudative diathesis in the chick, but not reproductive abnormalities in the rat nor encephalomalacia in the chick.

Vitamin K_1 (phylloquinone) occurs naturally in cereals and green fodder but the synthetic vitamin K_3 (menadione) is cheaper and therefore the more usual form for supplementation in laboratory diets. Vitamin K_2 (menaquinone) is synthesized by bacteria, and micro-organisms in the gut produce considerable amounts that are available to coprophagous animals when they recycle their faeces. Germfree animals require a higher dietary supplement of vitamin K to compensate for the lack of microbial synthesis and specified pathogen free animals or others with a limited gut microflora may also need more. If animals depleted of vitamin K are required, some device such as a tail-cup must be used to prevent coprophagy. For germfree animals vitamin K_1 is many times more effective than menadione in preventing signs of deficiency (Gustafsson et al., 1962). It is assumed that the conversion of menadione to the physiologically active form of vitamin K is facilitated by bacterial action. In very high doses menadione, but not vitamin K_1, is toxic (Owen, 1971).

The water-soluble vitamins, with the exception of vitamin B_{12}, are not well stored in the tissues. In the absence of a dietary source reserves are rapidly exhausted and acute signs of deprivation appear. Like vitamin K, vitamins of the B complex are synthesized in significant amounts by bacteria in the gastrointestinal tract. Although little if any of the products are absorbed directly from the gut, rodents and lagomorphs obtain an adequate supply when they practise coprophagy. Gnotobiotic animals need ample amounts in the diet to compensate for the lack of a microbial source of these vitamins. A suitable supplement for an autoclavable stock diet has been recommended by Kellogg and Wostmann (1969).

Vitamin C (ascorbic acid) is synthesized in adequate quantity in the tissues of most animal species, with the exception of the guinea pig and all primates. It is an unstable substance, and should be incorporated into the diet with sufficient excess to allow for losses during preparation and storage. As much as 50 per cent may be lost from the diet during the first six weeks after manufacture (Eva et al., 1976). It may be given in the drinking water, but the solution (0.5 g/l) must be fresh each day. Fruit and vegetables are unreliable sources of vitamin C as their content can vary widely. Furthermore, they rapidly become stale, which accelerates loss of vitamin C and provides a substrate for microbial contaminants.

Other accessory food factors

Other compounds required by most animal species in small amounts are choline, myo-inositol and the essential fatty acids (EFA). They are usually present in adequate amounts in diets of natural ingredients, but it may be necessary to add them to purified diets. Choline, an important source of methyl groups and a component of lecithin, is synthesized in animal tissues but not in sufficient quantity to satisfy requirements, and some must be provided in the diet. Amounts between 1 and 2 g per kg have been recommended for addition to purified diets. The presence of other methylating compounds such as methionine or betaine obviates the need for choline.

Myo-inositol is an important constituent of cell membranes. Although its addition to purified diets has been recommended, it is questionable whether the recommendation is justified, since inositol is synthesized in animal tissues and also by microbes in the alimentary tract.

Linoleic acid (C18:2, n–6) is an important dietary fatty acid since it can be converted in animal tissues to arachidonic acid (C20:4, n–6), the metabolically active compound. It has been estimated that an amount of linoleate equivalent to 1–2 per cent of the total dietary energy is adequate for several species of animal. This requirement would be met by inclusion of 5–10 g linoleate, e.g. 1–2 per cent soya or maize oil, in a diet providing 4 kcal (17 kJ)/kg. There is some evidence that fatty acids of the linolenic series (C18:3, n–3) may also be of physiological importance. Soya bean oil, which contains both linoleic and linolenic acids, is recommended for inclusion in purified diets.

Minerals

The list of inorganic elements essential to animal metabolism is very long, and all must be provided from an exogenous source. Many are required in minute amounts which are present in the drinking water or as contaminants in the major ingredients of larger mineral supplements. These so-called 'trace' elements need not be added to diets of natural ingredients but a complete supplement of all essential elements must be included in purified diets. Some of the trace elements, especially the heavy metals, are toxic in excess and care should be taken against oversupplementation.

Electrolytes including Na, K, Mg and Cl as well as Ca and P are needed to maintain osmotic balance in cells and tissue fluids. Dietary intake may vary over a wide range without harmful effects, because tissue concentrations are maintained by homeostatic mechanisms. The minerals needed in greatest quantity, especially during growth and lactation, are those concerned in bone formation, mainly Ca, P and Mg. The ratio of Ca to P is important, because a gross excess of one can interfere with proper utilization of the other. In most species about 5–10 g Ca/kg diet satisfies the requirement for growth and reproduction, with an optimal Ca:P ratio between 1:1 and 1:2. Much of the P in cereals is in the form of phytate, an organic compound which is poorly utilized as a source of P and which combines with other inorganic elements such as Fe, Zn and Mg, and prevents their absorption.

In designing a mineral supplement account must be taken of the quantities already present in the other dietary ingredients (for example, casein contributes appreciable amounts of P). Various combinations of salts can be used, but all elements must be included in the calculation and allowance made for water of crystallization.

Water

Water may be considered as the most essential nutrient, since it is the medium in which most metabolic reactions proceed and is itself an important component of many of them. Animals can survive deprivation of food much longer than deprivation of water. Although some desert species have become adapted to scarcity of water most laboratory animals require a constant supply.

Water may be provided in troughs, bottles or through an automatic watering device. The practice of providing at least part of the animal's water requirement in the form of fresh vegetables or wet mash should be avoided from considerations of hygiene. It is, however, permissible to give a gel diet (see p. 214) to animals being transported. Open troughs encourage spillage and are readily fouled; they should only be used where a drip supply is unacceptable to the animals. The standard system of water supply for small mammals consists of a bottle with a well-fitting bung through which a short metal or glass tube is passes. When the full bottle is inverted over the cage a drip of water forms on the end of the tube and is constantly replaced as the animal drinks. Several variations on this type of watering system are available commercially. The bottles and tubes must be regularly cleaned to prevent growth of algae and bacteria. The closures must be tightly fitting otherwise leaks occur.

Automatic devices usually comprise a storage tank from which narrow pipes lead to each individual cage. At the end of each pipe is a nozzle with a loose nipple. When the animal presses the nipple with its nose water is released. These devices effect considerable saving of labour, but they must be constantly checked. Deposits can form in the nozzle, particularly with hard water, so that the nipples become stuck either in the open position causing flooding or, more dangerously, in the closed position so that the animal is deprived of water. The storage tanks and supply pipes are excellent sites for microbial growth, so it is advisable to flush the whole system at regular intervals with, for example, a strong detergent or a dilute acid solution.

Tap water is a source of trace elements and may contribute a significant proportion of the animal's requirement for calcium. It may also carry undesirable contaminants such as lead or high concentrations of fluorine. The conditions will vary in different areas, according to the geological source of the water, its hardness and the type of piping through which it is supplied. Information on its mineral composition should be sought from the local water authority.

If an experiment requires measurement of fluid intake the bottle system, with a suitably calibrated bottle, should be used. In these circumstances special care must be taken to prevent leaks. For some experimental purposes it may be necessary to administer drugs or other test materials in the drinking water. The substance should be dissolved in distilled water and the control animals should also receive distilled water. Due attention should be paid to the stability of the test substance in solution, which may necessitate inclusion of a buffer or other stabilizing agent. In this event the control animals should be given distilled water containing all the additives except the test material. Daily fluid intakes should be measured to determine the total intake of the test substance.

PREPARATION AND PRESENTATION OF DIETS

The diet must be presented as a uniform mixture so that it is impossible for the animal to select some ingredients and ignore others. To achieve this, cereal grains and other particulate components must be ground or milled to a roughly similar particle size before mixing. A mechanical mixer is the best means of achieving uniform distribution of ingredients. A domestic food mixer is suitable for small quantities of diet and for larger amounts the type of machine used in the baking industry has been found satisfactory. Before addition of ingredients that are present in a very low concentration, for example vitamins, a premix must be prepared by distributing the substance in part of one of the major ingredients, either by hand with a pestle and mortar or in a small mixing machine. Test substances such as drugs or food additives should be treated in the same way. The labour of mixing individual batches of diet can be reduced by keeping a stock premix of all the vitamins and a salt mixture containing most of the mineral ingredients. The two should never be mixed directly together because in close contact components of the mineral mixture could bring about oxidative destruction of some of the vitamins. Fat-soluble substances should be dissolved in the oily constituents of the diet.

The length of the mixing period depends on various factors, including particle size and density, number of ingredients and amounts of each, and performance of the mixer. After excessive mixing times, separation of some particles can occur. To determine the optimal period for a particular diet the degree of uniformity can be checked by assaying one of the minor ingredients in samples taken from the mixer at various times.

Since errors in composition of the diet can profoundly affect the performance of the animals and hence the results of an experiment, every effort must be made to avoid mistakes during its preparation. The formula should be recorded on a master sheet from which photocopies can be made to provide a checklist every time a batch of the diet is mixed. Each item is ticked as it is added, and the checklist retained among the other experimental records. As a further precaution the identity and weight of each ingredient should be independently verified by a second person, and the total weight of the final mixture checked.

Diets may be presented as mixed, i.e. as mash or powder, but in this form have several disadvantages. They are very prone to wastage, being easily scattered, and accurate food consumption records are therefore hard to obtain. In powder form purified diets in particular are unpalatable, tending to cake inside the animal's mouth and prevent it achieving its normal food intake. They generate dust, which can be hazardous if toxic substances are incorporated into the diet. Some of these problems can be overcome by feeding the diet in gel form. Knapka (1983) recommends a mixture of equal parts of a 3 per cent agar solution and the powdered diet which, when cool, sets into a gel that can be cut into blocks of suitable size for feeding. This form of diet readily develops moulds, and must be stored in a refrigerator and fed frequently in small batches.

Crumbs or pellets are more acceptable to the animals. Crumbs are easy to prepare in the laboratory by incorporating a binding agent such as methyl cellulose (10 g/kg diet), moistening the diet and pressing it through a coarse sieve. The crumbs are then dried in a current of warm air, not exceeding about 40°C. During this process there may be small losses of some of the less stable ingredients such as thiamin for which allowance should be made in the original formula.

In many ways pellets or cubes offer the most satisfactory form of diet. They are generally more acceptable to the animals, less easily wasted, simple to handle and have a good storage life. The optimal size and hardness of the pellet is important and differs between species. Rats, which gnaw their food, prefer a larger pellet than those usually supplied to guinea pigs and rabbits. Soft pellets encourage waste, as they easily crumble in the storage bags or the hoppers; very hard pellets are unacceptable, particularly for young animals, which may find them too difficult to chew. Special processes are necessary for the manufacture of pellets so they are normally obtained from a commercial supplier. They are in general use for breeding and stock colonies, but commercial pellets are impractical for some experimental work since test substances cannot be incorporated into the final product. Small batches of pelleted diet can be prepared in the laboratory by a process similar to that described for crumbs. The diet, incorporating a binding agent and moistened with water, is forced through a mincing machine and the cylinders of damp diet so formed are dried in a current of warm air. The die of the mincer can be varied to produce pellets of the desired size. Purified diets can readily be made into satisfactory pellets. Diets of natural ingredients should be finely milled before pelleting is attempted.

Liquid diets are necessary for immunological studies, when all antigens must be excluded. They contain only water-soluble chemically defined ingredients and are ultrafiltered to remove microbial contaminants and other impurities likely to have antigenic or mitogenic properties. Their preparation and uses have been fully described by Pleasants (1984).

STERILIZATION OF DIETS

Diets intended for gnotobiotic animals must be sterilized (see review by Coates, 1984). Those for specified pathogen free (SPF) animals must be free from pests and pathogens, so that it is preferable though not essential for them also to be sterilized. Since sterilization processes are designed to destroy microbes, it is inevitable that they also inflict some destruction of nutrients. The methods most commonly employed involve heat, ionizing radiation and fumigation, and in all instances a balance must be struck so that sterility of the diet is achieved without drastically impairing its nutritive value.

Vegetative organisms and most of the common pests and pathogens likely to occur in laboratory animal diets are destroyed by pasteurization, i.e. application of mild heat (about 80°C) for a relatively short time. This type of treatment is suitable for diets for SPF animals but not for gnotobiotes, since much higher temperatures are necessary to destroy spores.

Autoclaving is a procedure in which the application of steam under pressure raises the temperature above 100°C. Treatment at 121°C for 10–15 min is sufficient to render a diet sterile without much loss of its nutrient value. In general, short periods at high temperature are less destructive of nutrients than longer periods at lower temperature. If a high vacuum autoclave is available treatment at 134°C for only 3 min is adequate; it is, however, essential that the times taken to reach the temperature and to cool the diet after treatment are minimized by drawing a high vacuum at the beginning and end of the period.

Because the effectiveness of autoclaving depends on the latent heat generated when steam condenses, steam must penetrate to all parts of the diet. This implies packaging that is pervious to steam, for example linen or paper bags. Nylon film or solid containers must be fitted with cotton plugs. Air pockets between particles of diet should be removed by drawing a vacuum before introducing the steam. If a large bulk of diet is autoclaved the necessary temperature will not be reached in the centre unless the treatment is very much prolonged, during which time the outer layers of diet will have been grossly overheated, to the detriment of its nutrient content. Ideally, diet should be treated in layers about 1–2 cm deep, although if it is intended for gnotobiotic animals this requirement may have to be modified to suit the dimensions of the entry port on the isolator.

Ionizing radiation is a very effective method of sterilization of laboratory animal diets (Ley et al., 1969). The source of radiation is usually ^{60}Co and although for reasons of safety and cost the process is not available in most laboratories, commercial facilities exist in many countries. Because of the intensely penetrating power of gamma rays, diets can be packed in much bulkier quantities than they can for autoclaving. There is little variation in dose throughout the package, and only a few degrees rise in temperature. A dose of 2.5 Mrad or even less is adequate for SPF animals, but it is usual to give a higher dose (3.5–5.0 Mrad) to diets for gnotobiotes. There is no danger of residual radiation in the treated diet.

Chemical sterilization by fumigation with ethylene oxide has been applied to diets with varying success, and there is some lack of agreement concerning its effect on nutrient quality (Coates, 1970). Ethylene oxide is explosive, and fumigation must be performed in a special pressure chamber. Its chief disadvantage for sterilization of diets lies in the difficulty of removing all traces of the fumigant, which remains dissolved in the lipid components.

The method of choice for diet sterilization depends on a number of factors including cost, availability and experimental purpose. Autoclaving requires a high initial outlay on equipment but little cost thereafter. Conversely, irradiation needs no capital investment but treatment, and possibly transport to and from the radiation plant, is expensive. In general radiation is less damaging than heat to the nutrient quality of the diet. Vitamin losses are small (Coates et al., 1969) and protein quality unaffected by irradiation whereas it can be seriously impaired by autoclaving (Eggum, 1969; Ford, 1976). Radiation sterilization has proved particularly valuable in some gnotobiotic experiments in which the factor under investigation would have been destroyed by heat, for example the antitrypsin activity of raw legumes (Coates et al., 1970) or the integrity of starch grains (Bewa et al., 1979).

Interactions between ingredients, especially of the inorganic components, may occur during sterilization of diets. For instance, a significant difference in retention of iron was observed between germfree rats given a purified diet autoclaved in pellet form compared with others given the same diet irradiated as a paste (Andrieux *et al.*, 1980). Maillard reaction products, well known to be formed when amino acids are heated in the presence of carbohydrates, caused decreases in retention of Ca, P, Mg and Cu in germfree but not in conventional rats (Andrieux *et al.*, 1980). In general, chemical reactions that occur during radiation are of a mildly oxidative nature, but when an antioxidant was added to a chick diet losses of thiamin, folic acid and pyridoxine were markedly increased (Coates *et al.*, 1969). It is clearly important, therefore, that in any experimental work involving sterilized diet the effects of the sterilization process should first be examined by comparing the performance of the experimental animals on both the treated and untreated diets, particularly when test substances are administered in the diet.

CONTAMINANTS IN DIETS

Chemical contaminants and additives

Chemical contaminants occur in animal diets either naturally, as constituents of the raw materials, or accidentally, having been introduced in the course of manufacture. It is self-evident that their presence must be avoided as far as possible. Even though they may not cause overtly toxic signs in the animals, they may have more subtle effects which influence their responses to the experimental treatment (Newberne, 1975). The stimulation of hepatic microsomal enzymes by some pesticide residues, for instance, could be of concern to the toxicologist.

Nevertheless, it is unrealistic to demand very stringent limits for contaminants, the concentration of which may very considerably with the season and the place of origin of the raw materials. This is clear from a five-year study in which successive batches of a commercial rodent diet were analysed for several pesticides and heavy metals. There was wide variation, and some values approached concentrations at which the contaminant might be expected to have biological effects (Greenman *et al.*, 1980).

As a general rule the permitted levels of contaminants laid down for human foods should also apply to those for laboratory animals, and materials rejected for human consumption cannot be considered suitable for use in diets for experimental animals. Although routine analysis of all batches of diet would be costly and impractical, occasional checks are advisable to monitor the concentration of contaminants likely to influence the parameters under investigation.

Non-nutritive additives are undesirable in laboratory animal diets, with the exception of binders used in pelleting; methyl cellulose is acceptable for this purpose. If antioxidants are deemed necessary preference should be given to natural products such as the tocopherols or ascorbic acid. Antibiotics and other growth promoters used in farm animal feeds must be avoided. There is no evidence that their addition to diets can improve the performance of laboratory species, and may indeed be deleterious through disturbance of the normal microecology of the gastrointestinal tract.

Microbiological contaminants

Bacteria, fungi, helminths and arthropods are potential contaminants of diets, and the types and numbers present are inversely related to the standards of hygiene maintained during harvesting of the ingredients and preparation and storage of the diet. Ingredients that have been subjected to heat in the course of their manufacture, for example dried milk, fish meal, usually have a lower bacterial count than that of cereals. Nevertheless, materials of animal origin may be a source of pathogens unless adequately heated. Poor quality meat and bone meal, for instance, may be contaminated with the anthrax bacillus. Fresh vegetables and fruit are not advisable contributions to laboratory diets, but if they must be used they should be thoroughly washed and no stale or wilted specimens should be included. Pelleted diets usually have a low microbial count because the heat generated during their manufacture is sufficient to destroy non-spore-forming organisms, which includes the salmonellae. Microbiological standards for diets for laboratory rats and mice, together with appropriate monitoring tests, have been proposed by the Laboratory Animals Centre Diets Advisory Committee (Clarke *et al.*, 1977). They require that salmonellae and *Escherichia coli* type 1 should be absent, and that the total count of tetrazolium-positive organisms should be less than 5000/g. They suggest that not more than ten presumptive coliform organisms should be present. Somewhat less stringent proposals have been made by the Gesellschaft für Versuchtierkunde (GV–Solas, 1980).

To avoid chemical or microbial contamination during preparation of diets a strict routine of cleanliness and hygiene must be followed in the factory or laboratory. A special area should be reserved for diet preparation, protected from invasion by wild rodents or birds. All equipment should be thoroughly cleaned between each batch of diet mixed, taking special care after preparation of a diet containing a drug or other non-nutrient additive. Floors and bench surfaces must be frequently swept to avoid collecting residues of diet liable to attract insect infestation. When mixing is complete the diet should immediately be packed into strong containers impervious to penetration by vermin.

STORAGE OF DIETS

The stability of nutrients in diets is adversely affected by light, heat and moisture. Riboflavin, folic acid and vitamin B_{12} are photolabile, and losses can be expected to occur if diets are exposed to strong light. Gradual changes due to oxidation and hydrolysis are inevitable in stored diets, and these are accelerated with increasing temperature and moisture content. Moist conditions encourage mould growth, and thus increase the danger of contamination with mycotoxins.

In general, diets of natural materials retain their nutrient quality on storage better than purified diets, since the more vulnerable nutrients such as vitamins occur inside coarse particles and may be combined with other constituents which afford some protection against chemical or microbiological attack. However, natural ingredients may retain some enzymes that could have deleterious effects during storage. All diets should be stored in a cool, dry and preferably dark environment and purified diets are best kept under refrigeration. The length of time they may be kept without serious loss of nutrient quality depends to some extent on their constitution, but periods longer than three months are inadvisable. The date of manufacture or of preparation in the laboratory should always be recorded on the package.

EXPERIMENTAL DIETS

For experimental purposes it is frequently necessary to incorporate test materials into diets. If the test material constitutes an appreciable proportion of the diet its inclusion inevitably distorts the balance of nutrients and some adjustment of the basic formula must be made. This is very much easier to do with a purified diet than one of natural ingredients. For example, if the test material has a high content of fat, the amount of oil in the diet fed to the control animals can be increased to match that in the test diet. Nutritionally inert materials are usually added at the expense of part of the carbohydrate component, but the total energy content of the diet must not be seriously altered. Where energy is a vital consideration the control and test diets must be equicaloric, i.e. the amounts of carbohydrate, fat and protein must be manipulated so that the energy concentration is the same in each diet.

The potential interaction between the test material and other components of the diet is an important consideration that has received too little attention. Some test materials may influence the nutritional quality of the diet, for example sulphonamides and some other antibacterials, which induce a deficiency of folate by suppressing the synthetic activities of the gut microflora. Conversely, certain ingredients in the diet may modify the effect of the test material on the animal, for example calcium phytate, which has been shown to reduce the toxicity of lead acetate to mice (Wise, 1981). Diet can have a profound effect on the physiological status of the animal receiving it, and hence on the outcome of an experimental treatment. The incidence of 'spontaneous' tumours is higher in mice and rats fed *ad libitum* than in their counterparts allowed only a restricted intake of the same diet (Tucker, 1979), and diets high in protein or fat, or both, also induce a high incidence of tumours (Visek *et al.*, 1978). In some circumstances it might be advantageous to design a diet that will increase the animal's susceptibility to the substance under test. Full discussion of the possible interactions between the diet and test substance is outside the scope of this chapter. Further examples have been given and commented on by Calabrese (1980), Newberne (1975) and Wise (1982).

STANDARDIZATION OF DIETS

In any experimental programme it is essential that all factors be kept constant except the one under test. Thus the diet must be strictly controlled: neither its composition nor its quality can be allowed to vary from batch to batch. It is virtually impossible to impose strict controls on diets of natural ingredients, even though they may be made to a fixed formula. The raw materials may be carefully chosen and of good quality, but their nutrient content will still vary according to factors such as season, variety and place of origin (Greenman et al., 1980; Topham and Eva, 1981). Wise and Gilburt (1980, 1981) have reported considerable variability in content of minerals, trace elements, phytate and crude fibre in commercial diets available in the United Kingdom, and also that the crude fibre analyses given in the manufacturers' catalogues did not reflect the true fibre content of the diets. Variation in fibre content may be particularly disadvantageous in toxicological testing. The toxicity of some compounds has been observed to increase if fibre is omitted (Ershoff, 1974), and inclusion of pectin in a diet for rats increased reduction of nitrate to nitrite, as evidenced by development of methaemoglobinaemia (Wise et al., 1982).

Although variation within reasonable limits may be tolerated in diets for stock colony and even for breeding animals, the use of purified diets is the only way of eliminating any potentially adverse influence of diet on the reproducibility of experimental results.

REFERENCES

Aitken, F. C. and Rankin, R. G. (1972). Tables of the amino acids in foods and feedingstuffs, Technical communication No. 19, Commonwealth Agricultural Bureaux, Farnham Royal, Bucks.

Andrieux, C., Sacquet, E. and Gueguen, L. (1980). Interaction between Maillard's reaction products, the microflora of the digestive tract and mineral metabolism, Reprod. Nutr. Développment, 20, 1061–9.

Bewa, H., Charlet-Lery, G. and Szylit, O. (1979). Role de la microflore digestive et de la structure crystalline de l'amidon dans la digestion et l'utilization des régimes chez le poulet, Ann. Nutr. Aliment., 33, 213–31.

Calabrese, E. J. (1980) Nutrition and Environmental Health: the Influence of Nutritional Status on Pollutant Toxicity and Carcinogencity, Vols I and II, John Wiley & Sons, New York.

Clarke, H. E., Coates, M. E., Eva, J. K., Ford, D. J., Milner, C. K., O'Donogue, P. N., Scott, P. P. and Ward, R. J. (1977). Dietary standards for laboratory animals: report of the Laboratory Animals Centre Diets Advisory Committee, Lab. Anim., 11, 1–28.

Coates, M. E. (1970). The sterilization of laboratory animal diets. In W. D. Tavernor (ed.) *Nutritional Diseases in Experimental Animals*, Baillere, Tindall and Cassell, London, pp. 38–48.

Coates, M. E. (1984). Diets for germ free animals. Part 1. Sterilization of diets. In M. E. Coates and B. E. Gustafsson (eds) *The Germ-free Animal in Biomedical Research*, Laboratory Animals Ltd, London, pp. 85–90.

Coates, M. E., Ford, J. E., Gregory, M. E. and Thompson, S. Y. (1969). Effects of gamma-radiation on the vitamin content of diets for laboratory animals, *Lab. Anim.*, **3**, 39–49.

Coates, M. E., Hewitt, D. and Golob, P. (1970). A comparison of the effects of raw and heated soya-bean meal in diets for germ-free and conventional chicks, *Br. J. Nutr.*, **24**, 213–25.

Cohlan, S. Q. (1953). Excessive intake of vitamin A as a cause of congenital anomalies in the rat, *Science, N.Y.*, **117**, 535–6.

Davidson, J. and Spreadbury, D. (1975). Nutrition of the New Zealand white rabbit, *Proc. Nutr. Soc.*, **34**, 75–83.

Deuel, H. J. Jr (1957a). In *The Lipids*, vol. 3, John Wiley & Sons, New York, p. 605.

Deuel, H. J. Jr (1957b). In *The Lipids*, Vol. 3, John Wiley & Sons, New York, p. 681.

Edwards, D. G. and Dean, J. (1985). The responses of rats to various combinations of energy and protein II. Diets made from natural ingredients, *Lab. Anim.*, **19**, 336–43.

Edwards, D. G., Porter, P. D. and Dean, J. (1985). The responses of rats to various combinations of energy and protein I. Diets made from purified ingredients, *Lab. Anim.*, **19**, 328–35.

Eggum, B. O. (1969). Der Einflusz der Sterilisation auf die Protein-qualität von Futtermischungen, *Z. Tierphysiol. Tierernähr., Futtermittelk.*, **25**, 204–10.

Ershoff, B. H. (1974). Antitoxic effects of plant fiber, *Am. J. clin. Nutr.*, **27**, 1395–8.

Eva, J. K., Fifield, R. and Ricket, M. (1976). Decomposition of supplementary vitamin C in diets compounded for laboratory animals, *Lab. Anim.*, **10**, 157–9.

Ford, D. J. (1976). The effect of methods of sterilization on the nutritive value of a commercial rat diet, *Br. J. Nutr.*, **35**, 267–76.

Ford, D. J. and Ward, R. J. (1983a). The effect on rats of practical diets containing different protein and energy levels, *Lab. Anim.*, **17**, 330–5.

Ford, D. J. and Ward, R. J. (1983b). The effect on mice of practical diets containing different protein and energy levels, *Lab. Anim.*, **17**, 336–9.

Greenman, D. L., Oller, W. L., Littlefield, N. A. and Nelson, C. J. (1980). Commercial laboratory animal diets: toxicant and nutrient variability, *J. Toxicol. Environ. Health*, **6**, 235–46.

Gustafsson, B. E., Daft, F. S., McDaniel, E. G., Smith, J. C. and Fitzgerald, R. J. (1962). Effects of vitamin K active compounds and intestinal microorganisms in vitamin K-deficient germfree rats, *J. Nutr.*, **78**, 461–8.

GV-Solas (1980). *Definitions of Terms and Designations in Laboratory Animal Nutrition (Part 1). Use of Feedstuffs and Bedding Materials for Nonclinical Laboratory Studies*, Publication No. 9, First English edition, GV–SOLAS, Basle, p. 84.

Harvey, D. (1970). Tables of the amino acids in foods and feedingstuffs, 2nd edn, Technical communication No. 19, Commonwealth Agricultural Bureaux, Farnham Royal, Bucks.

Hunt, R. D., Garcia, F. G. and Hegsted, D. M. (1967). A comparison of vitamin D_2 and D_3 in New World primates. 1. Production and regression of osteodystrophia fibrosa, *Lab. Anim. Care*, **17**, 222–34.

Kellogg, T. F. and Wostmann, B. S. (1969). Stock diet for colony production of germ-free rats and mice, *Lab. Anim. Care*, **19**, 812–14.

Knapka, J. J. (1983) Nutrition. In *The Mouse in Biomedical Research*, Vol. iii, Academic Press, New York, p. 64.

Ley, F. J., Bleby, J., Coates, M. E. and Paterson, J. S. (1969). Sterilization of laboratory animal diets using gamma radiation, *Lab. Anim.*, **3**, 221-54.

Molanaar, I., Vos, J. and Hommes, F. A. (1972). The effect of vitamin E deficiency on cellular membranes, *Vitams Horm.*, **30**, 45-82.

Moore, T. and Wang, Y. L. (1945). Hypervitaminosis A, *Biochem J.*, **39**, 222-8.

National Research Council (1974). Nutrient requirements of domestic animals, No. 8, Nutrient requirements of dogs, National Academy of Sciences, Washington DC.

National Research Council (1977). Nutrient requirements of domestic animals, No. 9, Nutrient requirements of rabbits, National Academy of Sciences, Washington DC.

National Research Council (1978a). Nutrient requirements of domestic animals, No. 10, Nutrient requirements of laboratory animals, National Academy of Sciences, Washington DC.

National Research Council (1978b). Nutrient requirements of domestic animals, No. 11, Nutrient requirements of cats, National Academy of Sciences, Washington DC.

National Research Council (1978c). Nutrient requirements of domestic animals, No. 12, Nutrient requirements of nonhuman primates, National Academy of Sciences, Washington DC.

National Research Council (1982). Committee on Animal Nutrition, United States-Canadian tables of feed composition, National Academy of Sciences, Washington DC.

Newberne, P. M. (1975). Influence on pharmacological experiments of chemicals and other factors in diets of laboratory animals, *Fed. Proc.*, **34**, 209-18.

Owen, C. A. (1971). Vitamin K group. XI. Pharmacology and toxicology. In W. H. Sebrell and R. S. Harris (eds) *The Vitamins*, Vol. 3, Academic Press, New York, pp. 492-509.

Parker, D. S. (1976). The measurement of production rates of volatile fatty acids in the caecum of the conscious rabbit, *Br. J. Nutr.*, **36**, 61-70.

Pleasants, J. R. (1984). Diets for germ-free animals. Part 2. The germ-free animal fed chemically defined ultrafiltered diets. In M. E. Coates and B. E. Gustafsson (eds), *The Germ-free Animal in Biomedical Research*, Laboratory Animals Ltd, London, pp. 91-109.

Rechigl, M. (1977). *CRC Handbook* series in nutrition and food, Vol. G/71, CRC Press, Cleveland.

Remp, D. G. and Marshall, I. I. (1938). The antirachitic activity of various forms of vitamin D in the chick, *J. Nutr.*, **15**, 525-37.

Roe, F. J. C. (1981). Are nutritionists worried about the epidemic of tumours in laboratory animals? *Proc. Nutr. Soc.*, **40**, 57-65.

Scott, M. K., Noguchi, T. and Combs, G. F. (1974). New evidence concerning the mechanisms of action of vitamin E and selenium, *Vitams Horm.*, **32**, 429-44.

Topham J. C. and Eva, J. K. (1981). A quality-controlled rodent diet, *Lab. Anim.*, **15**, 97-100.

Tucker, M. J. (1979). The effect of long-term food restriction on tumours in rodents, *Int. J. Cancer*, **23**, 803-7.

Visek, W. J., Clinton, S. K. and Truex, C. R. (1978). Nutrition and experimental carcinogenesis, *Cornell Veterinarian*, **68**, 1-39.

Wise, A. (1981). The protective action of calcium phytate against acute lead toxicity in mice, *Bull, environ. Contam. Toxicol.*, **18**, 643-8.

Wise, A. (1982). Interaction of diet and toxicity — the future role of purified diets in toxicological research, *Arch. Toxicol.*, **50**, 287-99.

Wise, A. and Gilburt, D. J. (1980). Variability of dietary fibre in laboratory animal diets and its relevance to the control of experimental conditions, *Food cosmet. Toxicol.*, **18**, 643–8.

Wise, A. and Gilburt, D. J. (1981). Variations in minerals and trace elements in laboratory animal diets, *Lab. Anim.*, **15**, 299–303.

Wise, A., Mallett, A. K. and Rowland, I. R. (1982). Dietary fibre, bacterial metabolism and toxicity of nitrate in the rat, *Xenobiotica*, 12, 111–18.

Laboratory Animals: An Introduction for New Experimenters
Edited by A. A. Tuffery
© 1987 John Wiley & Sons Ltd

CHAPTER 13

Non-Surgical Experimental Procedures

P. A. FLECKNELL
Comparative Biology Centre, Medical School, Newcastle upon Tyne

INTRODUCTION

The scientific literature contains numerous descriptions of methods for sampling body fluids from laboratory animals, or for the administration of drugs or other substances. In view of this plethora of techniques, some selection is necessary. The criteria used for their inclusion in this chapter are that they should cause a minimum of discomfort and distress to the animal whilst being simple and comparatively easy to apply. Stress and pain are known to alter body responses to a wide range of stimuli, hence in addition to important humanitarian considerations, efforts to reduce pain or discomfort to a minimum should also be considered good scientific practice.

ADMINISTRATION OF DRUGS AND OTHER SUBSTANCES

Recommended sizes of needles and volumes for injection are listed in Table 13.1. This table is intended only as a general guide — it is preferable to use the smallest gauge needle compatible with the viscosity of the material to be injected since this will minimize any discomfort to the animal. The total volume of material that can be administered will vary with the physicochemical characteristics of the substance (see Chapter 11). The type of syringe used will be determined largely by personal preference; it will usually be found most convenient to select a disposable plastic syringe rather than a reusable one of glass or metal. A range of different designs of plastic disposable syringes is available — particularly useful for the administration of small volumes of material are some types of disposable insulin syringes.

Table 13.1 For each species, site of injection, maximum normally accepted volume (see text) and needle size (see text)

Species	Subcutaneous	Intramuscular	Intraperitoneal	Intravenous
Mouse	Scruff, 2–3 ml, <20 G	Quadriceps/posterior thigh, 0.05 ml, <23 G	2–3 ml, <21 G	Lateral tail vein, 0.2 ml, <25 G
Rat	Scruff, back, 5–10 ml, <20 G	Quadriceps/posterior thigh, 0.3 ml, <21 G	5–10 ml, <21 G	Lateral tail vein, sublingual vein, penile vein (jugular vein, femoral vein — cut down), 0.5 ml, <23 G
Hamster	Scruff, 3–4 ml. <20 G	Quadriceps/posterior thigh, 0.1 ml, <23 G	3–4 ml, <21 G	Femoral or jugular vein (cut down), 0.3 ml, <25 G
Guinea pig	Scruff, back, 5–10 ml, <20 G	Quadriceps/posterior thigh, 0.3 ml, <21 G	10–15 ml, <21 G	Ear vein, saphenous vein, dorsal penile vein, 0.5 ml, <23 G
Rabbit	Scruff, flank, 30–50 ml, <20 G	Quadriceps/posterior thigh, lumbar muscles, 0.5–1.0 ml, <20 G	50–100 ml, <20 G	Marginal ear vein, 1–5 ml. (slowly) <21 G
Cat	Scruff, back, 50–100 ml, <20 G	Quadriceps/posterior thigh, 1.0 ml, <20 G	50–100 ml, <20 G	Cephalic vien, 2–5 ml, (slowly), <23 G
Dog	Scruff, back, 100–200 ml, <20 G	Quadriceps/posterior thigh, 2–5.0 ml, <20 G	200–500 ml, <20 G	Cephalic vein, 10–15 ml (slowly), <21 G
Bird (domestic fowl)	—	Pectoral muscles, 1–2 ml, <21 G	Midline, halfway between cloaca and sternum, 10–15 ml, <21 G	Brachial (wing) vein, 2–3 ml, <21 G

The rat

Oral dosing

Whilst some substances are highly palatable and will be readily consumed when mixed with the drinking water or food, or even licked from the end of a syringe, unpalatable materials must be administered by stomach tube. Administration of accurate volumes of material will almost invariably require the use of a stomach tube. Suitable tubing can be made from 15 or 16 gauge needles, about 10–12 cm in length. The needle should have the sharp tip removed and a smooth ball of solder attached to its distal end. Alternatively, a rubber or polyethylene catheter, 3–4 f.g., may be used. A 16 gauge 'Medicut' cannula (Sherwood Medical Industries Ltd, England) may also provide a suitable tube. Oral dosing needles are also available commercially. If using a plastic or rubber tube, a small gag may be necessary to prevent the rat biting through the catheter. To pass the tube the rat should be firmly restrained by grasping it by the skin of the back and neck; this ensures that the head and neck are extended and in line with the back (Fig. 13.1). The tube can then be passed into the

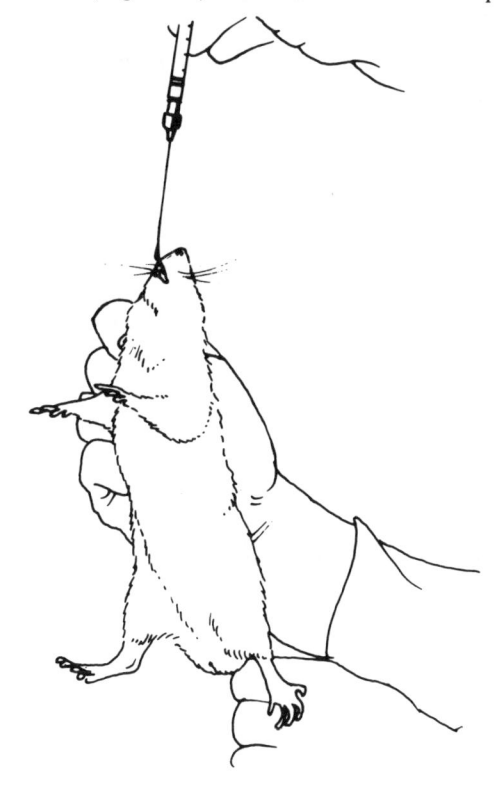

Fig. 13.1 The rat — oral dosing

mouth via the interdental space and advanced gently, rotating the tube slightly to ease its passage into the oesophagus and/or into the stomach if necessary. In a conscious rat it is unlikely that the tube will pass into the trachea. Some rats will resent being restrained by the method described above, and may struggle violently during the procedure. In these cases it may be found simpler and more humane to anaesthetize the animal prior to passing the tube, especially if the procedure is being undertaken by one person. Following induction of anaesthesia the tube can be passed easily, but when the rat is deeply anaesthetized it will be found that the tube occasionally enters the trachea. By keeping the animal at a light plane of anaesthesia the cough reflex is maintained and hence an indication will be given if the trachea is inadvertently intubated. An alternative hold for oral dosing is described in Chapter 9 (Figs. 9.6 and 9.7).

Subcutaneous injection

The rat should be restrained by grasping by the skin of the scruff, and the injection made into the area of skin 'tented' by this procedure (Fig. 13.2). The needle should be positioned almost parallel to the skin and introduced through the skin and then on subcutaneously for most of its length. If assistance is available, the rat can be restrained by placing one hand around the shoulders and the other holding the hindlimbs. The operator can then tent a section of skin on the flank or back and carry out the injection (Fig. 13.3). This latter method of restraint is to be preferred since it causes least discomfort to the rat.

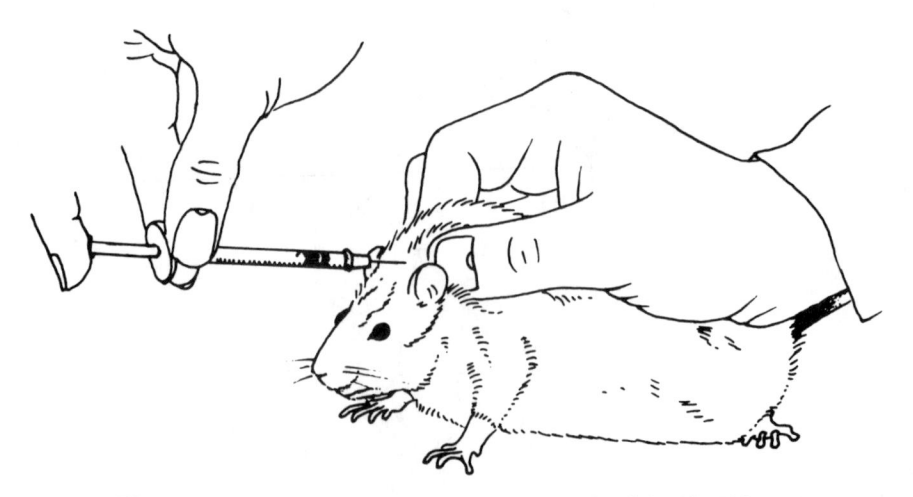

Fig 13.2 The rat — subcutaneous injection into the back of the shoulder

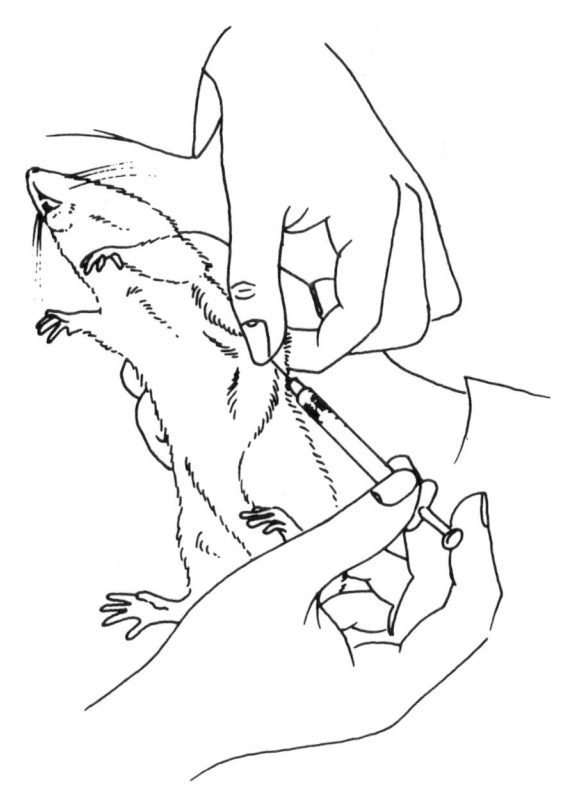

Fig. 13.3 The rat — subcutaneous injection into the flank or back

Intradermal injection

The skin over the injection site (usually the back) should be shaved, and the animal restrained around the shoulders by an assistant. A 26 gauge needle should be used, and held almost parallel to the skin surface. The needle can then be advanced carefully a few millimetres into the skin. If there is a sudden loss of resistance to passage of the needle, this is usually due to it passing subcutaneously, and it must be withdrawn and reintroduced. Following injection of material (up to 100 µl) a small bleb should be visible within the skin.

Intraperitoneal injection

The rat is held by an assistant, as shown in Fig. 13.4, with one leg held by the operator. The needle is advanced parallel to the line of the leg, and pushed through the abdominal wall into the peritoneal cavity; a loss of resistance to needle passage will be noted as the needle passes through into the abdomen. By advancing the needle along the line of the leg, injection is made in an area that

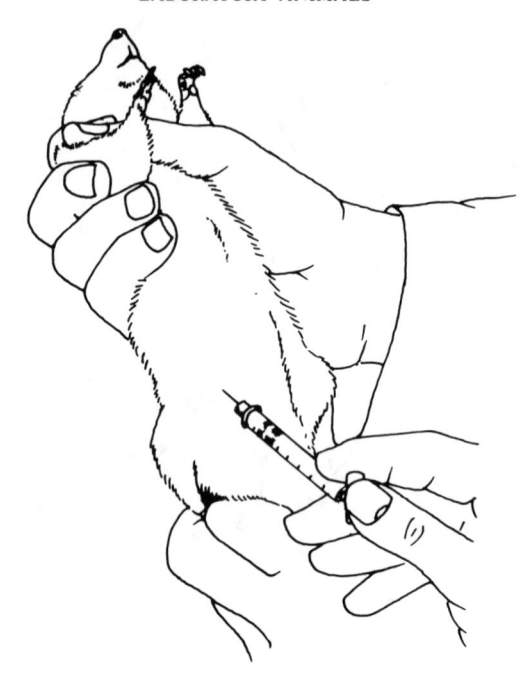

Fig. 13.4 The rat — intraperitoneal injection

avoids the urinary bladder in the posterior abdomen, and the liver anteriorly. Although it is possible to inject into the gastrointestinal tract, this is a rare complication of intraperitoneal injection. It should be noted that most rats will urinate when restrained, and if this coincides with the injection of material intraperitoneally it can alarm an inexperienced operator!

Intramuscular injection

Suitable sites for injection include the quadriceps, the posterior thigh and the triceps. Injection into the posterior thigh will cause least discomfort to the animal, but injection into the fascial planes between the muscle bellies may be a problem when using this site. Care must be taken to inject into the most posterior part of the muscle mass, to avoid damage to the sciatic nerve. Injection into the quadriceps (Fig. 13.5) often appears to cause more pain to the animal, although the pain can be minimized by slow injection of relatively small volumes (about 50 per cent of the volumes quoted in Table 13.1) of material. This latter route has the advantage that no major blood vessels or nerves are present in the main muscle bellies.

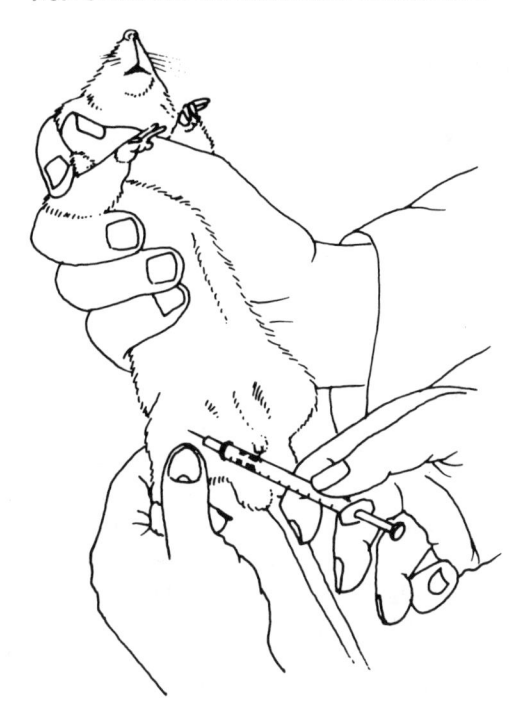

Fig. 13.5 The rat — intramuscular injection

Intravenous injection

The most frequently used vessels are the lateral tail veins, which can be readily seen in young rats. In older animals thickening of the skin over the tail makes it difficult to detect the vessel. Warming the tail by placing the rat under a heating lamp will dilate the vessel and facilitate venepuncture but great care must be taken to avoid excessive heat being applied (see comments on bleeding, p. 250, for more details). To inject into a tail vein the rat should be held by an assistant and the tail held by the operator. The vein can then be located and the needle advanced at a shallow angle. A loss of resistance will be felt as the needle enters the vein. To be certain that the needle is lying in the vein, a small quantity of material must be injected — if the needle tip is not lying within the vein a small bleb will be produced, and the needle must be repositioned. It is often easier to place the rat in a suitable restraining device, since any sudden movements by the animal make injection difficult. Injections into other veins can only be undertaken after first inducing general anaesthesia. Full details of use of the sublingual, penile, femoral, jugular and lateral marginal veins can be found in Waynforth (1980).

Injection of neonatal rats

A range of specialized techniques have been devised for injection of material into neonatal rats; these are described by Waynforth (1980), Bader and Klinger (1974) and Gibson and Becker (1967).

The Mouse

Oral dosing

A similar technique to that described for the rat is used. A 2–3 f.g. polyethylene catheter, about 2–3 cm in length, provides a suitable tube (an 18 gauge 'Medicut' is a suitable size). Alternatively, metal tubes can be made by blunting a 19–20 gauge needle, as described for the rat, or purchased from a commercial supplier. The mouse should be grasped by the skin of the neck and back, and the tube passed into the mouth via the interdental space. After passing the tube gently down into the oesophagus, the fluid can be administered. Considerable struggling may ensue during this procedure, and it is often preferable to lightly anaesthetize the mouse (for example by inhalation of methoxyflurane) before attempting to pass the tube. If undertaken in conscious mice, the plastic tubing will require frequent replacement due to biting by the animal, and it will usually be preferable to use metal dosing needles.

Subcutaneous injection

The mouse should be restrained as described previously, and injection made into the skin of the back using the technique as described for the rat. An alternative method of restraint is to grasp the mouse by the scruff using the thumb and forefinger, and allow it to continue to grasp a suitable rough surface. The injection can then be made into the skin of the neck which is tented by this procedure. The injection is made directly into the raised skin, parallel to the back (Fig. 13.6).

Intradermal injection

The method described for the rat can be applied to the mouse; however, the thinner skin in this species makes the technique more difficult. A very fine needle (30 g or less) should be used, and the skin over the back shaved carefully prior to injection. Because of the fine control needed when making the injections, it is often easier first to anaesthetize the mouse.

Intraperitoneal injection

The mouse should be restrained by the skin overlying the neck and back, and

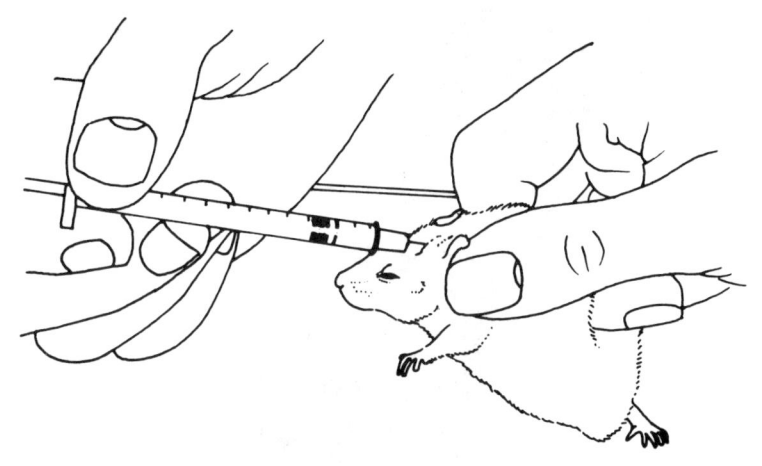

Fig. 13.6 The mouse — subcutaneous injection

the needle advanced along the line of one leg, as described for the rat. The needle should be passed through the body wall in the middle of the posterior quadrant of the abdomen, so avoiding the bladder and liver.

Mice, like rats, frequently urinate when restrained, and this should not be thought to be a result of the injection.

Intramuscular injection

The small muscle mass of mice allows injection of only relatively low volumes of material. The muscles of the posterior or anterior thigh are the sites most frequently used. The animal is restrained by an assistant by the method described for i.p. injection and one leg held by the operator. It may be found easier to hold the quadriceps (anterior thigh) between the forefinger and thumb, and inject down into the immobilized muscle mass.

Intravenous injection

Intravenous injection is carried out via the lateral tail vein. Either the whole mouse, or only its tail, should be warmed by placing in a suitable container (see section on bleeding, p. 251). Once the tail vessels have dilated the animal should be transferred to a restraining device. The tail should be grasped firmly, and rotated so that one of the veins (which are easily visible) is uppermost (Fig. 13.7). The needle should be advanced almost parallel to the tail, directly over the vein, and pushed gently through the skin. A loss of resistance should be noted as the needle enters the vein. Injection of a small volume of fluid will indicate whether the needle is correctly positioned. Upon withdrawal of the needle, firm pressure on the puncture wound for about a minute should be sufficient to prevent any haemorrhage.

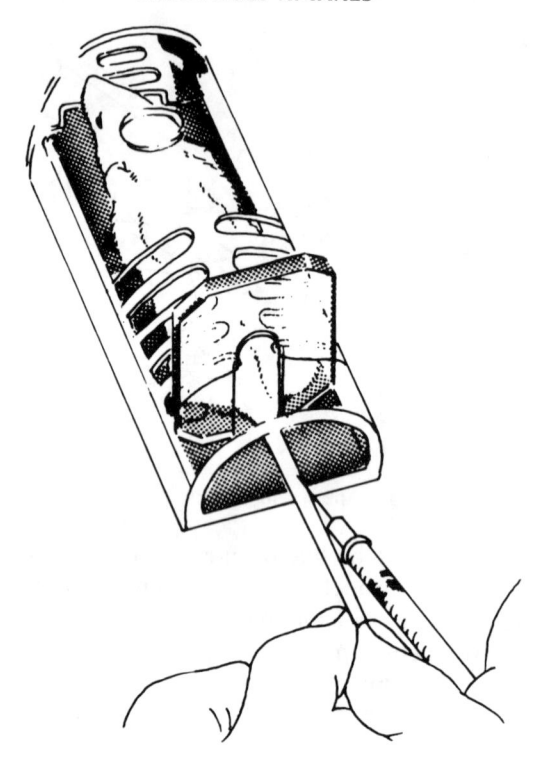

Fig. 13.7 The mouse — intravenous injection

The guinea pig

Oral dosing

Guinea pigs can be stomach tubed using a similar technique to the rat. An assistant restrains the animal by grasping it around the shoulders and supporting the hindquarters to prevent undue struggling (see Chapter 9). A blunted 15–16 hypodermic needle, polyethylene catheter (3–4 f.g.) or commercially manufactured dosing needle, as used for rats, is introduced into the mouth through the interdental space and advanced gently into the oesophagus. A small gag made from a solid plastic rod with a hole drilled centrally may be used to prevent the animal biting a plastic catheter.

Subcutaneous injection

The animal should be restrained by an assistant as described above, and a small area of skin on the flank tented by the operator. The needle is introduced into the raised skin, parallel to the body wall. The skin of the guinea pig is thicker than in smaller rodents and provides more resistance to needle passage, hence injection is easier if a short (1–2 cm) 21–23 gauge needle is used.

Intradermal injection

This is carried out as described for rats. The thicker skin of the guinea pig makes the technique relatively easy in this species. As described previously, the presence of a small bleb of material indicates successful intradermal rather than subcutaneous injection.

Footpad injection

The footpads are occasionally used as injection sites, particularly of material intended to act as an antigen for antisera preparation. The animal should be restrained by an assistant as described above and injection made into the large central pad of the foot. In view of the considerable swelling that often ensues, only one pad should be inoculated, so that the animal can avoid placing full weight on that limb. Since there is little evidence to suggest that footpad inoculation of antigen results in any better antibody production than does inoculation at other sites, it is preferable to avoid using this technique.

Intraperitoneal injection

A similar technique to that described in the rat and the mouse is used for intraperitoneal injection of guinea pigs. The animal should be restrained by an assistant as shown in Fig. 9.17. The operator extends one of the animal's legs and introduces the needle along the line of the thigh, into the centre of the posterior quadrant of the abdomen.

Intramuscular injection

As with other rodents, the anterior and posterior thigh are the sites most frequently used for intramuscular injection, although it is also possible to inject into the triceps muscles on the anterior aspect of the shoulder. To inject into the thigh, the animal should be restrained by an assistant as for intraperitoneal injection, and one leg held firmly by the operator. If the quadriceps muscles are to be used, they should be held between the thumb and forefinger and the needle introduced at right angles to the skin into the centre of the muscle mass.

Intravenous injection

Guinea pigs have few superficial veins; those which are reasonably accessible are the ear veins and the penile vein (in males). These veins are small and

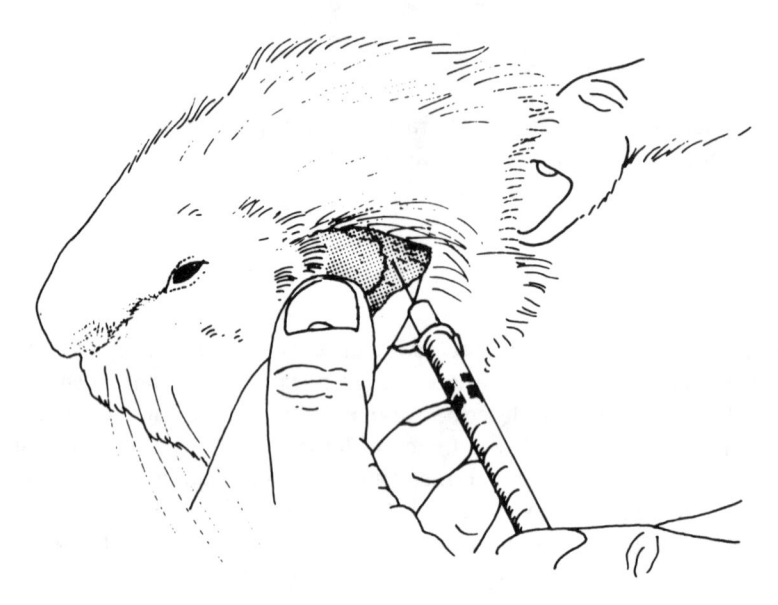

Fig. 13.8 The guinea pig — intravenous injection

fragile, and hence intravenous injection is difficult in this species. In large (>500 g) guinea pigs, the ear veins should be used. The ear should first be swabbed with a *small* quantity of xylene to dilate the vessels, and the animal restrained by an assistant on a firm surface.

The ear should be held firmly at one edge, and a suitable vein selected. A very fine needle (29–30 gauge) should be used, and once positioned in the vein, the syringe can be steadied using the remaining fingers of the hand restraining the ear (Fig. 13.8). Any movement of the animal during this procedure is likely to result in damage to the vein, and to avoid this it may be preferable to anaesthetize the animal. Following successful venepuncture, the xylene should be removed from the ear using a wet swab.

Penile vein injections should always be carried out in anaesthetized animals, since the procedure may cause considerable discomfort. The penis is extruded from beneath the inguinal skin by pressure at either side of the genital opening. The tip should then be grasped between the thumb and forefinger and the organ extended and rotated so that the dorsal penile vein lies uppermost. When introducing the needle, care should be taken to avoid damaging the fragile vein.

The rabbit

Oral dosing

The technique adopted for oral dosing of rabbits will be dependent upon whether administration of accurate volumes of fluid is required. It is possible to administer fluids quite simply by restraining the rabbit on a firm surface and placing the nozzle of a syringe in the corner of the mouth. The material should be injected slowly, in 0.25–0.5 ml boluses, allowing time for swallowing between each mouthful. Some material may be spilt during this procedure, but provided it is relatively bland most will be taken quite readily. Accurate volumes can be administered using a stomach tube, but a gag must be used in this species unless the animal is anaesthetized. A suitable design is illustrated in Fig. 13.10. To pass a stomach tube the animal should be restrained on a firm surface, and a gag introduced into the interdental space. The catheter is

Fig. 13.9 The rabbit — method of restraint for an intraperitoneal injection

threaded through the central hole and gently passed into the oesophagus. Although larger tubing can be used, an 8 f.g. soft polythene catheter is adequate for most materials. To check that the tube is in the oesophagus, and not in the trachea, it should be examined for signs of condensation in the lumen which would indicate air passage. Inadvertent tracheal intubation normally results in violent coughing, and if this is noted the tube should be withdrawn and repositioned. Once the catheter is correctly placed, material can be instilled slowly into the stomach or oesophagus and the tube withdrawn.

Subcutaneous injection

Subcutaneous injections are easily carried out in rabbits and often require only minimal restraint in animals of good temperament. The rabbit should be placed in a suitable transport box or restraining device to limit its movements and the skin over the neck grasped with one hand. Material can then be injected into the tented skin with little difficulty. If large volumes (>5 ml) are to be injected, it is advisable to have an assistant to restrain the animal on a firm surface. Following injection under the skin of the neck or back, the needle can be withdrawn slightly and redirected to spread the fluid over a wider area.

Intradermal injection

The usual sites for intradermal injection are the flanks or back. The area should be carefully shaved and the rabbit held by an assistant on a firm surface, or placed in a suitable restraining box. The skin should be gently stretched over the underlying tissues and the injection made using a 25–27 gauge needle. As with other species, raising a small bleb indicates successful intradermal injection.

Intraperitoneal injection

If assistance is available, the rabbit should be restrained as shown in Fig. 13.9 or Fig. 9.15. The xyphisternum and pubis should be palpated at the anterior and posterior extremes of the abdomen, the needle being introduced at right angles to the abdominal wall, just lateral to the midline, at a point equidistant between the xiphisternum and pubis. This positioning should avoid accidental puncture of the bladder or stomach. A short (2.5 cm) needle should be used. Puncture of the viscera is a rare complication. If no assistance is available, the rabbit should be placed on a firm surface and restrained by grasping the skin over the neck. The injection should be made at a point just anterior to the hind limb in the lower third of the abdomen. This is a less satisfactory technique since sudden movement by the rabbit could lead to incorrect positioning of the needle, possibly damaging the abdominal viscera.

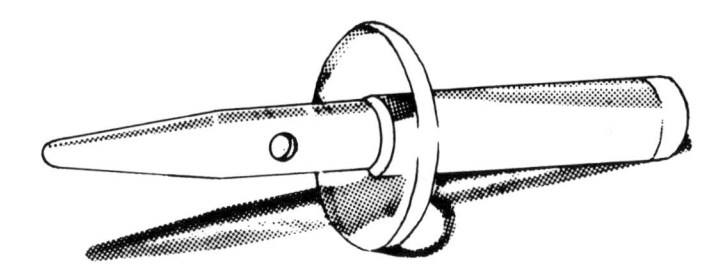

Fig. 13.10 The rabbit — plastic gag for oral closing

Intramuscular injection

The most usual sites for intramuscular injection are, as with other species, the posterior and anterior thigh. The larger size of the blood vessels within the muscle masses in this species compared to small rodents makes inadvertent intravenous injection a possibility. To avoid this potential complication, following introduction of the needle into the muscle mass, the plunger of the syringe should be drawn back to confirm that no blood can be obtained.

To inject into the hind limb muscles, an assistant should restrain the rabbit on a firm surface and the operator should grip the muscles of the anterior or posterior thigh between the thumb and forefinger. The needle should then be introduced at right angles to the skin surface into the centre of the muscle mass.

Intravenous injection

Intravenous injections are easily carried out in rabbits, the marginal ear veins being the vessels of choice. An assistant should restrain the rabbit, or alternatively it can be placed in a restraining box (see Chapter 9). The hair overlying the vein should be shaved with a scalpel or razor, so that the vessel becomes clearly visible. If assistance is available, then the vein should be compressed at the base of the ear, so that it dilates along the remainder of its length. The ear should be held firmly in one hand and the needle held almost parallel to the vein and directly above it. It can then be gently introduced into the vein and advanced about 1–2 cm along the lumen. It may be found easier to use a 'butterfly' infusion set for intravenous injection, rather than a needle fixed directly to the syringe. Many rabbits will twitch or jump as the needle is pushed into the ear, and the flexible coupling between syringe and needle provided by a 'butterfly' infusion set enables the needle to be maintained in its position in the ear without pulling on the syringe. If repeated injections are to be made, an indwelling catheter (e.g. Abbocath, Quickcath) should be taped in place.

The hamster

Oral dosing

The technique for oral dosing described for the mouse can be applied to the hamster.

Subcutaneous injection

Subcutaneous injection is made into the scruff as for the mouse; however, the very loose skin enables larger volumes to be administered by this route than is possible in other species of similar body size.

Intradermal injection

As for the mouse.

Intraperitoneal injection

Intraperitoneal injections can be made as described for the mouse; however, a large area of loose skin must be held to provide adequate restraint (see Chapter 9).

Intramuscular injection

As for the mouse.

Intravenous injection

There are no superficial veins that can be readily used for intravenous injection in hamsters. The most reliable technique is to anaesthetize the animal and to expose the jugular vein through a skin incision.

The cat

Oral dosing

Most cats will tolerate administration of liquids directly into the mouth using a syringe. The animal should be restrained as shown in Fig. 13.11, and fluid administered in 0.5–1.0 ml boluses, removing the syringe nozzle from the mouth between each mouthful to allow the animal to swallow. Stomach tubing is relatively simple, using a soft rubber or polyethylene 10–12 f.g. catheter which should be lubricated before use (e.g. using 'K–Y Gel'). The mouth is

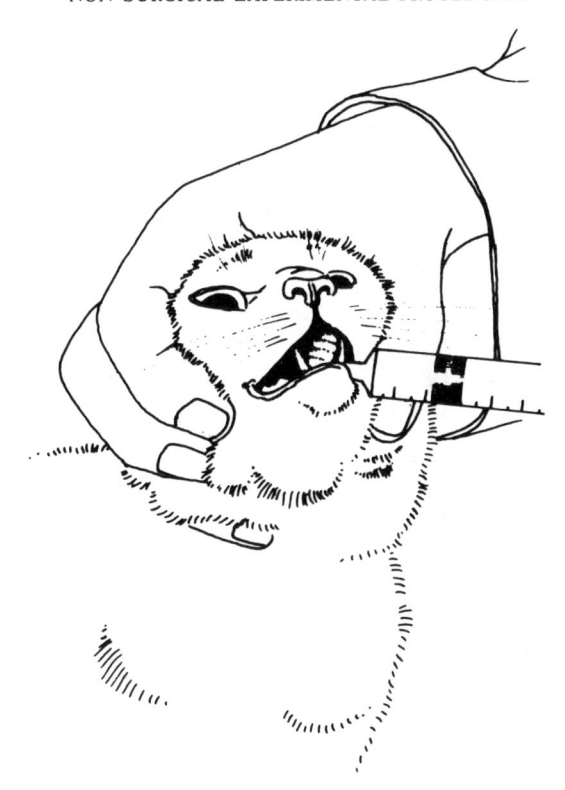

Fig. 13.11 The cat — oral dosing. Note that finger V is on the same side of the head as the thumb, i.e. the neck is between fingers IV and V

held open by pressure of the thumb and forefinger in the angle of the jaw, and the lubricated tube passed down into the oesophagus. It is virtually impossible to introduce the tube into the trachea in a conscious cat. If the cat becomes distressed during any of these procedures, then further attempts to administer fluids should be abandoned until the animal can be calmed and reassured.

Subcutaneous injection

Subcutaneous injection is made into the skin overlying the neck and shoulders. Larger volumes should be administered under the skin overlying the dorsal chest area. If several sites are used, large volumes (up to 100 ml) can be instilled by this route.

Intradermal injection

The skin must be carefully shaved and the technique described for the rabbit applied.

Fig. 13.12   The cat — intraperitoneal injection

Intraperitoneal injection

The cat should be restrained as shown in Fig. 13.12. The hair should be clipped along the midline of the abdomen, and the area just posterior of the umbilicus swabbed with a suitable antiseptic. The needle should be introduced subcutaneously, then redirected to pass into the abdomen. This technique is suitable only for use in tractable animals, and, as with all procedures involving cats, the assistance of an expert handler is essential.

Intramuscular injection

Intramuscular injections are usually made into the posterior thigh, ensuring that the femoral vessels and the sciatic nerve are avoided. Alternative sites are the quadriceps (anterior thigh) and the triceps muscles, although injection into the quadriceps is often painful and so is best avoided if possible.

NON-SURGICAL EXPERIMENTAL PROCEDURES

Fig. 13.13 The cat — method of restraint for an intravenous injection into the cephalic vein on the forelimb

Intravenous injection

The most convenient site for intravenous injection is the cephalic vein on the dorsal aspect of the forelimb. The cat is restrained as shown in Fig. 13.13, the hair over the area shaved or clipped and then swabbed with a suitable antiseptic. Pressure around the elbow will distend the vein, which can then be easily located beneath the skin. Once the skin has been punctured the needle should be advanced into the vein and successful positioning can be confirmed by withdrawing blood. If unsuccessful, the needle should be redirected, preferably without withdrawing completely and reintroducing through the skin, since repeated puncture will be strongly resented by the cat. Once the needle has been correctly positioned, pressure over the vessel should be released but firm support of the elbow joint and carpus maintained. Use of a 'butterfly' infusion set may be found advantageous in minimizing the effects of movement of the cat during venepuncture (see section on intravenous injection of the rabbit).

The dog

Oral dosing

Oral administration of fluid is easily achieved in this species. If the fluid is not unpalatable it can be administered gradually using a syringe as described for the cat. The head should be held with the nose tilted upwards to prevent fluid running out of the mouth. Passing a stomach tube usually presents no difficulty. The mouth is first opened by pressing the lips against the junction of upper and lower molars and premolars, then, as the mouth opens, the assistant should slide the fingers of one hand into the animal's mouth and press them against the hard palate. The lower jaw should be gently retracted using the other hand. The operator can then introduce a rubber or soft polyethylene lubricated tube over the tongue and on into the oesophagus. As the tube passes over the tongue the animal should be allowed time to swallow, ensuring that the tube passes into the oesophagus and not into the trachea.

Subcutaneous injection

Subcutaneous injections are normally made into the skin overlying the neck or over the back, as described for the cat.

Intradermal injection

The thicker skin in this species facilitates intradermal injection. As with other species, the skin must be carefully shaved and cleaned prior to injection.

Intraperitoneal injection

The animal should be restrained on its back, the abdomen being shaved along the midline. Injection is made just posterior to the umbilicus, on the midline, as described for the cat.

Intramuscular injection

Injection into the posterior thigh is well tolerated in the dog. The muscle should be held firmly in one hand, and the most posterior muscle belly palpated. Once this has been located the injection can be made with minimal risk of damage to vessels or nerves. An assistant will be needed to restrain the animal during this procedure. Alternative sites include the quadriceps and the triceps. As with the cat, injection into the quadriceps is often painful.

Intravenous injection

The cephalic vein is most frequently used, employing the technique as described for the cat. The cephalic vein is a more mobile structure in the dog, however, and venepuncture is facilitated by gently tensing the skin overlying the vein. The vein bifurcates just above the carpus, and is more firmly anchored at this point, hence injection just above the bifurcation is often easier. The recurrent tarsal vein which runs from the medial aspect of the hock is also a useful site for injection.

The domestic fowl

Oral dosing

Fluid should be administered using either blunt metal or flexible polythene tubing. An assistant is needed to restrain the bird. The operator should hold the head so that finger and thumb are placed at the angle of the beak. The neck is straightened by gently pulling the head and the jaw opened by pressing in the angle of the beak. The tubing can then be passed over the tongue and into the oesophagus. Accidental tracheal intubation rarely occurs.

Subcutaneous injection

With the bird restrained on its back, the skin at the junction of the thigh and abdomen can be tented and subcutaneous injection carried out.

Intradermal injection

The thin, fragile skin of this species makes intradermal injection difficult. The thickest skin is to be found in the lumbar region and, following plucking of some feathers, injection should be made in this area.

Intraperitoneal injection

Care must be taken to avoid damage to the abdominal air sacs when attempting intraperitoneal injection, and only small volumes of fluid (1–2 ml) should be administered.

Intramuscular injection

The feathers overlying a small area of the pectoral muscles should be removed and the injection made at this site.

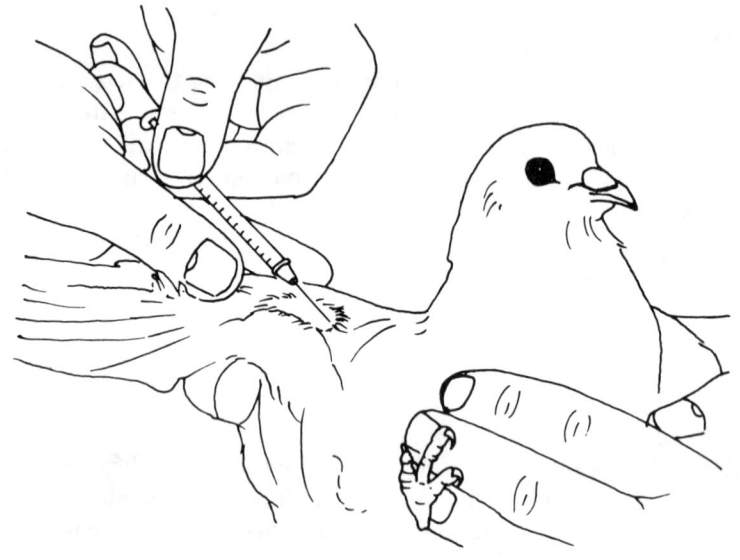

Fig. 13.14 The pigeon — intravenous injection into the alar vein; the bird is lying on its back on a cloth

Intravenous injection

The alar vein, which runs along the ventral aspect of the wing (Fig. 13.14), is a suitable site for injection. Considerable practice is needed, however, since the fragile vein is easily damaged and, due to the thin poorly anchored skin, a large haematoma is rapidly produced, obscuring the vein and making further injections almost impossible. The jugular vein can be used but the bird must first be anaesthetized, since puncture is most reliably achieved following surgical exposure of the vein. The jugular veins run subcutaneously, embedded in fat, parallel and lateral to the trachea.

COLLECTION OF BODY FLUIDS

Blood

The collection techniques used will vary depending upon the species, volume required and purpose for which the blood is being collected. In general, a sample which is free from haemolysis is to be preferred. If plasma or cellular constituents are required, clotting must be prevented by use of an appropriate anticoagulant. In general, EDTA (ethylene diaminetetra-acetic acid) is the most suitable agent if haematological parameters are to be measured. Heparin is suitable for most biochemical estimations, but since it interferes with the staining of leucocytes it is unsuitable for haematological examinations. Sodium fluoride (often combined with oxalate in commercially prepared sample tubes) is used both as an anticoagulant and to prevent glucose utilization by red cells during storage of the blood sample. Whichever anticoagulant is used, haemolysis should be avoided, since this may interfere with many biochemical assays. It can usually be prevented by avoiding excessive suction when collecting the sample, removing the needle before transferring the blood from the syringe to the sample tube, and by *gentle* mixing of blood and anticoagulant in the sample tube. Blood may be collected using a syringe or dripped directly into a suitable container. If a syringe is used it must be clean and dry, and this is best ensured by using plastic disposable syringes. If it is anticipated that collection will be slow, or that the sample may be contaminated with tissue fluid (which appears to increase the clotting time), then the addition of anticoagulant to the syringe may be found useful.

A range of disposable needles is widely available, and these are to be preferred to 'reusable' needles which require resharpening. It is often convenient to use 'butterfly' infusion sets rather than a simple needle, since the flexible coupling between needle and syringe provided by this apparatus enables more control over the needle position, particularly if the animal moves, or if a fresh syringe is to be attached. Although blood can be dripped directly into a sample tube, it may be found difficult to position the tube close to the site of venesection in some species. Small volumes of blood may be conveniently obtained by use of a capillary tube to collect the blood as it flows from the needle or from a small incision in the vein. Capillary tubes may be purchased ready coated with anticoagulant, which facilitates sample collection and processing. Alternative techniques include the use of vacuum tubes, precoated with anticoagulant if this is required, and a double-ended needle (e.g. Vacutainer). The maximum volume of blood usually obtainable from each species is listed in Table 13.2.

If the animal is required to survive the procedure, then the sample volume should not normally exceed 10 per cent of the animal's blood volume (Table 13.2). In practice, as much as 30–40 per cent of blood volume may be removed from some species; however, there will be a considerable risk of producing serious hypovolaemia and cardiovascular failure ('shock'). Repeated blood samples of 5–10 per cent of total blood volume can usually be obtained at 2–3 week intervals without normally causing any ill-effects. If this procedure is continued for some months then it is advisable to monitor the red blood cell count, or more simply to monitor the packed cell volume (haematocrit). It may also be useful to examine blood smears in order to detect early changes associated with anaemia, for example polychromasia of the red cells. Normal haematological values for the common laboratory species are listed in Table 13.2. In selecting a method for obtaining blood samples, it should be remembered that both stress and anaesthetics may radically alter haematological and

Table 13.2 Haematological parameters of laboratory animals

Species	Haemoglobin (g/dl)	PCV (%)	RBC count (× 10⁶/ml)	Average adult blood volume (ml)	Maximum sample volume (ml)	Expected volume on exsanguination (ml)	Adult body weight (g)
Rat	11.1–18.0	36–52	5.0–12.0	30	2.5	12	300–500
Mouse	10.2–16.6	32–54	6.7–12.5	2.5	0.3	1.2	25–40
Guinea pig	11.2–18.1	37–51	3.0–7.0	60	5	30	700–1200
Rabbit	9.9–19.3	30–53	4.0–8.6	250	50	150	2000–6000
Hamster	10.0–20.2	36–59	3.0–10.0	9	0.5	3	85–150
Cat	8.0–15.0	24–45	5.0–10.0	200	30	120	2500–4000
Dog (20 kg)	12.0–18.0	37–55	3.5–8.5	1600	200	1000	—
Domestic fowl	7.0–18.6	23–55	1.25–4.5	240	20	160	1500–2500

Species	WBC count (× 10³/ml)	Neutrophils (%)	Lymphocytes (%)	Monocytes (%)	Eosinophils (%)	Basophils (%)
Rat	3.0–15.0	4–50	40–95	0–8	0–4	0–2
Mouse	5.4–16.0	5–16	8–43	0–8	0–3	0–1
Guinea pig	5.0–18.6	5–18	20–60	1–9	0–8	0–2
Rabbit	2.0–15.0	10–85	25–95	0–16	0–8	0–8
Hamster	2.6–11.6	3–42	50–96	0–5	0–4	0–1
Cat	10.6–20.2	40–74	20–44	0–1	2–10	0–1
Dog (20 kg)	7.0–14.6	60–88	9–31	1–7	0–4	0–2
Domestic fowl	9.0–32.0	15–50	29–84	0–7	0–16	0–8

Data adapted from Canadian Council on Animal Care (1980), Mitruku and Rownsley (1977), Sanderson and Phillips (1982) and Coles (1980).

biochemical parameters. Stress caused by handling and physical restraint commonly results in an increase in haematocrit, and alterations in the white blood cell count. Concentrations of other plasma constituents, for example glucose and some hormones, may also be affected. This may be overcome to some extent by familiarizing the animals to the technique of venepuncture, and training them to accept the degree of restraint required. This is particularly effective with dogs and cats, but can also be achieved with rabbits and, to a lesser extent, with rodents. If anaesthetics or tranquillizers are administered to provide restraint and/or analgesia, these may also produce alterations in haematological and biochemical parameters. Often more predictable, and quite acceptable, control values and 'normal ranges' can be obtained under standard methods of anaesthesia for many parameters. An alternative, particularly when repeated blood samples are required, is to implant indwelling arterial or venous cannulae. This will often enable repeated samples to be obtained in an apparently stress-free animal. Cannulae may be implanted with the animal anaesthetized and utilized at a later date following full recovery from the anaesthetic, or they may be introduced into superficial veins without the use of an anaesthetic. This avoids the necessity for repeated venepuncture, which is resented by many animals.

A suitable anaesthetic should be used whenever possible before undertaking any procedure likely to cause pain or distress to an animal. In all species, following superficial venepuncture, it is important to re-examine the animal 30–60 min after the procedure to ensure that all haemorrhage has been successfully controlled.

The rat

(a) Exsanguination. To obtain blood by exsanguination, without the need for anaesthesia, the rat can be decapitated. This can be achieved by stunning the rat by a blow to the back of the head followed by cutting through the neck with strong scissors or by use of a purpose-made guillotine. The operator must be well trained and confident in the technique, and must be thoroughly experienced in animal-handling procedures. The rat should be handled gently and allowed to recover from any excitement caused by removal from its cage. It should then be stunned and rapidly decapitated. The neck should be immediately placed over a suitable container. The volume of blood collected will be low compared with other techniques, and the blood will be contaminated with tissue fluids. The instrument and surrounding area should be cleaned before a further animal is killed. Stunning followed by decapitation is a procedure which many workers find distasteful, although when carried out expertly it is an extremely rapid and presumably humane method of killing the animal.

(b) Cardiac puncture. The rat should be anaesthetized and placed in right lateral recumbency. The heartbeat can be palpated with the finger and thumb placed on either side of the chest, just posterior to the rat's elbows. If the animal is to be allowed to recover, a 2 ml syringe and 25 gauge needle should be used, to minimize damage to the myocardium. The needle should be introduced into the left side of the chest, perpendicular to the chest wall, immediately over the area in which the heartbeat is most easily palpable. As the needle is introduced, gentle suction should be applied with the syringe, and the needle advanced slowly until blood is obtained. Once in position, it is important to avoid moving the needle tip whilst continuing to aspirate blood. An alternative approach is to place the rat in dorsal recumbency, locate the xyphisternum by palpation and introduce the needle beneath this into the chest. The needle should be inserted at an angle of 25–30 degrees relative to the animal's body wall, and gentle suction applied as soon as the body wall has been punctured. The needle is then advanced until blood is aspirated. If the animal is not to recover, a 23 or 21 gauge needle may be used, which considerably facilitates blood sampling and minimizes the risk of producing haemolysis. If blood is not obtained, the needle should be withdrawn, redirected slightly and the procedure repeated. Repeated puncture of the myocardium often produces serious damage, haemorrhage and death of the animal.

(c) Jugular puncture. The rat should be anaesthetized and the hair on the ventral aspect of the neck shaved. A skin incision is made parallel to the midline and the jugular vein exposed; in obese animals it is necessary to first clear the overlying fat by blunt dissection. The vein should not be handled since this often causes the smooth muscle in its wall to go into spasm, although this can often be prevented by the local application of procaine hydrochloride. The vein should be entered using a 25 gauge needle. If the needle is introduced through the belly of the pectoral muscle which overlies part of the vein, then this will help prevent haemorrhage once the needle has been withdrawn.

(d) Tail vein. The rat may be anaesthetized, or placed in a suitable restraining apparatus. Venepuncture is made considerably easier if the tail is first gently warmed. This can be achieved either by placing the rat under a warming lamp or by dipping the tail in warm (40–45°C) water. The lateral tail veins are readily visible in young rats, but in older animals the thickened skin on the tail makes localization difficult. The vein can be entered using a 23–25 gauge needle and small volumes of blood (0.2–0.5 ml) gently aspirated. It may be found easier to either cut off the tip of the tail (in an anaesthetized rat) to obtain a small blood sample (0.1–0.2 ml) or to incise the vein using a scalpel blade. Gentle pressure is usually sufficient to produce haemostasis, although overenthusiastic incision may require chemical cauterization (using potassium permanganate or ferric chloride). Repeated sampling from the tail tip will cause significant damage to the tail and so should be avoided.

The mouse

(a) Exsanguination. The animal should be anaesthetized or stunned by a blow to the head, and the neck completely severed using a pair of scissors. The body should then be immediately placed over a suitable container for collection of blood. The problems associated with this method are as described for the rat.

(b) Cardiac puncture. The mouse should be anaesthetized and placed in dorsal recumbency. It is often convenient to immobilize the animal using a suitable restraining board. A 25 gauge needle and a 1 or 2 ml syringe should be used. The needle should be introduced beneath the xyphisternum as described for the rat. The small size of the heart makes needle positioning critical. Small movements will result in passage of the needle tip out of the ventricle and into the myocardium.

(c) Tail vein. This technique is the preferred method for blood collection, and allows repeated venepuncture over long periods. The animal should be warmed to dilate the vessels, and should either be anaesthetized or placed in a suitable restraining device. Rather than introduce a needle into the vein, the tail should be cleaned with a disinfectant and a small incision made over one of the lateral tain veins. Blood can then be dripped directly into a sample tube. Haemostasis is readily achieved by gentle manual pressure. Small volumes of blood may also be obtained by snipping off the end of the tail of an anaesthetized mouse. Since the tail veins are easily located even in old animals, this method of sampling by tail-snipping may be thought an unnecessary mutilation of the animal.

The hamster

(a) Exsanguination. The method as described for the mouse should be used.

(b) Cardiac puncture. The hamster should be anaesthetized and the procedure followed as described for the mouse.

(c) Retro-orbital plexus. The retro-orbital venous plexus provides a convenient source of venous blood in many species. Since the hamster lacks other suitable superficial veins, this plexus is often considered the best site for collection of small blood samples. It is, however, a controversial method of sampling, and is aesthetically unacceptable to many people. It is also discouraged by the Home Office Inspectorate in the United Kingdom. If the technique is to be employed, the animal should be anaesthetized and then held firmly by the skin of the back and nape, which causes the eyeball to protrude. A fine glass tube (previously prepared by drawing out capillary tubing to an outside diameter of 0.5 mm) is introduced into the orbit. The tube is placed at the inner canthus of the eye and advanced gently alongside the globe into the plexus. The tube ruptures the fine vessels of the plexus and blood is withdrawn by capillary action. Gentle rotation of the tube may be required to maintain the flow of blood. Great care must be taken not to snap the tube and leave a portion embedded in the eye. Following removal of the tube, and relaxation of the grip on the scruff of the animal (which causes the eye to retract), bleeding stops immediately. Although no damage to the eye results when the technique is carried out competently, the unpleasant nature of the method makes it unacceptable to many workers.

The guinea pig

(a) Exsanguination. Anaesthetized guinea pigs may be decapitated using a guillotine or large scissors. However, the poor yield obtained when using this method, coupled with the relative ease of cardiac puncture in this species, limits its usefulness.

(b) Cardiac puncture. The animal should be anaesthetized, and a 21–23 gauge needle used (23 gauge if the animal is to recover). The animal is placed in right lateral recumbency, the heartbeat palpated and the needle introduced over the point of the strongest beat, as described for the rat.

(c) Ear veins. Small blood samples (0.1 ml) can be obtained from the ear veins of larger guinea pigs (>300 g). The vessels should first be dilated by warming the animal, or by swabbing with a few drops of xylol. A large vein is selected and a small incision made using a scalpel blade. Although this technique can be undertaken in conscious guinea pigs, restraint is often a problem and so anaesthesia is desirable.

(d) Jugular vein. Surgical exposure of the jugular vein may be undertaken in anaesthetized guinea pigs, the technique described for the rat being used to obtain blood samples.

The rabbit

(a) Cardiac puncture. The rabbit should be anaesthetized and placed in right lateral recumbency. The heartbeat is palpated and a 19–21 gauge needle used for cardiac puncture as described for the rat and guinea pig. Alternatively, the animal may be placed in dorsal recumbency, the chest entered under the xyphisternum and the needle advanced slowly until the heart is punctured. This technique is usually employed only for exsanguination, since large volumes of blood are readily obtained from the marginal ear veins.

(b) Ear vein. The vein is easily located on the caudal margin of the ear. The overlying skin should be shaved and the area swabbed with disinfectant to allow the clearest view of the vessel. The vein can be dilated by applying a few drops of xylol to the tip of the ear. The rabbit need not be anaesthetized, but should be well restrained by placing in a suitable restraining box or wrapping in a towel. The ear is held between thumb and forefinger, the forefinger providing a firm support for the edge of the ear. A 23–21 gauge needle can be introduced into the vein and blood gently aspirated. The suction should be reduced if the vein is seen to collapse. A 'butterfly' (scalp vein) infusion set may be found convenient, as it allows small movements of the syringe without dislodging the needle. An alternative technique is to make a small (2–3 mm) longitudinal incision over the vein and collect the blood as it drips from the incision. Both techniques enable volumes of up to 50 ml to be collected. A number of variations of these methods have been described, including techniques to apply slight vacuum to the ear to speed bleeding from an incision in the vein. These methods allow more rapid bleeding and are useful if large numbers of rabbits are to be sampled. Following successful completion of sampling, any xylol used should be cleaned from the ear and haemostasis produced by manual pressure using a clean swab. If haemorrhage continues, pressure can be maintained by holding the swab in position using a paper clip. It is essential to remove this within the next 30 min or so, since continued pressure will result in necrosis and sloughing of that part of the ear. If repeated blood samples are required, a flexible catheter can be introduced into the vein and taped in place, thus avoiding repeated puncture of the vessel. Blood can also be sampled from the central artery of the ear, but since inadvertent damage to this vessel may severely impair the circulation to the ear, it is best avoided and the marginal vein used whenever possible.

The dog

(a) Cardiac puncture. The larger size of this animal enables a needle to be introduced through an intercostal space into the heart without the necessity of anaesthetizing the animal. Sudden movements of the animal can cause unnecessary trauma, however, and hence it is usually preferable to sedate or anaesthetize it. A 23–21 gauge needle (or larger, 19–16 gauge, if the animal is to be exsanguinated) is introduced into the chest between the fourth or fifth intercostal space, over the area of strongest heartbeat. Due to the relatively large size of the animal, a 4–5 cm needle should be used.

(b) Cephalic vein. The technique of venepuncture is as described on p. 245. Once the needle has been introduced into the vein, the operator's assistant should continue to maintain pressure around the dog's elbow to dilate the vein. Small volumes of blood (1–5 ml) are readily collected. If large volumes are required, gentle intermittent pressure around the foot may increase the rate of venous return and so speed the collection of blood. Alternatively, blood can be dripped from a needle directly into sample tubes. The tarsal vein on the medial aspect of the hock may also be used for blood sampling.

(c) Jugular vein. Large volumes of blood (10–100 ml or more) are most conveniently obtained from the jugular vein. Anaesthesia is not necessary in good-tempered animals. The neck should be shaved, the overlying skin swabbed with alcohol or a suitable disinfectant, and the jugular vein located in the jugular groove. Pressure with the thumb at the vein's point of entry to the thorax will distend the vessel and facilitate venepuncture. A 21 gauge needle should be introduced under the skin overlying the vein, almost parallel to the vessel. It should then be advanced further to puncture the vein. Haemostasis is readily achieved by applying firm digital pressure over the puncture site for a minute or two.

The cat

(a) Cardiac puncture. The cat should be anaesthetized and placed in right lateral recumbency. Cardiac puncture can then be undertaken as described for the rabbit.

(b) Cephalic vein. The technique described on p. 243 should be used. 1–2 ml of blood can be collected fairly readily by this route. Larger volumes (5–10 ml) can be obtained, but since collection may be slow in some individuals, it is advisable to add anticoagulant to the syringes to be used.

(c) Jugular vein. The jugular vein is easily located following shaving of the neck and pressure with the thumb at the point of entry to the thorax. A 21 gauge needle should be used, and 20–30 ml of blood can easily be obtained by this method. It is usually preferable to anaesthetize the cat, unless an experienced animal handler is available and the animal has a tractable nature.

The domestic fowl

(a) Cardiac puncture. Although this technique can be undertaken in conscious birds, anaesthesia may be needed to provide good restraint. The bird is restrained in dorsal recumbency and the needle (23–21 gauge) introduced either through the anterior thoracic inlet or through the left thoracic wall, as described in other species. It is not usually possible to localize the heartbeat as accurately as in laboratory mammals, and hence considerable practice may be necessary before consistently good results are obtained.

(b) Wing vein. Smaller volumes of blood can be obtained from the wing vein, either following venepuncture or by making a small incision over the vein.

(c) Jugular vein. The bird should be anaesthetized and placed in dorsal recumbency. The feathers should be plucked from the ventral aspect of the neck and a skin incision made parallel to the midline, 1–2 cm posterior to the head. The jugular vein is usually easily visible, although in some birds it may be embedded in fat and require blunt dissection prior to puncture with a 23–21 gauge needle.

Urine

Urine samples may be obtained either by sampling during voluntary urination, by catheterization, by direct puncture of the bladder or, if some contamination of the sample with faecal and other material is acceptable, by use of a metabolic cage. Metabolic cages designed for all of the common laboratory species are available commercially, or may be manufactured 'on site' to meet particular requirements. If large volumes of an animal's urine output are to be collected, a metabolic cage of the appropriate design should be used. It is the only reliable method of urine collection from small laboratory rodents.

Sampling during voluntary urination is occasionally useful in laboratory rodents and dogs and cats. When rats and mice are picked up and restrained, they frequently urinate, and this may be taken as an opportunity to collect small volumes of urine for analysis. Samples will be contaminated with small amounts of faecal material, cells and bacteria from the genital tract, skin, hair and other debris, but may be useful for some purposes. Cats may often be persuaded to urinate by applying gentle manual pressure to the bladder through the body wall. Once again, the sample will be contaminated with debris, but prior cleaning of the perineum may minimize this. Dogs will often urinate when removed from their pens to an exercise area, although as with other species, samples obtained during voluntary urination usually have some degree of contamination.

Urethral catheterization is applicable routinely only in the larger species — the rabbit, cat and dog. In all species care must be taken to minimize the risk of introducing infection into the urinary tract during the procedure. Catheters must be sterilized before use, as should any lubricating gels and speculums.

Cystocentesis, or direct puncture of the bladder through the body wall, is possible in most species, but practicable as a routine procedure only in larger animals. The bladder must be palpable to enable accurate insertion of the sampling needle, limiting this procedure to rabbits and larger species.

The rabbit

(a) Urethral catheterization. Male — The procedure is made easier if the animal is sedated or anaesthetized. The animal should be restrained in dorsal recumbency and the penis extruded. A suitable catheter (cat urethral catheters, available commercially) is introduced into the urethral opening and advanced gently into the urethra and up into the bladder. The catheter should not be completely removed from its plastic container, as this enables it to be held without contaminating it. Successful catheterization is indicated by flow of urine from the end of the catheter. The plastic container can then be removed and the urine it contains retained for analysis. A sterile syringe can then be attached to the catheter and the urine remaining in the bladder aspirated. The presence of large quantities of sediment in the bladder may block the catheter. If this occurs a small volume of urine should be reinjected and then further urine can usually be aspirated.

Female — Catheterization is impracticable as a routine method of urine collection.

(b) Cystocentesis. The animal should be sedated ('Hypnorm', 0.3 ml/kg i.m.) and the skin shaved in the midline in the inguinal region. The bladder is normally palpable in the posterior abdomen and can be firmly held through the body wall. A 23 gauge needle can then be introduced through the body wall in the midline at an angle of about 45°. The needle should be directed to puncture the posterior section of the bladder, so that it will remain in the bladder lumen as urine is expelled. Urine can be withdrawn by suction, using a syringe, or expelled through the needle into a collection tube.

The cat

(a) Urethral catheterization. Male — Anaesthesia is not necessary, but if the animal resents handling light sedation (e.g. with Ketamine HC1) is helpful. The animal can remain standing if conscious, or be restrained in lateral recumbency. The penis is extruded by manual pressure either side of the prepuce and a urethral catheter introduced into the urethral opening. The catheter should be advanced until at the level of the posterior edge of the pelvis, when resistance to its passage will be encountered. The catheter should then be withdrawn slightly, redirected so that it lies parallel to the line of the pelvis and then advanced again into the bladder. Alternatively, the penis should be extended caudally by gentle traction, and elevated so that the catheter lies parallel to the line of the pelvis. Relaxation of the traction and advancing the catheter allows it to pass on into the bladder.

Female — Light anaesthesia is helpful in providing good restraint, but the procedure can be carried out in a conscious animal. The animal is restrained in dorsal recumbency or allowed to remain standing. The vulva should be cleansed using a suitable disinfectant and then dilated using a small illuminated speculum. The urethral opening can then be located on the ventral surface of the vagina. The urethra opens in the centre of a small mound, and gentle probing with the catheter will usually distinguish the opening. The catheter can then be gently threaded into the bladder.

(b) Cystocentesis. Anaesthesia is not necessary. The animal should be restrained in dorsal recumbency and the technique for cystocentesis is as described for the rabbit.

The dog

(a) Urethral catheterization. Catheterization can usually be carried out in unanaesthetized animals, but nervous or aggressive animals will require sedation.

Male — The animal is restrained in lateral recumbency and the penis extruded. A human urethral catheter can be used (4–10 f.g.). It should be lubricated with xylocaine gel and introduced gently into the urethral opening and threaded up into the bladder. Usually no difficulties are experienced in advancing the catheter into the bladder, provided the catheter is flexible and of suitable size.

Female — The animal is restrained in dorsal recumbency and the vulva cleaned using a suitable disinfectant. A lubricated (e.g. using 'K–Y jelly') speculum should then be introduced into the vagina and the urethral opening located. As with the cat, the urethra opens in the centre of a small raised area on the ventral floor of the vagina. The catheter can then be advanced under direct vision into the urethral opening and on into the bladder.

(b) Cystocentesis. The technique is as described for the rabbit, but a 4–6 cm needle should be used.

Faeces

Faecal samples may be collected using a metabolic cage, or, if fresh material is required, the animal should be removed from its cage or pen and, in the case of larger species (cat, rabbit and dog), small quantities removed from the rectum using a swab. Use of a small speculum may enable larger quantities of material to be removed. Small rodents will almost invariably pass one or more faecal pellets when manually restrained for a few minutes. Collection of total faecal output is complicated by the occurrence of coprophagy in rodents and rabbits. If total faecal output is required, special apparatus is needed to prevent the animal ingesting faeces as they are passed. It must also be remembered that prevention of this activity may interfere with normal growth and metabolism. Suitable apparatus includes restraining harnesses that prevent access to faeces as they are voided, coupled with a grid floor to prevent accumulation of faeces in the cage. An alternative technique is to attach an anal cup, and a variety of designs of these have been described (Kraus, 1980).

Saliva

Mixed salivary secretions can be obtained from anaesthetized animals by placing small cotton swabs or filter paper discs in the mouth and then eluting the contents with known volumes of water. With larger species (dog and cat) sedation with Ketamine (without premedication with atropine) results in fairly copious salivation, enabling small quantities to be aspirated from the mouth. It is possible to catheterize the salivary ducts in some species to collect saliva from a specific gland. Anaesthesia is necessary and in most instances catheters must be specially constructed by drawing out heated nylon tubing (Portex) or glass capillary tubing. The submandibular duct in the rabbit can be localized on a small papilla close to the rostral attachment of the phrenulum and a catheter threaded into it following dilation of the opening using a piece of monofilament nylon (7 or 8/0 initially, followed by progressively thicker material). A similar technique can be employed in the cat. In the dog, the larger size of the duct enables larger cannulae to be inserted. The parotid duct may be cannulated in the dog in a similar manner to the submandibular gland after locating its opening dorsolateral to the upper premolars. Methods for cannulating the salivary ducts in the rat have been described (Baker et al., 1980) using purpose-made cannulae.

Vaginal fluid; vaginal cells

Collection of materials for the preparation of vaginal smears for the determination of the stage of the oestrus cycle is frequently required in laboratory species. Several techniques have been described. These involve either introduction of a sterile swab or glass rod into the vagina and then smearing the adhering material onto a glass slide, or instilling a small volume of normal saline into the vagina using a blunt-ended pipette, following by aspiration of the liquid which will then contain a suspension of cellular debris. In the bitch, larger volumes (>1 ml) of fluid can be aspirated directly during the oestrus period, using a blunt-ended glass pipette. Details of the cytology of vaginal smears at various stages of the oestrus cycle are given by Hafez (1970).

Peritoneal fluid

The simplest method for collection of peritoneal fluid is by aspiration through a surgical incision at postmortem. A method for collection of peritoneal cells from conscious rats, mice and hamsters has been described (Baker et al., 1980). The technique requires the use of a specially constructed micropipette, which is used to instil warmed Hanks' solution into the peritoneal cavity followed by aspiration of the fluid which contains suspended cells.

Milk

Milk can be collected from most species, although specially constructed apparatus is often necessary to obtain a reasonable sample volume. Manual stimulation may be sufficient to produce milk secretion from lactating females; however, in some instances an injection of oxytocin may be required to stimulate mammary gland activity. A range of 'milking machines' for small rodents has been described (see Baker et al., 1980, for a review); these involve application of pulsatile suction to the gland using a small vacuum pump. Milk samples can be obtained from dogs and cats by manual manipulation of the teats following injection of oxytocin, or by milking from an unused teat during suckling by the animal's offspring. With all species, removal of the litter for several hours prior to attempting milk collection will tend to produce engorgement of the glands and aid collection. It should be noted that such disturbances to the mother may lead to problems of neglect and rejection of her litter.

Semen

The range of techniques for collecting semen in rodents is reviewed by Bennett and Vickery (1970). Collection of semen in rats can be achieved using electroejaculation. The technique has been described in detail by Waynforth (1980). An electroejaculation technique for use in guinea pigs has been described (Freund, 1969). Semen can be obtained from dogs following manual manipulation whilst exposing the animal to a bitch in oestrus (Kirk and Bistner, 1981).

Cerebrospinal fluid

Collection of cerebrospinal fluid in most species may be achieved by puncture of the fourth ventricle at the cysterna magna. The technique is basically similar in all species, involving anaesthesia of the animal, shaving of the dorsal cervical and occipital region, positioning the animal with its head in flexion followed by puncture of the ventricle using a suitable sized needle (26–23 g). Details of the technique in individual species are described in the rat (Waynforth, 1980), the rabbit (Kusumi and Plouffe, 1979), the guinea pig (Reiber and Schunck, 1983), the cat (Parker and Small, 1975) and the dog (Kirk and Bistner, 1981). Repeated puncture is feasible in most species; however, a common problem is contamination with blood due to inadvertent puncture of small dural blood vessels. Gross contamination results from deviation of the needle from the midline and puncture of the venous sinuses. To avoid significant red cell contamination about 3–5 days should be allowed to elapse between successive punctures. Care must be taken to avoid collection of an excessive volume if the animal is to survive; about 0.05–0.1 ml from rats and guinea pigs, 0.5 ml from cats and 2.0 ml from dogs is considered a safe volume.

REFERENCES

Bader, M. and Klinger, W. (1974). Intragastric and intracardiac injections in newborn rats. Methodological investigation, *Z. Versuchstierkd*, **16**, 40–2.
Baker, H. J., Lindsey, R. J. and Weisbroth, S. H. (1980). *The Laboratory Rat*, Vol. II, *Research Applications*, Academic Press, New York.
Bennett, J. P. and Vickery, B. H. (1970). Rats and mice. In E. S. E. Hafez (ed.) *Reproduction and Breeding Techniques for Laboratory Animals*, Lea and Febiger, Philadelphia.
Canadian Council on Animal Care (1980). *Guide to the Care and Use of Experimental Animals*, Vol. 1, Canadian Council on Animal Care, Ottawa.
Coles, E. H. (1980). *Veterinary Clinical Pathology*, Saunders, Eastbourne.
Freund, M. (1969). Interrelationship among the characteristics of guinea pig semen collected by electro-ejaculation, *J. Reprod. Fert*, **19**, 393–403.
Gibson, J. E. and Becker, B. A. (1967). The administration of drugs to one day old animals, *Lab. Animal Care*, **17**, 524–7.
Hafez, E. S. E. (1970). *Reproduction and Breeding Techniques for Laboratory Animals*, Lea and Febiger, Philadelphia.
Kirk, R. W. and Bistner, S. I. (1981). *Handbook of Veterinary Procedures and Emergency Treatment*, W. B. Saunders, Philadelphia.
Kraus, A. L. (1980). Methodology. In H. J. Baker, J. R. Lindsey and S. H. Weisbroth, (eds) *The Laboratory Rat*, Vol. II, *Research Applications*, Academic Press, New York.
Kusumi, R. K. and Plouffe, J. E. (1979). A safe and simple technique for obtaining cerebrospinal fluid from rabbits, *Laboratory Animal Science*, **29**, 681–2.
Mitruka, B. M. and Rawnsley, H. M. (1977). *Clinical, Biochemical and Haematological Reference Values in Normal Experimental Animals*, Masson Publishing USA Inc. New York.
Parker, A. J. and Small, E. (1975). The nervous system. In E. J. Catcott (ed.) *Feline Medicine and Surgery*, American Veterinary Publications, Santa Barbara.
Reiber, H. and Schunck, O. (1983). Suboccipital puncture of guinea pigs, *Laboratory Animals*, **17**, 25–7.
Sanderson, J. H. and Phillips, C. E. (1982). *An Atlas of Laboratory Animal Haematology*, Oxford University Press, Oxford.
Waynforth, H. B. (1980). *Experimental and Surgical Techniques in the Rat*, Academic Press, London.

Laboratory Animals: An Introduction for New Experimenters
Edited by A. A. Tuffery
© 1987 John Wiley & Sons Ltd

CHAPTER 14

Anaesthesia and Analgesia

C. J. GREEN

Division of Comparative Medicine, Clinical Research Centre, Harrow

CHOICE OF ANAESTHETIC METHOD

Welfare of the animal and safety of personnel

The humane treatment of experimental animals and the safety of personnel involved in their management take precedence over all other considerations in selecting the most suitable anaesthetic protocol. Above all it is mandatory to provide adequate analgesia, using drugs which suppress the perception of painful stimuli at all stages of the experiment including postoperative recovery. Indeed, it is essential to avoid *any* suffering and this necessitates provision of humane restraint by gentle handling and using carefully selected drugs after environmental adaptation and caring conditioning. Among the hazards to personnel which should be avoided are: kicks, bites or scratches; sparks or flames in the presence of flammable volatile anaesthetics like ether; possible risks of volatile agents building up in poorly ventilated rooms; and self-injection with potent drugs of addiction.

Available anaesthetic techniques

Surgical anaesthesia has four components, comprising analgesia, decreased perception of other external stimuli, suppression of reflex activity and loss of skeletal muscular tone. These may for example during general anaesthesia be accompanied by loss of consciousness. It must be understood that ideal surgical conditions can be achieved without a general anaesthetic; for instance a sedative agent can be used to depress central sensory and motor activity, and a local analgesic agent can be injected adjacent to the operative site to prevent transmission of painful stimuli. As a general proposition, analgesia and humane restraint are always essential whereas muscular relaxation is often unnecessary. The aim should be to maintain the lightest level of central nervous system (CNS) depression possible.

Many drugs are available to satisfy these requirements and can be used singly or in combination. The concept of administering a combination or succession of drugs which may act synergistically or additively on the CNS is known as *balanced anaesthesia*. However, no technique can be entirely satisfactory. Interaction with several bodily systems is inevitably unpredictable and even hazardous. The anaesthetist can merely minimize the resultant physiological deviations by exercise of skill and an appreciation of the agent's pharmacology.

Choice can be made from several techniques.

Local analgesics (local anaesthetics)

Surface application by spray, drops or ointments.

Infiltration by injection around a discrete area or by injection around the whole surgical field.

Regional nerve block where the drug is injected immediately around a specific nerve trunk supplying the operative area.

Spinal nerve block where the drug is deposited into the vertebral canal to block post-thoracic nerve supply.

Centrally acting analgesics

These drugs are given parenterally and are potent suppressors of pain perception in the CNS.

General central depressants

Sedatives, tranquillizers and hypnotics which do not in safe doses produce really profound depression of the CNS. These agents may be used either alone, in conjunction with any of the above techniques or as aids to general anaesthesia.

General anaesthetics

These may involve: inhalation of gas or volatile liquid vapour; parenteral administration by the oral, rectal, intravenous, intraperitoneal or subcutaneous routes of non-volatile agents; or a combination of inhalation and parenterally administered agents.

Muscle relaxants

These agents act either centrally or peripherally to block motor tone and, because they neither depress the CNS nor act as analgesics, should not be used in experimental animals except in exceptional circumstances. If used at all, administration must be accompanied by analgesics and general anaesthetics.

Variation in response to anaesthetic agents

Animals may vary both in their qualitative and quantitative response to a drug. Species, strain, weight, obesity, age, sex, health, nutritional status, prior exposure and adaptation, body temperature, pulmonary and cardiovascular function and seasonal or circadian biorhythms are but some of the factors which have been shown to modulate extremely complex biological events during anaesthesia. In general, most variation is related to differences in metabolic rate (and associated differences in uptake, distribution and elimination of the drugs) and in the health status of the animal.

Species variations usually correlate with the ratio of basal metabolic rate (BMR) to the body surface area. Small mammals, such as mice, having a high ratio need relatively large doses of anaesthetic for an equivalent degree of CNS depression. Birds have a higher BMR than mammals of comparable mass and this is reflected in their high body temperature and rapid heart rate. In their case, anaesthetic management can be further complicated by rapidly exhausted responses to stress, by hypothermia, particularly in small species, if the feathered integument is impaired and by cardiac arrhythmias. Some species differences can be explained by the presence of high concentrations of hydrolysing enzymes which prevent the agent reaching the CNS in sufficient concentration for anaesthesia. Others, however, defy simple explanation. For example it is still not known why morphine-like drugs should act as powerful CNS depressants in rats yet in equianalgesic doses produce tremors and other signs of CNS stimulation in the closely related mouse.

Within a given species, genetic variance exhibited in strain and breed differences is also reflected in variations in dose response to anaesthetics. These probably reflect liver and plasma enzyme differences. Age-associated variations also reflect changes in liver enzyme activity and the laying down of fat deposits in older animals. During periods of rapid growth, enzyme activity alters in relation to organ weight and altering function. In general, young animals are far more sensitive to anaesthetics because of their limited ability to metabolize drugs to glucuronides. This is true of rodents and rabbits in the first three weeks of life and of cats and dogs in the first six weeks. Drug metabolism will be slowest in the young, the aged and the obese subject, so less anaesthetic agent will be needed to obtain the desired level of CNS depression. Sex-linked differences are less important and are unlikely to be of much account to the scientist seeking effective anaesthesia.

Conversely, the health status of the animals has the greatest bearing on results. Most laboratory animals reared under conventional conditions are subject to low-grade, subclinical infections. The resultant cardiopulmonary, hepatic or renal dysfunction may profoundly influence the uptake and elimination of anaesthetics. Chronic respiratory disease (CRD) is endemic in most conventional colonies of rodents and lagomorphs and is probably the single most common cause of anaesthetic failure other than technical incompetence. The best solution to this problem is to use only animals recently issued from specified pathogen free (SPF) breeding colonies, ventilate them well during anaesthesia with high partial pressures of oxygen and ensure that the airway is not obstructed by catarrhal exudates. Hepatic and renal insufficiency are less frequently encountered. Whatever the species, the animals should be carefully selected before inclusion in an experiment. If it is impossible to use SPF stock then choose young adults before they become chronically diseased.

Nutritional composition and the quantity of diet may also affect drug responses. It is dubious whether rodents or lagomorphs need be fasted prior to anaesthesia, and in small birds with their high BMR starvation is rapidly lethal — only six hours without food prior to anaesthesia can produce hypoglycaemia and a high risk of fatal CNS depression.

In our experience, prior handling and adaptation to the laboratory reduce extremes in response to drug administration in rodents and rabbits. However, repeated administration of barbiturates, phenothiazines and morphine-like drugs leads to non-specific tolerance which may last up to fourteen days because enzyme induction in hepatocytes accelerates degradation of the drug.

Biological rhythms are a minor consideration. Nevertheless, cyclic rises and fall in BMR associated with circadian, hibernatory or seasonal sexual activity can affect the outcome.

Environmental factors such as abnormal temperature, humidity, light, noise and changes in social conditions can each induce neuroendocrinological disturbances in the animal and make anaesthetic management more hazardous. Animals should be allowed time for conditioning in a stable environment before using them in experiments, and repeated handling by gentle personnel will minimize fear and stress when anaesthesia is induced.

Maintenance and homeostasis

For many experiments the anaesthetized animal must be kept in as near a physiological state as possible. This entails maintaining homeostasis, especially the thermal, fluid and electrolyte balance. Of these variables, temperature is perhaps the most important.

Postanaesthetic management

However much skill and care has been devoted to the animal during the operation, it will be set at nought if the poor beast is simply dumped in a cold environment afterwards and left to fend for itself. The partially recovered animal needs protection from hypothermia, respiratory obstruction, cardiovascular failure, self-inflicted injury and from mutilation by peers if returned to a box of conscious animals. Under no circumstances must it be allowed to suffer pain.

The experimental requirements

Having taken all the above considerations into account, it is now time to fit them into the experimental protocol before making a final decision on the anaesthetic regimen. Clearly, the nature of the surgical interference is important. For example, many procedures such as skin grafting performed on mice are brief, but it may be necessary to anaesthetize a whole group of animals at a time. Injectable agents will obviously be more convenient in this situation and, because these drugs are inexpensive, need no special apparatus for administration and masks around the head are avoided, they are most commonly used in small animals.

It is also essential to consider if, and to what extent, the anaesthetic drugs and techniques used will affect the validity of experimental results, and how they interact with other drugs being used. This is really a vexed question because we often do not know the answer and can therefore make no predictions beyond the realms of intelligent guesswork. At least it is essential that the animal is maintained in as physiological a state as possible. At least, too, we should make every effort to understand the pharmacology of the drugs involved. Concurrent drug administration may interfere with metabolic rate, with the uptake and distribution of anaesthetic agents and with their mode of action by competing for receptors in the CNS. For example, hyperthyroidism increases metabolic rate and therefore higher dosages are needed, whereas totally thyroidectomized animals need markedly reduced doses of anaesthetic. Several antibiotics interfere with anaesthesia degradation: for instance, chloramphenicol can prolong pentobarbitone narcosis, and neomycin, streptomycin, tetracyclines, sulphonamides and polymyxin B may interact with several anaesthetic agents.

Recommendations

1. Ensure that the animals are healthy before using them in an experiment.
2. Consider if and how much the anaesthetic drugs are likely to affect the validity of the study, and whether they will interact with other drugs.

3. In choosing an anaesthetic technique, aim for the minimum degree of CNS depression compatible with complete analgesia and the animal's welfare.
4. Whenever possible, assay the anaesthetic method in a limited trial before using it in the experiment.
5. Regard airway patency and pulmonary ventilation as the prime responsibility of the anaesthetist. An endotracheal tube should be passed whenever practicable, a regulated source of O_2 should be supplied whether the animals are anaesthetized with injectable or inhalational agents, and aspiration suction should always be available.
6. Regard the conservation of heat as an integral part of anaesthetic management in all species, but particularly in young and small mammals and birds. External sources of heat are essential and body temperature should be monitored continuously.
7. Administer warm, balanced salt solutions by continuous i.v. infusion whenever practicable and keep plasma expanders available in case of extensive haemorrhage.
8. Pay particular attention to postanaesthetic nursing. Allow animals to recover in an environment approaching the normal body temperature of the species, maintain i.v. infusion, leave an endotracheal tube in place until the swallowing reflex is fully recovered, and finally ensure that postoperative pain is eliminated by the judicious use of analgesics.
9. Consider the safety of laboratory personnel.

PHARMACOLOGY OF DRUGS ACTING ON THE CNS

Introduction

Agents which modify neural function in the CNS can also affect other systems in the body either *indirectly* via both the somatic and autonomic divisions of the nervous system or *directly* by their actions on, for example, cardiac and smooth muscle. The drugs are broadly classified as CNS stimulants, non-selective CNS depressants or partially selective depressants depending on their peak activity at safe dosages. A classification is given in Table 14.1 (after Lees, 1977) and a list of definitions later in this section. However, because drugs are categorized in rather arbitrary fashion and usually by their effect on human beings, they often do not fit neatly into these compartments in all species. Nor can the CNS be compartmentalized into different states. It is always in dynamic flux. Anaesthesia is not a steady state. It is more satisfactory to think of a continuum of pharmacological activity stretching from stimulation exhibited clinically by tremors or even convulsions at one end of the spectrum, thence through sedation, sleep, anaesthesia and coma at the other. Even this description is not characteristic of all CNS depressants. It is typical of barbiturates but not of tranquillizers or dissociative anaesthetics.

Table 14.1 Classification of Drugs Acting on the CNS (after Lees, 1977; see Green, 1979)

1. *CNS stimulants*
   Three groups may be listed according to the main site of action — spinal, medullary and cortical. None are further discussed in this chapter.

2. *Non-selective CNS depressants* — divided into groups according to the nature of the final response. Drugs recommended for use in this chapter include:
   (i) Sedatives — some barbiturates, sedative analgesics of the morphine type, xylazine
   (ii) Hypnotics — short-acting barbiturates (thiopentone, methohexitone), alphaxolone–alphadolone
   (iii) General anaesthetics:
       (a) Gases — nitrous oxide
       (b) Volatile liquids — diethyl ether, methoxyflurane, halothane, chloroform, enflurane, isoflurane
       (c) Solids (water soluble) — some barbiturates (pentobarbitone)
           Solids (water insoluble) — alphaxolone–alphadolone

3. *Partially selective CNS depressants*
   (i) Tranquillizers (ataractics)
       (a) Major tranquillizers (neuroleptics) — chlorpromazine, acepromazine, droperidol, fluanisone
       (b) Minor tranquillizers — diazepam, midazolam
   (ii) Centrally acting muscle relaxants
   (iii) Anticonvulsants — phenobarbitone
   (iv) Analgesics
       (a) Narcotic or sedative analgesics — morphine, pethidine, fentanyl, pentazocine, buprenorphine
       (b) Antipyretic analgesics — acetylsalicylic acid
   (v) Neuroleptanalgesic combinations such as fentanyl–fluanisone, fentanyl–droperidol, etorphine–methotrimeprazine

Definitions and basic properties

Medullary stimulants or analeptics (e.g. bemegride, doxapram) stimulate some medullary centres, particularly the respiratory rate.

Narcotic is a general term for *any* CNS depressant including sedatives, hypnotics and anaesthetics.

Sedatives (e.g. xylazine) are drugs which depress the CNS sufficiently to produce lethargy, drowsiness, indifference to the surroundings and decreased motor activity. They allay fear and apprehension but the animal remains conscious.

Hypnotics (e.g. thiopentone) are drugs which induce deep sleep from which the animal can only be aroused with difficulty and depress the CNS to the level of basal narcosis or even light general anaesthesia. They may have some analgesic properties but by no means all do.

Tranquillizers or ataractics (e.g. acepromazine, diazepam) exert a quietening effect, lessening anxiety and calming naturally vicious animals. They differ from sedatives in that the animals do not become very drowsy and at high dosage they do not produce hypnosis or general anaesthesia. They do not themselves have analgesic properties.

Analgesics are CNS depressants whose main action is to diminish the perception of pain. They are divided into two main groups on the basis of their chemical structure and potency: (a) the 'strong' narcotic or sedative analgesics (e.g. morphine, fentanyl, buprenorphine) and (b) the 'weak' antipyretic analgesics (e.g. acetylsalicylic acid).

Neuroleptanalgesics (e.g. fentanyl–fluanisone) depress the CNS by the combined administration of a neuroleptic and sedative–analgesic. Most combinations available commercially produce a state resembling light anaesthesia but the animals may respond reflexly to noise and light, and muscular tone is usually retained at normal levels.

Dissociative anaesthetics (e.g. ketamine) produce unconsciousness, catalepsy and some analgesia (depending on species) but muscular tone is often enhanced.

General anaesthetics (e.g. ether, halothane, methoxyflurane and pentobarbitone) produce a general but reversible depression of the CNS involving hypnosis, unconsciousness, analgesia, relaxation of voluntary muscles and suppression of reflex activity. Sensory perception and motor activity are both diminished and the overall effect depends upon route and rate of administration as well as the total dosage and concentration achieved in the CNS.

PREANAESTHETIC MEDICATION, CHEMICAL RESTRAINT

Introduction

The main aims of preanaesthetic medication are to: reduce or abolish perception of pain before *and* after the operation; reduce apprehension in a frightened subject; permit quiet recovery, ideally eliminating struggling and hence danger to the animal and anaesthetist alike; reduce the amount of general anaesthetic needed; reduce salivary and bronchial secretions; reduce gastric and intestinal motility, and protect the heart from vagal inhibition.

Anticholinergics

Atropine sulphate is the only agent commonly used (Table 14.2).

Atropine is used to diminish salivary and bronchial secretions, to protect the heart from vagal inhibition and to prevent the muscarinic action of anticholinesterases such as neostigmine when given to reverse non-depolarizing muscular relaxants. It should be given routinely whenever irritant inhalational

Table 14.2 Dosage and administration (i.m. or s.c.) of atropine in some species

Species	Dosage (mg/kg)	Time to effect (min)	Duration of effect (min)[b]
Dogs	0.05	30–40[a]	20
Cats	0.1	30–40[a]	20
Rabbits	0.2	20	15
Guinea pigs	0.05	10	15
Rats	0.05	10	15
Mice	0.05	10	25

[a] Effects produced in 2–3 min after slow i.v. injection.
[b] These times are very approximate.

Table 14.3 Dosage and administration of acepromazine maleate in some species

Species	Dosage (mg/kg)	Route	Time to effect (min)	Duration of effect (hr)
Dogs	0.5	i.m. (or slow i.v.)	10–15 (1–2)	3–5
Cats	0.5	i.m. (or slow i.v.)	5–10 (1–2)	3–5
Rats	1.0	i.m.	5–10	2

agents or ketamine are being used. It should not be used where a marked tachycardia is already established, and its value in ruminants is debatable.

Tranquillizers (ataractics and neuroleptics)

The most commonly used agents in animals are: the *phenothiazine* derivatives promazine, chlorpromazine, acepromazine and methotrimeprazine; the *butyrophenones* droperidol, fluanisone and azaperone; and the *benzodiazepines* diazepam and midazolam. In some species these drugs do cause drowsiness but the animals are easily aroused.

Acepromazine maleate (ACP)

Acepromazine is useful in providing a stress-free subject for induction and emergence from anaesthesia and in enhancing the potency of anaesthetics and analgesics. It is rapidly absorbed from i.m. sites and reaches peak effect in 15–20 min. It is effective in dogs, cats and rats. It is also surprisingly free of toxic side-effects at clinical dosages. (See Table 14.3)

Table 14.4 Dosage and administration of diazepam in some species

Species	Dosage (mg/kg)	Route	Time to effect (min)	Duration of effect (hr)
Rabbits	1.0	i.m. (i.v.)	5–10	1–2
Guinea pigs	2.5	i.p. (i.m.)	2–5	1–2
Rats	2.5	i.p.	2–3	1–2
Hamsters	5.0	i.p.	2–3	1–2
Mice	5.0	i.p.	1–2	1–2

Droperidol (Droleptan®)

Droperidol is more potent than acepromazine and also produces mental calm and indifference. It is often given in combination with a sedative–analgesic like fentanyl (the combination marketed as Innovar–Vet®) and has been widely accepted for neuroleptanalgesia in rabbits, guinea pigs, rats and mice. Fluanisone is a similar butyrophenone which is also used in combination with fentanyl (the commercial combination Hypnorm®) for neuroleptanalgesia. Dosages of these combinations are given in Table 14.7. Droperidol alone is a very useful tranquillizer for dogs at a dosage of 0.4 mg/kg i.m.

Diazepam (Valium®)

Diazepam exerts potent tranquillizing, muscle relaxant and anticonvulsant effects in animals, and potentiates barbiturates, analgesics and inhalational anaesthetics. It is an extremely valuable agent in the anaesthetic management of most species, particularly for enhancing the sedative–analgesic properties of morphine-like analgesics, and for use with the dissociative anaesthetic ketamine in primates, rabbits, guinea pigs, rats, mice and birds (Green *et al.*, 1981). (See Table 14.4)

Midazolam (Hypnovel®)

Midazolam is a water-soluble benzodiazepine with similar pharmacological activity to diazepam. At suitable dilutions it can be mixed with other water-soluble agents such as fentanyl–fluanisone and is extremely valuable for producing surgical anaesthesia in laboratory species (Flecknell and Mitchell, 1984). (See Table 14.5)

Xylazine hydrochloride (Rompun ®)

Xylazine is a potent hypnotic with marked central muscle relaxant and modest

Table 14.5 Dosage and administration of midazolam–fetanyl–fluanisone in some species

Species	Dosage (ml/kg)[a]	Route	Time to effect (min)	Duration of effect (hr)
Mice	13.0	i.p.	3–6	35–90
Rats	3.0	i.p.	4–5	55–160
Gerbils	8.0	i.p.	3–5	30–60
Hamsters	4.0	i.p.	4–8	15–90
Guinea pigs	8.0	i.p.	5–10	30–90
Rabbits[b]	—	—	5–10	60–120

[a] Stock solution prepared 2 parts water for injection, 1 part Hypnovel®, 1 part Hypnorm® (2.5 mg fluanisone, 0.097 mg fentanyl and 1.25 mg midazolam per ml).

[b] For rabbits, it is more convenient to inject midazolam at 2 mg/kg i.v. or i.p. and Hypnorm® at 0.3 ml/kg i.m. separately.

Table 14.6 Dosage and administration of xylazine in some species

Species	Dosage (mg/kg)	Route	Time to effect (min)	Duration of effect (min)
Primates	1.0–2.0	i.m.	5–10	30–60
Dogs	1.0–2.0	i.m.	3–10	30–90
Cats[a]	1.0–2.0	i.m.	3–5	20–60
Rabbits[a]	3.0	i.v.	Rapid	20–60

[a] Surgical conditions are provided if given concurrently with ketamine (10 mg/kg i.m.)

analgesic properties. It is the sedative of choice in cattle and sheep, and is extremely useful in horses, cats, dogs and primates. Its pronounced muscle relaxant properties render it useful in combination with ketamine in several species but particularly cats and primates. (See Table 14.6)

Fentanyl citrate (Sublimaze®); *Fentanyl citrate–fluanisone* (Hypnorm®); *Fentanyl citrate–droperidol* (Innovar-Vet®)

Fentanyl is a potent analgesic with marked central depressive properties and is particularly useful in laboratory animals either alone or in neuroleptanalgesic combination with fluanisone, droperidol or diazepam (Green, 1975; 1979). Because of its depressive effects on the cardiovascular and respiratory systems, it should be given with care during i.v. administration or when used as a supplement for other respiratory depressants such as pentobarbitone. It can be antagonized and reversed very rapidly by i.v. administration of naloxone. (See Table 14.7)

LABORATORY ANIMALS

Table 14.7 Dosage and administration of fentanyl–fluanisone in some species

Species	Dosage (ml/kg)	Route	Time to effect (min)	Duration of effect (min)
Dogs	0.1–0.4	i.m.	10	20–40
Primates	0.2	i.m.	10	20–40
Rabbits	0.3	i.m	6–10	20–30
Rats	0.3–0.4	i.m., i.p.	8.3	15–30
Guinea pigs[a]	1.3	i.m., i.p.	15.5	20–40
Hamsters[b] Mice	0.3	i.p.	2	15–20

[a] In guinea pigs, neurolepsis does not develop with this combination but analgesia does. For neuroleptanalgesia it is necessary to use fentanyl–fluanisone (0.5 ml/kg i.p.) concurrently with diazepam (2.5 mg/kg i.p.)
[b] In mice and hamsters, it is necessary to inject diazepam (5 mg/kg i.p.) beforehand and to use 0.1 ml of a 1/10 diluted solution per animal.

Table 14.8 Dosage and administration of ketamine hydrochloride in some species

Species	Dosage (mg/kg)	Route	Time to effect (min)	Duration of effect (min)	Time to full recovery (min)
Cats	10–30	i.m.	1–5	30–40	100–150
Primates	15–30	i.m.	3–5	30–45	100–150
Rabbits[a]	15–25	i.m.	3–5	20–40	100
Rats[a]	40–60	i.m.	5–10	45–70	100

[a] To provide peaceful sedation, diazepam (1.0 mg/kg i.m.) or xylazine (1.0 mg/kg i.m.) should also be given.

Ketamine hydrochloride (Vetalar®)

Ketamine alone produces profound analgesia and catalepsis in cats and primates, but in other species the effects are unpredictable. However, if used in combination with xylazine or diazepam its disadvantages of enhanced muscle tone and tonic–clonic convulsions can be overcome and it is then valuable in cats, rats and guinea pigs. Ketamine has certain advantages for the inexperienced anaesthetist: it tends to stimulate both the cardiovascular and respiratory systems; it is relatively safe if given by the i.m. route; and it has a wide safety margin. (See Table 14.8)

Alphaxalone–alphadolone (Saffan®)

This is a combination of anaesthetic steroids (alphaxalone 9 mg/ml: alphadolone acetate 3 mg/ml). Good CNS depression and muscle relaxation are

ANAESTHESIA AND ANALGESIA

Table 14.9 Dosage and administration of alphaxalone–alphadolone acetate in some species

Species	Dosage (mg/kg)	Route	Time to effect (min)	Duration of effect (min)[a]	Time to full recovery (min)
Cats	9	i.v.	Rapid	10	30–90
	12	i.m.[b]	7–8	15	60–90
Rabbits	6–9	i.v.	Rapid	3–7	15
Guinea pigs	10–20	i.v.	Rapid	4–8	15
Rats	10–20	i.v.	Rapid	4–6	10
Mice	10–20	i.v.	Rapid	4–6	10
	90	i.p.	3	15–20	30

[a] Light surgical anaesthesia in most species after i.v. administration. May be continued for many hours by incremental or continuous infusion in all these species.
[b] By deep i.m. injection into anterior thigh muscles.

produced but the degree of analgesia depends on the route of administration and species: for example, it is good by the i.v. route in primates, domestic cats, rats and mice, but poor in rabbits (Green *et al.*, 1978); by the i.m. route, sedation and hypnosis are produced in cats and primates. The steroids are short acting but are valuable for long periods of anaesthesia if given as a continuous infusion or by sequential i.v. injections. We regard alphaxalone–alphadolone as the drug of choice in cats, marmosets and neonatal piglets. It should never be used in dogs. (See Table 14.9)

GENERAL ANAESTHESIA

Introduction

General anaesthesia is a state of general depression of the CNS involving hypnosis, unconsciousness, analgesia, suppression of reflex activity and relaxation of voluntary muscles. It can be achieved in several different individual ways or by combining different techniques to suit the surgical requirements:

1. Parenteral administration of:
   (a) Short-acting water-soluble barbiturates
   (b) Anaesthetic steroids
   (c) Dissociative anaesthetic agents
   (d) Neuroleptanalgesic combinations
   (e) Miscellaneous drugs
2. Inhalation of:
   (a) Volatilized liquids such as ether, methoxyflurane, enflurane or halothane
   (b) Gaseous agents such as nitrous oxide (N_2O)

3. Combinations of any of these with muscle relaxant drugs.

Signs of anaesthesia

So long as excitement and motor activity during induction are suppressed by suitable preanaesthetic medication, the course of narcosis deepening to surgical anaesthesia follows a fairly standard pattern in mammals. The animal becomes progressively ataxic before laying down with its head held up without support. As narcosis deepens, the *righting reflex* is lost and the animal makes no attempt to turn over to the prone position when placed on its back. The whole body then relaxes although limited movement may still be provoked by painful stimuli and the *swallowing reflex* may be evoked by opening the jaws and retracting the tongue. The latter is lost as muscle relaxation develops. Reflex withdrawal of a limb after the toe web is pinched (the *pedal reflex*) weakens and slows until it disappears when light to medium surgical anaesthesia has been attained. The disappearance of head shake in response to an ear pinch is a surer guide to surgical anaesthesia in rabbits, guinea pigs and pigs, whilst failure to respond to tail pinching is a good indication in rats and mice. Ocular signs vary but provide a useful index in many species. As surgical depths of anaesthesia are approached, the blink response to gently touching the inner corner of the eye (the *palpebral reflex*) disappears. Respiratory signs are also valuable in dogs; when deliberate dilatation of the anal sphincter fails to elicit an increased rate of breathing (the *respiratory reflex*) surgical anaesthesia is nigh. When the animal fails to react to *any* external stimuli, surgical anaesthesia has been attained.

The respiratory pattern provides another index of anaesthesia. The rate and quality are each important. During induction or recovery, breathing may range from rapid irregular movements to temporary cessation (*apnoea*). Breathing generally becomes slow but regular during medium surgical anaesthesia. If the CNS is depressed still further, thoracic movements may stop between inspiration and expiration (*delayed thoracic respiration*). If then anaesthesia is deepened to dangerous levels, inspiration becomes laboured and involves diaphragmatic and abdominal muscles rather than thoracic muscles.

Inhalational anaesthesia

Anaesthetic chambers

For small laboratory species, anaesthesia can be induced in a chamber. At its simplest this can be a large glass jar containing a pad of cotton wool impregnated with a volatile anaesthetic. However, because these agents are irritant if the animal comes into direct contact with them, the cotton wool should be taped to the side of the bottle or, better still, attached inside the lid; a simple arrangement is to bolt a metal gauze (e.g. a tea strainer) over a pad of cotton wool to the lid of the jar. A measured volume of anaesthetic liquid should then be used and time allowed for it to vapourize in the closed container before introducing the animals. The main problem with this simple technique is that the concentration of anaesthetic is uncontrolled so it is only really suitable for ether or for methoxyflurane; it must not be used for halothane or enflurane because dangerous concentrations of vapour can build up very quickly inside the jar.

It is preferable, though, to induce anaesthesia in a clear perspex box in which the animal can be observed and to which known concentrations of anaesthetic agent in oxygen can be piped. The gases can then be vented from an outlet and ducted to safety outside the room.

Anaesthetic machines and circuits

The only pieces of equipment which are essential and common to all controlled methods of inhalational anaesthesia are: (1) the anaesthetic machine, consisting of a source of oxygen (O_2) delivered from its cylinder through a reducing valve and a flowmeter calibrated for low and high flow rates and a system for vapourizing volatile drugs, and; (2) an anaesthetic circuit to deliver the resultant gas mixtures to the animal's respiratory system.

The basic anaesthetic machine is shown schematically in Fig. 14.1. It consists

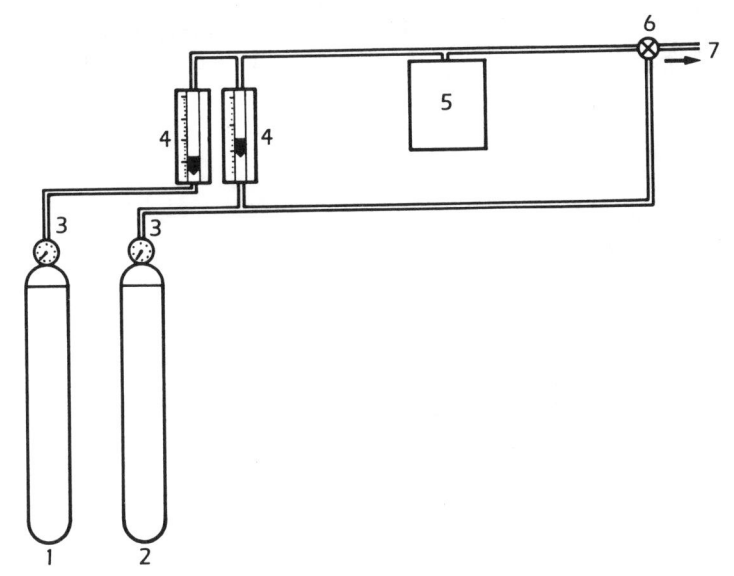

Fig. 14.1 Basic components of an anaesthetic machine. 1. N_2O cylinder; 2. O_2 cylinder; 3, pressure reducing valves; 4, flowmeters for each gas; 5, calibrated vapouriser; 6, emergency O_2 flush; 7, ducting to animal

276   LABORATORY ANIMALS

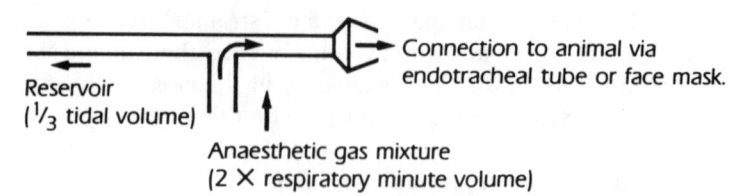

Fig. 14.2 Ayres T-piece

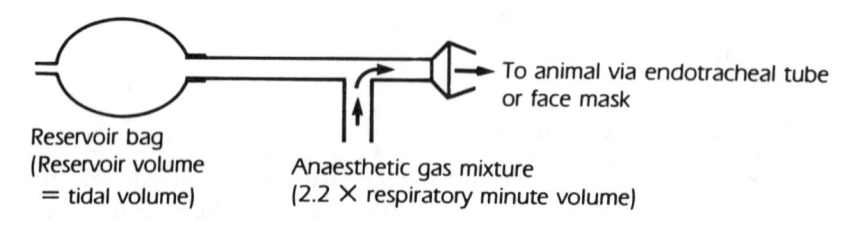

Fig. 14.3 Rees-modified T-piece

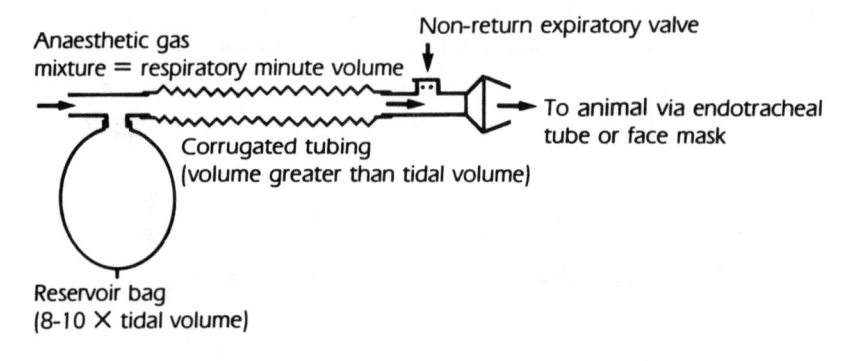

Fig. 14.4 McGill attachment

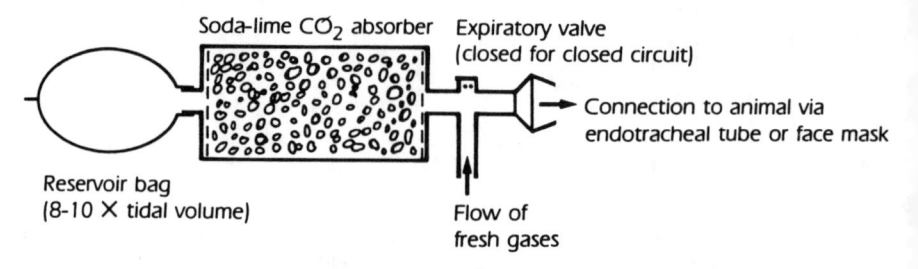

Fig. 14.5 Waters 'to and fro' system

of: gas cylinders, usually nitrous oxide (1) and oxygen (2); pressure-reducing valves and gauges indicating the volume of gas remaining in cylinders (3); flow meters for each gas (4); a vapourizer for each anaesthetic in use (calibrated, e.g., for halothane or methoxyflurane) (5); an emergency oxygen system (6); and ducting to the animal via the chosen circuit (7).

Many circuits have been described for conducting anaesthetic agents to the animal's respiratory system and are classified at their simplest into open, semi-open and closed systems. In open systems no rebreathing of exhaled gases occurs, but high-flow fresh gases must be supplied continuously and this is both wasteful and pollutes the atmosphere; however, simple and inexpensive apparatus such as the Ayre's T-piece (Fig. 14.2) provides the greatest safety to the animal. Semi-open systems have a reservoir bag and do not allow rebreathing of exhaled gases, but gas flows are still high even though lower than in open systems; the Rees modified T-piece (Fig. 14.3) and McGill circuits (Fig. 14.4) are widely used. In closed systems rebreathing of exhaled gases is permitted after CO_2 has been removed in a soda-lime absorber; the Water's to-and-fro (Fig. 14.5) and circle systems (Fig. 14.6) are each economical in gases and anaesthetic agents, and reduce the risk of atmospheric pollution significantly, but in practice present problems which put the animals at risk. For dogs, cats, non-human primates, rabbits and guinea pigs, mobile trolleys equipped with flowmeters, vapourizers designed for ether, halothane, enflurane or methoxyflurane, and emergency oxygen-flushing facilities are basic requirements. Suitable units which are available commercially are the Penlon 'SAM', the Cyprane Small Animal Unit and the MIE 'Casualty and Out-Patients Anaesthetic Unit'. Because the vapourizers operate at rather high flow rates on these machines they are not suitable for rodents, and small vapourizers which are efficient at gas flow rates ranging from 200 to 600 ml/min should be specially constructed. The anaesthetic units described by Sebesteny (1971), Carvell and Stoward (1975) and Norris (1981) are each suitable for small laboratory animals.

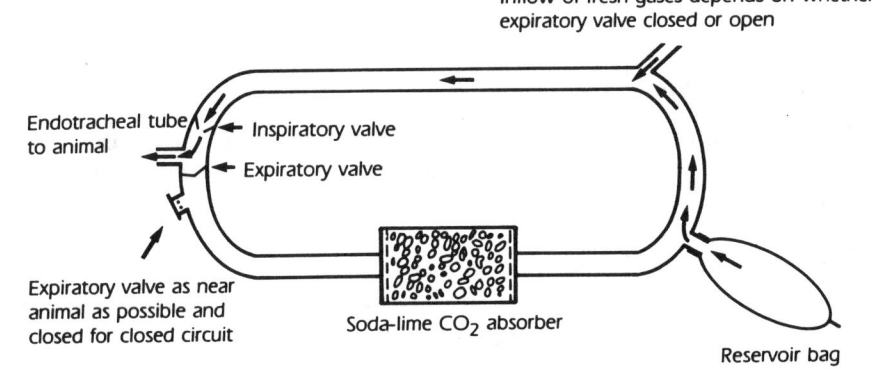

Fig. 14.6 Circle system (can be used closed or semi-closed)

These circuits can be connected to the animal by a close-fitting face mask but it is usually better to pass an endotracheal tube and connect them to that. Endotracheal intubation is relatively simple in dogs, sheep, primates and birds greater than 1 kg in size so long as a suitable laryngoscope is available. In cats, laryngospasm is easily induced and the larynx should be sprayed with lignocaine before attempting to pass the 3 mm tube. In rabbits, a size 1 Wisconsin laryngoscope blade (Penlon Ltd) or a purpose-built blade described by Brown (1983) are essential aids and the vocal cords should be sprayed with lignocaine to prevent laryngospasm before passing a 0.3 mm diameter tube. Intubation of rats is not difficult after practice and if a suitable mini-laryngoscope is built (Medd and Heywood, 1970). However, intubation of smaller rodents is not usually realistic and requires purpose-designed blades to visualize the larynx whilst passing plastic catheters. Intubation of pigs can be reliably achieved with practice so long as relatively small tubes with inflatable cuffs are used.

The main advantages of inhalational anaesthesia are that the depth of anaesthesia can be altered rapidly, the animals recover quickly and the agents are either exhaled unchanged or are metabolized by the liver such that minimal interference with experimental results is likely to be incurred. However, there are disadvantages to be considered. Although both ether and methoxyflurane can be administered by simple open techniques, halothane and enflurane should only be given through carefully calibrated apparatus and this is often inconvenient when anaesthetizing several mice at a time. Then the agents may be unpleasant for the animals during induction and recovery. Finally, these agents are potentially dangerous in a confined space: for example, ether forms explosive mixtures with air, oxygen and nitrous oxide, and its flammability presents a considerable fire hazard whenever it is used; other agents pollute the atmosphere and present a more subtle hazard so that appropriate scavenging measures should be taken (Green, 1981 for review; Hunter et al., 1984).

Table 14.10 summarizes the main properties of five inhalational anaesthetics. Further details are given when the anaesthesia of individual species is discussed.

Anaesthesia with injectable agents

The only equipment essential for parenteral administration of drugs is a suitable hypodermic needle and syringe; but butterfly infusion sets, Medicut® intravenous cannulae, Bardic® I-catheters and burettes infusion sets are all valuable.

Choice of agent and route of administration will be governed by the size of animal, accessibility or otherwise of superficial veins and ease of restraint. Intravenous injection is generally preferable as it allows controlled administration of the anaesthetic and enables the dosage to be judged according to effect. Uptake and distribution of the drug and hence induction of anaesthesia will

Table 14.10 Major properties of inhalational anaesthetic agents

	Ether	Halothane	Methoxyflurane	Enflurane	Nitrous oxide
Commercial name	—	Fluothane®	Penthrane® Metofane®	Enthrane®	—
Potency	Fairly high	High	Very high	High	Low
Speed of induction and recovery	Slow	Rapid	Slow	Rapid	Rapid
Concentrations (a) induction	10–20%	3–4%	2–3%	0.5–2.0% in $N_2O:O_2$	Use at 50% with O_2 and other agents
(b) maintenance	4–5%	0.5–2.0%	0.4–1.0%	0.5–1.5% in $N_2O:O_2$	
Effect on respiratory rate	Does not fall until deep anaesthesia attained	Depressed	Depressed only after long period of use	Does not fall until deep anaesthesia attained	Does not fall
Irritancy	V. irritant	Nil	Nil	Nil	Nil
Stimulations of secretions	Profuse, esp. in guinea pigs	Nil	Nil	Nil	Nil
Effect on cardiovascular system	Depressed only at deepest levels of anaesthesia	Depressed progressively	Less than halothane	Usually v. little effect	Nil
Analgesic value	Good	Fair/good	Very good	Good	Very little
Safety	Flammable, explosive	Non-flammable	<4% is non-flammable	Non-flammable	Non-flammable

Table 14.11 Major properties of injectable anaesthetics — short-acting barbiturates

	Pentobarbitone sodium	Thiopentone sodium	Inactin sodium	Methohexitone sodium
Commercial name	Sagatal®	Intraval® Sodium pentothal®	Inactin®	Brevane®, Brietal® Sodium
Effect on respiratory system	Fails at close to gen. anaes. doses	Depressed	Very slight depression	Depressed
Effect on cardiovascular system	Depressed at anaes. doses	Varies with speed injected	Very little effect	Some depression and fall in BP
Route of administration	i.p., i.m., s.c. Best = i.v.	i.v. only	i.v. and i.p.	i.v.
Analgesic value	Poor	Poor	Poor	Poor
Hypnotic effect	Good	Good	Good	Good
Muscle relaxation	Poor	Not very good	Fair	Poor
Dosage	Up to 30 mg/kg in cat, dog, rabbit; 30–60 mg/kg in other spp. as 3 or 6% soln	20–30 mg/kg as 1% soln in cat, dog; as 2.5% soln in larger spp.	150 mg/kg i.p.	4–10 mg/kg as 1% soln in small spp. or 2.5% soln in large spp.
Major uses	For induction and maintenance of light anaesthesia	For induction, followed by inhalational anaesthesia	For rats only in acute experiments	For rapid induction followed by inhalational anaesthesia
Special notes	Better agents are available	Cumulative	Not for other spp.	Ultra-short-acting barbiturate

Table 14.12 Major properties of other injectable anaesthetics

	Neurolept-analgesic combinations	Ketamine	Alphaxalone–alphadolone
Commercial name	Hypnorm® Innovar-Vet® Immobilon-SA®	Vetalar®	Saffan®
Effect on respiratory system	Depressive — especially Immobilon-SA®	Slight and variable	Slightly depressed
Effect on cardiovascular system	Bradycardia	Slight and variable	Slightly depressed
Analgesic value	Excellent	Poor to good (good in primates and birds)	Good
Hypnotic effect	Fair but easily	Good	V. good
Muscle relaxation	Poor	Poor	V. good
Dosage	See species tables	15–60 mg/kg i.m. (see species tables)	6–17 mg/kg
Route of administration	i.m., i.p., s.c.	i.m., i.v., s.c.	i.m., i.v.
Duration of main effects	15–30 min	30–60 min	10–15 min
Major uses	Deep sedation/analgesia alone; anaesthesia if given with diazepam or midazolam	Deep sedation primates, cats and birds	G.a. in cats, primates, pigs
Special notes	Unless given with benzodiazepines in mice they produce excitation	Better if given with diazepam or xylazine	Not to be used in dogs. Most useful if given i.v. in other species

necessarily be slower if it is injected by the intramuscular, intraperitoneal or subcutaneous routes and it is then necessary to give a single bolus to the animal at a calculated dosage. Clearly, given the known wide variations in response to all these agents, the results will not always be satisfactory; it is, too, more important to select agents with a wide safety margin. The volume of solution to be injected must also be considered. Intramuscular injections greater than 0.05 ml in an adult mouse hind leg or greater than 0.3 ml in an adult rat cause considerable pain and are less likely to be absorbed as quickly as smaller volumes.

When administering drugs intravenously, it is best to have butterfly infusion sets or indwelling flexible cannulae taped in place to allow easy access for additional anaesthetic agent and rapid administration of anaesthetic antagonists, respiratory stimulants and fluids if emergencies arise.

The main injectable agents are listed in Tables 14.11 and 14.12 and recommendations for their use in different species are given in the following section.

RECOMMENDED TECHNIQUES: INDIVIDUAL SPECIES

Inhalational anaesthesia can be used for *all* species for short and prolonged periods. Methoxyflurane is the safest agent for small laboratory species and for neonates of larger animals. Ether is commonly used but is irritant to mucous membranes and exacerbates chronic respiratory disease. If injectable anaesthetics are selected then it is best to keep the animal as lightly anaesthetized as possible using balanced anaesthesia and select agents which can be easily reversed in an emergency. Remember that local analgesics are useful in many species.

Mice

Analgesia: (a) buprenorphine 1–25 mg/kg i.p. every eight hours for acute postoperative pain; (b) aspirin (100 mg/kg by mouth every four hours or in drinking water for mild chronic pain).

Sedation/light anaesthesia: (a) fentanyl–fluanisone (0.1 ml/30g of a 1 in 10 dilution Hypnorm® i.p.); (b) metomidate 60 mg/kg s.c.; (c) pentobarbitone 25 mg/kg i.p.

Surgical anaesthesia (SA): (a) fentanyl–fluanisone (0.1 ml/30g of a 1 in 10 dilution Hypnorm ® i.p.) + diazepam (5 mg/kg i.p.) gives 60–90 min SA; (b) fentanyl–fluanisone + midazolam (0.2–0.3 ml/30g i.p. of mixture one part Hypnorm®, one part Hypnovel® and two parts water (FFM mixture)) gives 20–40 min SA; (c) metomidate (60 mg/kg) + fentanyl (0.06 mg/kg) premixed and injected s.c. provides 2–3 hours SA; (d) alphaxalone–alphadolone (15–20 mg/kg i.v.) provides 5–10 min SA and can give 7–10 mg/kg i.v. every 15 min or continuous infusion of 0.25 mg/kg/min for prolonged SA, but i.v. route a disadvantage; (e) inhalational anaesthesia with methoxyflurane (2 per cent concentration) is best by far — alternative ether (10 per cent concentration).

General comments: Keep warm and supply O_2 to recovery chambers. Give 1.0 ml of warmed dextrose–saline s.c. at end of surgery. Atropine 0.04 mg/kg s.c. 30 min before anaesthesia i.v. useful.

Rats

Analgesia: (a) buprenorphine 0.1–0.5 mg/kg s.c. every eight hours for acute postoperative pain; (b) aspirin 120 mg/kg by mouth every four hours for mild chronic pain.

Sedation/light anaesthesia: (a) fentanyl–fluanisone (0.2 ml/kg Hypnorm® i.m.); (b) pentobarbitone at 30 mg/kg i.p. — narcosis lasts 25–40 min; (c) inactin sodium at 150 mg/kg provides deep sedation for 6–8 hr.

Surgical anaesthesia: (a) fentanyl–fluanisone (0.3 ml/kg i.m.) + diazepam (5 mg/kg i.p.) or midazolam (5 mg/kg i.p.) provide 30–50 min SA; (b) fentanyl–fluanisone–midazolam (FFM mixture as above) at 3.3 ml/kg i.p. or 1.6 ml/kg s.c. provides about 30 min SA; (c) alphaxalone–alphadolone 10–12 mg/kg i.v. provides 10–15 min SA — can be maintained by 5–10 mg/kg i.v. every 20 min or continuous infusion of 0.5 mg/kg/min; (d) methohexitone (1 per cent soln) 5–8 mg/kg i.v. provides 5–10 min SA; (e) ketamine 60 mg/kg i.p. followed by pentobarbitone at 20 mg/kg i.p. produces SA for 60 min; (f) ketamine–acepromazine (dosage of 15 mg/kg + 1 mg/kg in mixture given as single i.p. injection); (g) inhalational anaesthesia with methoxyflurane, halothane, enflurane or ether in that order of preference.

Hamsters

Analgesia: buprenorphine 0.5 mg/kg s.c. every eight hours.

Sedation/light anaesthesia: (a) alphaxolone–alphadolone (15 mg/kg i.p.) produces deep sedation for 20–60 min; (b) pentobarbitone (36 mg/kg i.p.) produces light anaesthesia for 20–50 min.

Surgical anaesthesia: (a) fentanyl–fluanisone (1 ml/kg i.p.) + diazepam (5 mg/kg i.p.) provides 60 min SA; (b) fentanyl–fluanisone–midazolam (4.0 ml/kg i.p. of FFM mix) provides 45–60 SA; (c) ketamine–xylazine at 200:10 mg/kg i.p. produces SA for up to 70 min; (d) methoxyflurane, ether, halothane and enflurane can be recommended in that order of preference.

Gerbils

Analgesia: buprenorphine 0.1–0.2 mg/kg s.c. every eight hours.

Sedation/light anaesthesia: fentanyl–fluanisone at 1.0 ml/kg i.m.

Surgical anaesthesia: (a) metomidate (50 mg/kg) + fentanyl (0.05 mg/kg) premixed and given as s.c. injection provides 30–60 min SA; (b) methoxyflurane inhalation.

General comments: Keep warm and supply O_2 during recovery. Very variable response to injectable anaesthetic agents (Flecknell et al., 1983).

Guinea pigs

Analgesia: pethidine at 20 mg/kg s.c. or i.m. every three hours.

Sedation/light anaesthesia: diazepam at 5 mg/kg i.p. is a good sedative — useful for combination with spinal epidural anaesthesia.

Surgical anaesthesia: (a) fentanyl–fluanisone (1 ml/kg i.m.) and diazepam (5 mg/kg i.p.) provides SA for 30–60 min; (b) fentanyl–fluanisone–midazolam (8.0 ml/kg i.p. of FFM mix) provides SA for 30–60 min; (c) methoxyflurane, halothane or enflurane are recommended in that order — ether should be avoided in guinea pigs; (d) spinal epidural anaesthesia is particularly valuable for abdominal and intrauterine surgery (Green, 1981).

General comments: Avoid pentobarbitone in this species. Give atropine (0.05 mg/kg s.c.) 30 min before other agents.

Rabbits

Analgesia: (a) buprenorphine 0.02–0.05 mg/kg s.c. or i.v. every 8–14 hr; (b) diazepam 0.5 mg/kg i.m. or i.v. is useful in addition to buprenorphine particularly after skeletal surgery.

Sedation/light anaesthesia: (a) fentanyl–fluanisone (0.1–0.5 ml/kg i.m.) produces sedation and enough analgesia for minor operations; (b) ketamine–xylazine (25:1.0 mg/kg i.m.) or ketamine–diazepam (25:5 mg/kg i.m.) produce deep sedation.

Surgical anaesthesia: The marginal ear vein is readily accessible for venepuncture with a butterfly infusion set and this allows several agents to be given to effect. The following agents are useful: (a) thiopentone (30 mg/kg of a 1.25 per cent solution i.v.) allows rapid induction of anaesthesia which then lasts only 5–10 min; (b) methohexitone (10 mg/kg of a 0.5 per cent solution i.v.) is also useful for rapid induction and this also lasts 5–10 min; (c) alphaxolone–alphadolone (5–7 mg/kg i.v.) is useful only for rapid induction in rabbits and lasts for 5–10 min; (d) fentanyl–fluanisone (0.3 ml/kg i.m.) followed 7 min later by diazepam or midazolam at 1.0 mg/kg i.v. provides SA lasting 30–50 min (can give increments of fentanyl–fluanisone at 0.01–0.05 ml/kg i.v. every 20–30 min to extend period of anaesthesia); (e) after induction by any of these methods, the rabbit can be intubated (3 mm diameter tube) and maintained on oxygen and low concentrations of methoxyflurane, halothane or enflurane for long periods (T-piece circuit best).

General comments: Reverse opiates with naloxone (0.1 mg/kg i.v.); keep warm and inject fluids i.v. and s.c. to ensure hydration after operation; pentobarbitone is a poor analgesic in rabbits and must be used with extreme caution; atropine at 0.2 mg/kg i.m. is useful in protecting heart from vagal inhibition.

Cats

Analgesia: (a) buprenorphine 0.01–0.01 mg/kg i.m. every 8–14 hr; (b) most mild analgesics are toxic in cats and should *not* be given.

Sedation/light anaesthesia: (a) xylazine 1 mg/kg s.c. or i.m. produces good sedation but the cats usually vomit; (b) alphaxolone–alphadolone 6–8 mg/kg i.m. produces sedation so long as it is injected properly into the belly of a muscle (e.g. quadriceps femoris); (c) ketamine (10–20 mg/kg i.m.) produces good sedation but enhances muscle tonus (atropine 0.05 mg/kg s.c. will reduce salivation with both (a) and (c)).

Surgical anaesthesia: Anaesthesia can be induced by i.v. injection of several agents into the accessible cephalic vein but if the agents above have been used for sedation, subsequent dosages of anaesthetics should be reduced by 50 per cent and, if alphaxolone–alphadolone has been used for sedation, then barbiturates should not be given. SA can be induced by: (a) i.v. administration of alphaxolone–alphadolone (9–12 mg/kg) which provides 15 min of SA — incremental doses of alphaxolone–alphadolone (3–5 mg/kg i.v. every 15 min) may be given to prolong anaesthesia; (b) i.v. administration of thiopentone (5–10 mg/kg i.v. using a 1.25 per cent solution) provides 10 min of SA; (c) i.v. administration of methohexitone (4–8 mg/kg) provides 5 min of SA; (d) ketamine–xylazine (20:1.0 mg/kg i.m.) is a simple way of producing SA lasting 30–40 min (beware respiratory depression and acidosis); (e) maintenance anaesthesia is best provided by inhalation via an endotracheal tube and T-piece of methoxyflurane (1.0 per cent) or halothane (1–2 per cent) in a 50:50 mixture of nitrous oxide and oxygen.

General comments: Methoxyflurane is the agent of choice in kittens under ten weeks of age.

Dogs

Analgesia: (a) buprenorphine (0.01–0.02 mg/kg) every 8–14 hr is best for acute postoperative pain; (b) pethidine is also useful at 10 mg/kg i.m. but should be given every 2–4 hr; (c) paracetamol (10–30 mg/kg) + codeine (0.25–0.5 mg/kg) every six hours by mouth is useful for mild to moderate pain.

Sedation/light anaesthesia: (a) acepromazine (0.5 mg/kg i.m.) provides light sedation to ease handling and venepuncture; (b) droperidol (0.4 mg/kg i.m.) provides light sedation and is useful in greyhounds; (c) fentanyl–fluanisone (0.1–0.2 ml/kg i.m.) or etorphine/methotrimeprazine (0.5 ml/4kg i.m.) provide deep sedation lasting 30 min and 2 hr respectively (still responsive to noise) — reverse with naloxone or diprenorphine; (d) thiopentone (10–20 mg/kg i.v. of a 2.5 per cent solution) is a safe induction agent in most breeds of dogs (not greyhounds) and produces anaesthesia lasting 7–10 min;

Table 14.13 Dosage and route of administration of some drugs used in the anaesthetic management of mice, rats, hamsters and gerbils (after Flecknell, 1984)

Drug	Mouse	Rat	Hamster	Gerbil
Methohexitone	6 mg/kg i.v. xxx	7–10 mg/kg i.v. xxx	—	—
Thiopentone	30–40 mg/kg i.v. xxx	30 mg/kg i.v. xxx	—	—
Pentobarbitone	40 mg/kg i.p. xx/xxx!	40 mg/kg i.p. xx/xxx!	50–90 mg/kg i.p. xx/xxx!	60–80 mg/kg i.p. xx/xxx!
Inactin®[a]	—	80 mg/kg i.p. xx/xxx	—	—
Ketamine	200 mg/kg i.m. xx	100 mg/kg i.m. xx	200 mg/kg i.m. xx	200 mg/kg i.m. xx
Ketamine + xylazine	200 mg/kg i.m. 10 mg/kg i.p. xx/xxx!	90 mg/kg i.m. 10 mg/kg i.p. xx/xxx	200 mg/kg i.p. 10 mg/kg i.p. xx/xxx	50 mg/kg i.m. 2 mg/kg i.m. xx
Ketamine + acepromazine	100 mg/kg i.m. 2.5 mg/kg i.m. xx/xxx	75 mg/kg i.m. 2.5 mg/kg i.m. xx/xxx	150 mg/kg i.m. 5 mg/kg i.m. xx/xxx	75 mg/kg i.m. 3 mg/kg i.m. xx/xxx
Ketamine + diazepam	200 mg/kg i.m. 5 mg/kg i.p. xx/xxx	—	—	50 mg/kg i.m. 5 mg/kg i.p. xx/xxx
Immobilon-SA® Etorphine + methotrimeprazine	—	0.05 ml/100 g i.m. xxx	—	—
Hypnorm® Fentanyl + fluanisone	0.01 ml/30 g i.p. x/xx	0.4 ml/kg i.m. or i.p. xx	1 ml/kg i.m. or i.p. xx	0.5–1.0 ml/kg i.m. or i.p. xx
Hypnorm® + diazepam	0.01 ml/30 g i.p. 5 mg/kg i.p. xxx	0.3 ml/kg i.m. 2.5 mg/kg i.p. xxx	1 ml/kg i.m. 5 mg/kg i.p. xxx	0.3 ml/kg 5 mg/kg i.p. xxx
Hypnorm® + midazolam[b]	10.0 ml/kg i.p.	2.7 ml/kg i.p.	4.0 ml/kg i.p.	8.0 ml/kg i.p.
Metomidate + fentanyl	60 mg/kg s.c. 0.06 mg/kg s.c. xxx	—	—	50 mg/kg s.c. 0.05 mg/kg s.c. xxx
Alphaxalone/ alphadolone	10–15 mg/kg i.v. xxx	10–12 mg/kg i.v. xxx	150 mg/kg i.p. xx/xxx!	80–120 mg/kg i.p. xx/xxx!
Tribromoethanol	125 mg/kg i.p. (0.25%) xx/xxx!	300 mg/kg i.p. xxx!	—	230–300 mg/kg i.p. (1.25%) xxx!

ANAESTHESIA AND ANALGESIA

Table 14.13 (contd.)

Drug	Mouse	Rat	Hamster	Gerbil
Atropine	0.04 mg/kg s.c., i.m.	0.05 mg/kg s.c., i.m.	0.04 mg/kg s.c., i.m.	0.04 mg/kg s.c., i.m.
Doxapram	5–10 mg/kg i.v.	5–10 mg/kg i.v.	5–10 mg/kg i.v.	5–10 mg/kg i.v.
Naloxone	0.01–0.1 mg/kg, i.p., i.v.	0.01–0.1 mg/kg, i.p., i.v.	0.01–0.1 mg/kg, i.p., i.v.	0.01–0.1 mg/kg, i.p., i.v.

[a] Inactin is only suitable for non-recovery experiments in rats.
[b] Dose rates in ml of a mixture of 1 part Hypnorm®, 1 part Hypnovel® and 2 parts water for injection.
x, sedation;   xx, immobilization;   xxx, anaesthesia/neuroleptanalgesia;   ! high mortality possible.

(e) methohexitone (4–8 mg/kg i.v. of a 1 per cent solution) is better for puppies and greyhounds and provides SA for 3–5 min but should only be given after administration of a suitable sedative (ACP or droperidol) otherwise excitement and convulsions may be a problem, particularly during recovery; (f) pentobarbitone (20–30 mg/kg i.v.) provides 45 min of light to medium anaesthesia but respiration is depressed and it is not a good analgesic; (g) it is best to induce anaesthesia with (d) or (e) then pass an endotracheal tube to maintain anaesthesia with halothane and a 50:50 mixture of nitrous oxide and oxygen (T-piece for puppies and small breeds and a McGill circuit for dogs > 10 kg).

ANAESTHETIC COMPLICATIONS AND EMERGENCIES

Anaesthesia is a pathological state, a departure from the normal physiology in any animal no matter how well controlled and benign the method used. Not surprisingly then, many complications may arise. These can usually be attributed to incompetence on the part of the anaesthetist — the most common emergency arises simply because of overdosage. The most important complications are:

1. Respiratory failure
2. Acute cardiovascular failure
3. Cardiac arrhythmias
4. Cardiac arrest
5. Hypothermia
6. Dehydration and hypovolaemia
7. Electrolyte imbalance
8. Regurgitation and vomiting
9. Anomalous anaesthetic responses and overdosage
10. Postanaesthetic failure

Table 14.14 Dosage and route of administration of some drugs used in the anaesthetic management of guinea pigs, rabbits, dogs and cats (after Flecknell, 1984)

Drug	Guinea pig	Rabbit	Dog	Cat
Methohexitone	31 mg/kg i.p. xx	10 mg/kg i.v. xxx	4–8 mg/kg i.v. xxx	4–8 mg/kg i.v. xxx
Thiopentone	—	30 mg/kg i.v. xxx	10–20 mg/kg i.v. xxx	10–15 mg/kg i.v. xxx
Pentobarbitone	37 mg/kg i.p. xx/xxx!	45 mg/kg i.v. xxx!	20–30 mg/kg i.v. xxx	25 mg/kg i.v. xxx
Ketamine	100–200 mg/kg i.m. xx	50 mg/kg i.m. xx		20 mg/kg i.m. xx
Ketamine + xylazine	40 mg/kg i.m. 5 mg/kg s.c. xx/xxx	10 mg/kg i.v. 3 mg/kg i.v. xxx	5 mg/kg i.v. 1.0–2.0 mg/kg i.v., i.m. xxx	15 mg/kg i.m. 1 mg/kg s.c., i.m. xxx
Ketamine + acepromazine	125 mg/kg i.m. 5 mg/kg i.m. xx/xxx	75 mg/kg i.m. 5 mg/kg i.m. xxx	—	—
Ketamine + diazepam	100 mg/kg i.m. 5 mg/kg i.m. xx/xxx	25 mg/kg i.m. 5 mg/kg i.m. xx	—	—
Immobilon-SA® Etorphine/methotrimeprazine	—	0.025–0.05 ml/kg i.m. xx/xxx	0.5 ml/4 kg i.m. xx/xxx	—
Hypnorm® Fentanyl/fluanisone	1.0 ml/kg i.m. x	0.5 ml/kg i.m. xx	0.1–0.2 ml/kg i.m. xx	—
Hypnorm + diazepam	1 ml/kg i.m. 2.5 mg/kg i.p. xxx	0.3 ml/kg i.m. 2.0 mg/kg i.p. or i.v. xxx	—	—
Alphaxalone/alphadolone	40 mg/kg i.p. xx	6–9 mg/kg i.v. xxx!	—	9–12 mg/kg i.v. xxx 12–18 mg/kg i.m. xx/xxx
Acepromazine	—	5.0 mg/kg i.m. x	0.1–0.25 mg/kg i.m. x	0.1–0.25 mg/kg i.m. x
Xylazine	—	3 mg/kg i.m. x	1–2.0 mg/kg i.m. x/xx	1–3.0 mg/kg i.m., s.c. x/xx

Table 14.14 (*contd.*)

Drug	Guinea pig	Rabbit	Dog	Cat
Atropine	0.05 mg/kg s.c., i.m.	0.2 mg/kg i.m.	0.05 mg/kg s.c., i.m.	0.05 mg/kg s.c., i.m.
Doxapram	5 mg/kg i.v.	2–5 mg/kg i.v.	2–5 mg/kg i.v.	2–5 mg/kg i.v.
Naloxone	0.01–0.1 mg/kg i.p., i.v.	0.01–0.1 mg/kg i.m., i.v.	0.01–0.05 mg/kg i.m., i.v.	0.01–0.05 mg/kg i.m., i.v.

x, tranquillization or sedation; xx, immobilization; xxx, anaesthesia; ! high mortality possible.

Table 14.15 Respiratory variables in some species of laboratory animals

Species	Respiratory rate (per minute)	Tidal volume (ml)	Minute volume (ml)
Mouse	180	0.15	24
Rat	90	1.6	220
Hamster	80	0.8	64
Gerbil	90	0.5	50
Guinea pig	100	1.5	150
Rabbit	55	20	1100
Cat	26	15	390
Dog (15 kg)	30	250	7500

Respiratory failure

Next to overdosage, respiratory failure is the commonest emergency during anaesthesia and can be defined as the inability to maintain arterial blood gases within normal limits. Failure usually results either from inadequate lung ventilation resulting in insufficient movement of gases in and out of the lungs, or less commonly, from poor gas exchange across the alveolar–blood barrier even though ventilation is adequate.

Inadequate ventilation can be caused by depression of the respiratory centres through anaesthetic overdosage or hypoxia; by upper airway obstruction; and by diminished thoracic and lung movements because of failure of muscle function.

Failure is signalled by alterations in the rate of breathing and in the nature of thoracic movements, by sounds indicating airway obstruction and by cyanosis of visible membranes. In rodents and rabbits, a fall to 40 per cent or below of normal respiratory rate (see Table 14.15) is likely to be followed by failure unless corrected. As soon as trouble is suspected, vigorous corrective measures should be adopted as first priority. First, the source of anaesthetic should be removed if it is an inhalational agent and the circuit should be flushed with a high flow of O_2; the airway should be checked to ensure patency and any obstructions cleared or fluids aspirated by suction; artificial respiration with O_2 should be initiated either by squeezing the rebreathing bag if an endotracheal tube is in place or by pressing rhythmically on the sternum to encourage to-and-fro movements of gas in the lungs; in dogs, an endotracheal tube should be passed if this is not already in place and in small mammals small plastic tubes passed into the mouth can be used for mouth-to-mouth resuscitation; finally, suitable drugs should be given i.v., using the analeptic doxapram at 2–10 mg/kg in all cases (repeating the dose at 15 min intervals) and the opiate antagonist naloxone (0.1 mg/kg) when sedative analgesics have been used for anaesthesia.

Laryngeal and bronchial spasm may be induced by inhalation of irritant agents or gastric fluids, and can easily be provoked in rabbits, cats and guinea pigs. In emergency, it may even be necessary to perform a tracheostomy and pass a catheter directly into the trachea. However, most cases can be relieved simply by spraying the larynx with 1.0 per cent lignocaine and administering O_2 by mask or tube after the cause has been removed.

Poor gas exchange may be due to: uneven perfusion of the lungs when the animal is lying on its back in one position for long periods; catarrhal exhudates and fibrous thickening of alveoli through disease; aspiration of blood, mucous or gastric content; and shunting of pulmonary blood through unventilated areas of alveolar collapse.

Treatment consists then of removing the cause of the problem if possible, changing the position of the animal and increasing inflation of the lungs.

It will be self-evident that preventive action will avoid most of these emergency situations. The lungs must be properly ventilated and an endotracheal tube should be passed whenever practicable. The position of the animals should be changed at regular intervals during long operations and they must be kept well hydrated with balanced salt solutions and plasma volume expanders to ensure the lungs are properly perfused with blood. Atropine should be given before induction of anaesthesia to reduce the risk of upper airway obstruction with excess bronchial and salivary secretion. The animal should be laid out with the head and neck extended to avoid partial obstruction of the larynx by the tongue or soft palate. The limbs must not be taped out and down so that thoracic movements are impeded and the surgeon must beware leaning on the chest during the operation.

Cardiovascular failure

Cardiovascular failure (CVF) is essentially a failure of blood flow through tissues resulting in cellular hypoxia. Four precipitating factors are well recognized:

1. Haemorrhage or water loss leading to reduced circulating volume and blood pressure.
2. Septic conditions, especially where endotoxins are released into the circulation.
3. Depressed cardiac function due to respiratory failure, anaesthetic overdosage or hypothermia.
4. Neurological factors stemming from spinal cord damage.

The clinical symptoms of a shocked animal are: pallor of skin and mucous membranes due to peripheral vasoconstriction; cyanosis of abdominal viscera due to hypoxia; a weak and rapid pulse due to a decreased blood volume and pressure; a fall in body temperature; and rapid and shallow breathing.

Emergency treatment consists of the following measures in order of priority:

(a) The precipitating factor and anaesthetic agents must be removed.
(b) Artificial respiration should be instituted immediately — an endotracheal tube is essential in this situation.
(c) Cardiac output should be assisted by expanding the circulating blood volume with Hartmann's solution.
(d) Diluted whole blood or plasma expanders should be transfused to maintain an adequate oncotic pressure in the vascular compartment and so minimize the risk of interstitial oedema.
(e) Further heat loss should be prevented by insulating the animal with cotton wool, blankets or aluminium foil depending on the species.
(f) Sodium bicarbonate should be injected at a rate of 500 mg/kg i.v. to counteract the inevitable acidosis.
(g) Doxapram should be injected at 2–10 mg/kg i.v. to increase cardiac output and hence increase tissue perfusion.

Cardiac arrhythmias

Most arrhythmias occurring during anaesthesia are caused by deviations in reflex sympathetic and parasympathetic arcs. The precipitating factors can be: the drugs in use; catecholamines released in response to apprehension; hypothermia; electrolyte imbalance, particularly if the serum concentrations of potassium rise significantly; and raised CO_2 in the blood associated with poor ventilation and hypoxia. Sedative-analgesics such as fentanyl all cause reflex slowing of the heart (bradycardia) but this can be prevented by administering atropine beforehand. Halothane and enflurane will also stimulate arrhythmias if administered in excessive concentrations. Perhaps most unpredictable of all the drugs in regular use is xylazine, which triggers arrhythmias in most species of animal.

Ventricular arrhythmias are dangerous and may proceed to cardiac arrest so vigerous corrective action should be taken to restore normal serum O_2, CO_2

potassium and bicarbonate concentrations. Anaesthetic administration should be stopped and steps taken to correct fluid and thermal imbalances.

Cardiac arrest

Arrest is diagnosed by ventricular fibrillation, dilation of the pupils, cyanosis of visible membranes, loss of sounds on auscultation and, almost invariably, respiratory arrest. The circulation must be restored within 2–3 min so instant action is required of the anaesthetist as follows:

1. The chest should be compressed manually by pressing on the sternum at a rate of 70–80 times/min with the animal lying on its back. This has the effect of massaging the heart externally.
2. Artificial ventilation should be started having ensured that there are no obstructions in the airway.
3. Ventricular fibrillation should be reversed with an electrical defibrillator or by injection of 0.5 mg/kg of 1.0 per cent lignocaine.
4. Adrenaline (1000 i.u.) injected into the heart is often useful in stimulating renewed breathing.
5. As a last resort, the animal should be picked up by its hind legs and swung back and forth head down. Alternatively, if the heart is still not beating after 2 min, it is worth opening the thorax, injecting calcium gluconate at 10 mg/kg of a 10 per cent solution directly into the ventricle and applying gentle massage.

Hypothermia

It is vitally important to maintain the core temperature of anaesthetized animals within normal limits, particularly when long periods of anaesthesia are anticipated. Hypothermia may affect the nature and duration of drug responses, it predisposes the heart to arrhythmias and it is an additional stressor to the pituitary–adrenal axis. It is probably the most common cause of anaesthetic failure in laboratory animals next to anaesthetic overdosage and respiratory obstruction.

Hypothermia results from several factors. Small animals, particularly when young, have a high surface to weight ratio so heat is lost to the environment over a relatively wide area. Peripheral vasodilation during anaesthesia significantly enhances this heat loss. Exposed moist surfaces and the lungs lose heat at an even greater rate due to latent heat of evaporation. The animal's temperature can be further depressed by i.v. administration of cold fluids.

Anaesthetic drugs and most CNS depressants override the thermoregulatory centre so that the animal is unable to respond to a fall in body temperature in such ways as shivering or increasing insulation by piloerection.

A few basic precautions should be taken to conserve body heat:

1. The operating environment should be as warm as convenient for personnel and should be free of draughts.
2. All i.v. infusions should be warmed to 38°C.
3. As little of the animal's hair as possible should be shaved during preparation for surgery.
4. Blankets, cotton-wool and aluminium foil should be used to wrap the anaesthetized animal before and after surgery commences.
5. External sources of heat including heated underblankets and electric light bulbs or infra-red heaters should be used sensibly to avoid overheating. Keep a thermometer next to the animal and do not allow the environmental temperature to rise above 32°C.
6. Core temperatures should be monitored at all times.

Dehydration and hypovolaemia

As discussed above, dehydration is often intimately related to hypothermia and the danger to small and young animals must again be stressed. The smaller the animal, the greater the surface area:mass ratio (both skin and lung) and the higher the basal metabolic rate. This in turn is directly related to the extracellular water 'reserve', so any deviation in fluid balance is more serious in these small beasts. Body fluid may be lost because of haemorrhage, as well as transudative and evaporative loss from the skin, lungs and open operative sites. A healthy animal can withstand rapid loss of >10 per cent of its circulating blood volume but during anaesthesia therapy should start if 5 per cent has been lost (estimate on basis of circulation = 75 ml/kg bodyweight).

Dehydration and hypovolaemia during anaesthesia will: prolong the effects of any drugs in use but particularly those excreted via the kidneys; increase the workload on the heart as it attempts to maintain normal central venous pressure; and allow the accumulation of toxic metabolites and abnormal concentrations of electrolytes in the extracellular space.

Dehydration should be prevented by commonsense measures:

1. Humidifying incoming dry anaesthetic gases.
2. Covering as much of the operation site as possible with moistened gauzes and irrigating the open wound at frequent intervals.
3. Infusing i.v. plasma expanders and balanced physiological salt solutions (e.g. Hartmann's) in appropriate volumes (at 5–10 per cent of circulating volume per hour).
4. Depositing isotonic or hypotonic salt solutions subcutaneously so that these can be taken into the circulation as required (far safer in small rodents than i.v. or i.p. injection) — a bolus of 15–25 per cent of the animal's calculated blood volume is adequate.

5. In the event of obvious haemorrhage or dehydration, it may be best to infuse citrated whole blood (one part acid citrate dextrose solution to 3.5 parts blood) at a rate of 10 per cent of the calculated blood volume per 30–60 min. Alternatively, Haemaccel® (Hoechst) in volumes not exceeding 10 per cent of circulating blood should be administered i.v. every 30–60 min.

Electrolyte and acid-base imbalance

The two ions sodium and chloride are responsible for over 90 per cent of the osmolality of extracellular fluid and the balance is maintained within fairly narrow limits under renal control. During anaesthesia, renal function may be impaired and any haemorrhage can deplete the sodium concentration in extracellular fluid. Excessive i.v. infusion of low sodium fluids can make this still worse. It is important then not to overload the animal with solutions but blood volume should be maintained with whole blood or plasma expanders.

The blood gas and hydrogen ion concentration are intimately linked and imbalances are far more likely to cause trouble in the anaesthetized subject than sodium, chloride or potassium ions. Anaesthesia and surgery can cause deviations in pH both from respiratory and non-respiratory (metabolic) dysfunction. In the laboratory situation a metabolic acidosis is only likely to occur after preoperative fasting and far the most likely problem to be encountered is respiratory acidosis. Poor ventilation and gas exchange allows CO_2 to build up rapidly and form excess carbonic acid in blood so that the pH falls from a normal of 7.4. If it falls below 6.8 (or in an alkalosis rises above 7.8) then the animal will almost certainly die. Modest alterations can usually be corrected by ensuring efficient ventilation of the lungs or even overventilating them artificially. In emergency, it is important to institute artificial respiration immediately and inject sodium bicarbonate at 500 mg/kg i.v.

Regurgitation and vomiting

Dogs, cats and primates vomit during anaesthesia and should be fasted beforehand. Preanaesthetic treatment with droperidol will usually prevent vomiting in cases which have to be anaesthetized in emergency and where they have been fed by accident.

It is essential to prevent gastric juices being aspirated into the bronchial tree, so if a small animal does vomit during induction of anaesthesia it should be picked up by the hind legs and vomitus cleared from the mouth with swabs and suction. Wherever possible, a cuffed endotracheal tube should be passed and inflated to protect the airway from irritant fluids.

Anomalous anaesthetic responses and overdosage

Most emergencies are due to overdosage. This can be reversed rapidly if

Table 14.16 Some analgesic drugs for use in animals to prevent pain during the postoperative period

	Morphine	Pethidine	Buprenorphine
Mouse	10 mg/kg s.c. 2–4 hourly	20 mg/kg s.c., i.m. 2–3 hourly	2.0 mg/kg s.c. 12 hourly
Rat	10 mg/kg s.c. 2–4 hourly	20 mg/kg s.c, i.m. 2–3 hourly	0.1–0.5 mg/kg s.c. 12 hourly
Rabbit	5 mg/kg s.c, i.m. 2–4 hourly	10 mg/kg s.c., i.m. 2–3 hourly	0.02–0.05 mg/kg s.c., i.v. 8–12 hourly
Guinea pig	10 mg/kg s.c., i.m. 2–4 hourly	20 mg/kg s.c., i.m. 2–3 hourly	0.05 mg/kg s.c. 8–12 hourly
Cat	0.1 mg s.c. 4 hourly	10 mg/kg s.c., i.m. 2 hourly	0.005–0.01 mg/kg 12 hourly
Dog	0.5–5.9 mg/kg s.c., i.m. 2–4 hourly	10 mg/kg s.c., i.m. 2–3 hourly	0.01–0.02 mg/kg i.m., s.c. 12 hourly

inhalational agents are in use but can be lethal where irreversible injectable agents, particularly barbiturates, have been employed. As emphasized above, respiratory depression and its reversal are central to the problem of management in these cases.

Treatment consists of:

1. Ensuring a patent airway and supplying the animal with high oxygen concentrations.
2. Injecting doxapram at 1–2 mg/kg i.v. (or i.p. in small rodents).
3. Where opiate drugs have been used, reversing them with the specific antagonist naloxone at 0.1 mg/kg i.v., i.p. or s.c.
4. Placing the animals in a warm environment and devoting constant attention to them until they have recovered.
5. Injecting warm 0.9 per cent sodium chloride solution i.v. or i.p. at approximately 10 ml/kg/hr and sodium bicarbonate i.v. at 500 mg/kg to all species.

Postanaesthetic failure

Unexplained deaths may occur in the next 12–24 hr even in cases where the animals appeared fully recovered from the anaesthetic. Failure may be due to endotoxins released into the circulation as a result of sepsis or extensively traumatized tissue. It is often associated with pulmonary oedema and progressively failing respiration. More obvious causes of failure are hypothermia, and mutilation by peers in the group if animals are put back together after

Table 14.17 Antimicrobial drugs for some laboratory species (after Flecknell, 1984)

	Mouse	Rat	Hamster	Gerbil
Ampicillin	50–150 mg/kg s.c. 200 mg/kg per os	50–150 mg/kg s.c. 200 mg/kg per os	Toxic	—
Amoxycillin	100 mg/kg s.c., i.m.	150 mg/kg s.c., i.m.	Toxic	—
Cephalexin	60 mg/kg per os	60 mg/kg per os	—	—
Cephaloridine	30 mg/kg i.m.	30 mg/kg i.m.	30 mg/kg i.m.	30 mg/kg i.m.
Neomycin	—	50 mg/kg s.c. per os	100 mg/kg per os	100 mg/kg
Trimethoprim 40 mg + sulphadiazine 200 mg/ml	0.5 ml/kg s.c.	0.5 ml/kg s.c.	—	—
Tylosin	10 mg/kg s.c.	10 mg/kg s.c.	100 mg/kg per os	—
Sulfamerazine	0.02% in drinking water	0.02% in drinking water	—	—
Tetracycline	5 mg/ml in drinking water 100 mg/kg s.c.	5 mg/ml in drinking water 100 mg/kg s.c.	Toxic	5 mg/ml in drinking water

operation. The following measures allied to close attention from interested personnel should minimize these risks:

1. Animals should recover in heated pens or boxes (25°C for cats, dogs and rabbits; 30°C for primates; and 35°C for mice, rats and neonates). Rodents should recover on tissue paper rather than in sawdust which may obstruct breathing.
2. An endotracheal tube should be left in place until chewing, swallowing and coughing reflexes are observed.
3. Animals should recover alone or in groups at the same stage of recovery.
4. Access to water should be allowed when recovery is obviously complete. Remember most animals require water at 40–80 ml/kg/24 hr so inject fluids if they will not drink voluntarily.
5. The animals should not lie in one position to one side but should be turned

Table 14.18 Antimicrobial drugs for some laboratory species (after Flecknell, 1984)

	Guinea-Pig	Rabbit	Dog	Cat
Ampicillin	Toxic	25 mg/kg i.m.	5 mg/kg s.c. 10 mg/kg per os, b.i.d.	5 mg/kg s.c. 10 mg/kg per os, b.i.d.
Amoxycillin	Toxic	—	7 mg/kg s.c. 5–10 mg/kg per os, b.i.d.	7 mg/kg s.c. 10 mg/kg per os, b.i.d.
Cephalexin	—	15–20 mg/kg per os, b.i.d.	10–15 mg/kg per os, b.i.d.	10–15 mg/kg per os, b.i.d.
Cephaloridine	25 mg/kg i.m.	15 mg/kg i.m.	—	10 mg/kg i.m.
Chloram-	20 mg/kg i.m.	15 mg/kg i.m.	10 mg/kg i.m. 50 mg/kg per os	10 mg/kg i.m. 50 mg/kg per os
Neomycin	10 mg/kg per os 30 mg/kg s.c.	—	5 mg/kg i.m., b.i.d. 10 mg/kg per os, b.i.d.	5.0 mg/kg i.m., b.i.d. 10 mg/kg per os, b.i.d.
Trimethoprim 40 g + sulphadizine 200 mg/ml	0.5 ml/kg s.c.	0.2 ml/kg s.c.	1 ml/8 kg s.c.	1 ml/8 kg s.c.
Oxytetra-cycline	Toxic	15 mg/kg s.c. or i.m.	10 mg/kg s.c. or i.m.	10 mg/kg s.c. or i.m.

frequently until they are capable of sitting up in the prone position. The head and neck should be extended to minimize the risk of airway obstruction. Pressure points should be massaged to prevent pressure sores developing. The animals must be kept clean and dry.

6. Cats and dogs may vomit during recovery so suction should be available (a 4 mm catheter attached to a 50 ml syringe is useful).
7. Analeptic agents such as doxapram are useful in the event of failing respiration.
8. If cats or dogs become constipated then administer an enema (e.g. Micolax).
9. *Always* assume that the animal is suffering *some* postoperative pain for at least 72 hr (Flecknell, 1984b; Morton and Griffiths, 1985; Taylor, 1985) and administer analgesics accordingly (Table 14.16).
10. Avoid giving antibiotics unless infection is suspected, remembering that most of these agents are toxic in rodents and may interact adversely with anaesthetic agents. However, if they are considered essential, select one appropriate to the species (Tables 14.17 and 14.18).

Table 14.19 Generic drugs with proprietary names and manufacturers

Acepromazine	'ACP injection', C-Vet Ltd
Alphaxalone/alphadolone	'Saffan', 'Glaxovet', 'Althesin'
Azaperone	'Suicalm', Crown Chemical Company
Buprenorphine	'Temgesic', Reckitt & Coleman Ltd
Codeine	
Diazepam	'Valium', Roche Products Ltd
Doxapram	'Dopram V', A.H. Robins
Droperidol	'Droleptan', Janssen
Enflurane	'Ethrane', Abbott Laboratories Ltd
Etorphine/methotrimeprazine	'Immobilon SA', C-Vet Ltd
Fentanyl	'Sublimaze', Janssen
Fentanyl–fluanisone	'Hypnorm', Crown Chemical Company
Halothane	'Fluothane', ICI, 'Halothane', May and Baker
Inactin	'Inactin', Byk Grulden Konstanz
Isoflurane	'Forane', Abbott Laboratories Ltd
Ketamine	'Vetalar', Parke Davis
Methohexitone	'Brietal', Elanco Products Ltd
Methoxyflurane	'Metofane', C-Vet Ltd, 'Penthrane', Abbott Laboratories Ltd
Metomidate	'Hypnodil', Crown Chemical Company
Midazolam	'Hypnovel', Roche Products Ltd
Naloxone	'Narcan', Winthrop
Pentobarbitone	'Sagatal', May and Baker Ltd
Thiopentone	'Intraval', May and Baker Ltd
Tribromoethanol	Aldrich Chemical Co, Milwaukee, Wisconsin, USA
Xylazine	'Rompun', Beyer UK Ltd

REFERENCES

Brown, P. M. (1983). A laryngoscope for use in rabbits, *Laboratory Animals,* **17**, 208–9.

Carvell, J. E. and Stoward, P. J. (1975). Halothane anaesthesia of normal and dystrophic hamsters, *Laboratory Animals,* **9**, 345–52.

Flecknell, P. A. (1984a). Notes on anaesthesia and analgesia in laboratory animals, Internal publication, MRC Clinical Research Centre.

Flecknell, P. A. (1984b). The relief of pain in laboratory animals, *Laboratory Animals,* **18**, 147–60.

Flecknell, P. A., John, M., Mitchell, M. and Shurey, C. (1983). Injectable anaesthetics in 2 species of gerbil (*Meriones libycus* and *Meriones unguiculatus*), *Laboratory Animals,* **17**, 118–22.

Flecknell, P. A. and Mitchell, M. (1984). Midazolam and fentanyl–fluanisone: assessment of anaesthetic effects in rodents and rabbits, *Laboratory Animals*, **18**, 143–6.

Green, C. J. (1975). Neuroleptanalgesic drug combinations in the anaesthetic management of small laboratory animals, *Laboratory Animals*, **9**, 161–78.

Green, C. J. (1979). *Animal Anaesthesia*, Laboratory Animals Ltd, London.

Green, C. J. (1981). Anaesthetic gases and health risks to laboratory personnel: a review, *Laboratory Animals*, **15**, 397–403.

Green, C. J., Halsey, M. J., Precious, S. and Wardley-Smith, B. (1978). Alphaxalone–alphadolone anaesthesia in laboratory animals, *Laboratory Animals*, **13**, 85–9.

Green, C. J., Knight, J., Precious, S. and Simpkin, S. (1981). Ketamine alone and combined with diazepam or xylazine in laboratory animals: a 10 year experience, *Laboratory Animals*, **15**, 163–70.

Hunter, S. C., Glen, J. B. and Butcher, C. J. (1984). A modified anaesthetic vapour extraction system, *Laboratory Animals*, **18**, 42–4.

Lees, P. (1977). Quoted in *Animal Anaesthesia*, Laboratory Animals Ltd, London.

Medd, R. K. and Heywood, R. (1970). A technique for intubation and repeated short-duration anaesthesia in the rat, *Laboratory Animals*, **4**, 75–8.

Morton, D. B. and Griffiths, P. H. M. (1985). Guidelines on the recognition of pain, distress and discomfort in experimental animals and an hypothesis for assessment, *Veterinary Record*, **116**, 431–6.

Norris, M. I. (1981). Portable anaesthetic apparatus designed to induce and maintain surgical anaesthesia by methoxyflurane inhalation in the Mongolian gerbil (*Meriones unguiculatus*), *Laboratory Animals*, **15**, 153–5.

Sebesteny, A. (1971). Fire-risk-free anaesthesia of rodents with halothane, *Laboratory Animals*, **5**, 225–31.

Taylor, P. (1985). Analgesia in the dog and the cat, *In Practice*, **7**, 5–13.

GLOSSARY OF TERMS, ABBREVIATIONS, SYMBOLS AND CONVENTIONS

Anaesthesia	— a state of controllable, reversible insensibility in which sensory reception and motor response are both markedly depressed
Analgesia	— the temporary abolition or diminution of pain perception
Anoxia	— complete deprivation of oxygen for tissue respiration
Apnoea	— temporary cessation of breathing
Ataxia	— incoordination, unsteady gait
b.i.d.	— twice daily
BMR	— basal metabolic rate
bradycardia	— slowing of the heart rate
CNS	— central nervous system

CNS depressant	— any agent which modifies function by depressing sensory or motor responses in the CNS
Cyanosis	— blue or purple colouring of the skin or visible membranes due to reduced haemoglobin in capillary blood, symptomatic of hypoxia
Dead space	— space in the respiratory system and/or anaesthetic equipment between the lung capillaries and fresh gas supply, where no gas exchange takes place
Dosages	— all dosages are expressed as mg of drug per kg body weight (mg/kg) except for the neuroleptanalgesic combinations which are more conveniently expressed as ml of commercial or diluted premixed solution per kg body weight (ml/kg)
Dyspnoea	— laboured breathing
Hypercapnia	— elevated blood carbon dioxide content
Hyperpnoea	— fast or deep breathing
Hypnotic	— a drug which induces a state resembling deep sleep, but usually with little analgesic effect
hypopnoea	— slow or shallow breathing
Hypotension	— a fall in (arterial) blood pressure
Hypothermia	— a fall in body temperature
Hypovolaemia	— a fall in circulating blood volume
Hypoxia	— depressed levels of oxygen

Injections	— routes of administration are abbreviated:
	    i.v., intravenous
	    i.m., intramuscular
	    i.p., intraperitoneal
	    s.c., subcutaneous
	    per os, by mouth

Laryngospasm	— spasm of the vocal cords, producing complete or partial obstruction of the airway
Minute volume	— the volume of gas breathed in one minute calculated from tidal volume × respiratory rate
Narcosis	— a state of insensibility or stupor from which it is difficult to arouse the animal

Polypnoea	— rapid, panting breathing
Pulmonary ventilation	— the mechanical expansion and contraction of the lungs in order to renew alveolar air with fresh atmospheric air
q.i.d.	— four times daily
Tachycardia	— a significant increase in heart rate
Tidal volume	— three times daily
u.i.d.	— once daily

Laboratory Animals: An Introduction for New Experimenters
Edited by A. A. Tuffery
© 1987 John Wiley & Sons Ltd

CHAPTER 15

Standards of Surgery for Experimental Animals

H. B. WAYNFORTH
Smith Kline & French, Welwyn, Herts

INTRODUCTION

The performance of surgery in all laboratory animals, regardless of species and size, is governed by the same principles as those for surgery of human beings. In practice, circumstances may occur or precedents may have been set which dictate that some experimental animals are treated differently, though this difference is usually only one of degree. Thus the standards of surgery for the small laboratory species (e.g. rats, mice, guinea pigs, rabbits) usually differ in several respects to those for the larger animals such as cats, dogs, pigs, sheep and goats. The opinion is widespread amongst scientists that rodents in particular are specially resistant to surgical infection, but evidence from work on intentional infection of these animals with pathogenic bacteria suggests that this may not be true. There is little doubt, however, that rodents rarely get noticeable diseases as a result of surgery — possible reasons are discussed later — and a dichotomy in surgical standards may be acceptable for laboratory animals and will be considered in the following pages.

CONSIDERATIONS REQUIRED BEFORE SURGERY

Planning

When a surgical procedure unfamiliar to the investigator is to be carried out, it is good practice for the work to be well planned. Planning together with practice (see below) are important to ensure that the investigator feels and acts calmly and confidently and that the animal gets the best possible treatment.

The result is a healthy and well-prepared animal and one which is likely to give the best response in the subsequent experimental study. Preparation of a checklist of requirements before surgery is begun is helpful both to the investigator and to any ancillary personnel who might be involved. Such a checklist might contain the following:

> Animals to be used
> Facilities and accommodation required
> Instruments required
> Apparatus and accessories required
> Type of pre- and postoperative care to be carried out
> The anaesthetic to be used
> Drugs, disinfectants and other chemicals required
> Any assistance needed
> An outline of the surgical technique.

An important consideration is when to carry out the operation. Surgery done late in the day may not allow time for good postoperative care to be given. Similarly, operations done at the end of the week may also result in animals being given less than proper attention over the weekend period. Consequently these two situations should be avoided, but to some extent this will be determined by what animals are being operated on, the length of the operation and the extent of the insult to the animal. For example, there would be little objection to carrying out ovariectomy on a rat in these periods since the animals are known to recover quickly from this operation with little subsequent morbidity or postoperative complications. On the other hand, a complicated operation on a dog, and even a rat, may necessitate overnight care if the animal is to be encouraged to survive.

Practice

The need to first practice an operative technique is a paramount requirement and failure to do so may jeopardize the experiment. In the United Kingdom, live animals cannot be legally used and practice should therefore be conducted on a freshly killed specimen. The use of cats, dogs and farm animals in this way may present a problem because of their reduced availability but small animals are readily available and easily obtained. It should be remembered, however, that dead animals do not show the effects of bleeding, respiratory movements and tissue tension on the operative method and this may very easily complicate the efforts of the novice surgeon when he comes to use a living animal. Experience will be needed to deal with this. Practice should include handling instruments, making incisions, tying knots and suturing wounds.

It should be readily evident that before practising, the investigator should gain a knowledge of the relevant anatomy and surgical technique. This is best achieved by discussions with an experienced colleague or alternatively from reading the available literature. This will prevent unnecessary manipulation of tissues which otherwise could result in unwanted haemorrhage and nerve and other tissue damage and in infection. Having an experienced colleague on hand to advise during the actual surgery until adequate experience is gained is worth many books and every attempt should be made to obtain such help.

Health and acclimatization

Animals are easily stressed and upset. Merely moving a cage of rats from one position to another within a room is known to markedly change the levels of various chemical and cellular constituents in the blood, which may then not return to normal for some time. Animals which have to undergo transport from a distant supplier to the place where they are to be used are particularly vulnerable to stress-induced effects and mice and rats, for example, can lose 10–20 per cent of their body weight by the time of their arrival. This may often take 48 hours or longer to return to normal, particularly as in their new environment they will encounter new stresses such as unfamiliar food, different husbandry practices, different cage sizes and different animal care personnel. Animals which are stressed in these ways make poor surgical and experimental subjects unless they are allowed to acclimatize to their new circumstances. The length of time allowed for such adaptation will vary with the species and the situation encountered. It may take several months for primates caught in the wild and transported to the laboratory, whereas animals moved from room to room will require only a few hours. Consideration for acclimatization may not be necessary where the stress is minimal and where the added stress of the subsequent surgical procedure is likely to be small (e.g. adrenalectomy in a rodent) but in other cases the need for acclimatization should always be seriously considered. Certainly animals brought in from outside should be allowed several days to adapt to their new surroundings before being used. This also has the added advantage that they can be observed for the development of any clinical symptoms of a disease which they may have been incubating and which will suggest that they are not in the best of health.

Immediately before surgery, all animals should be examined for signs of ill-health. With larger animals such as dogs or pigs, a more thorough clinical examination is generally advocated in the veterinary textbooks even to the extent of biochemical analyses of blood and urine samples. However, in many establishments, such facilities are not available but it should nevertheless be determined that overt signs of ill-health are absent. Such animals should be active, behave normally, be well fed and adequately hydrated. It is good practice to keep an individual record file for each dog or other large animal and details such as those shown in Figs 15.1 and 15.2 are well worth recording. It is almost invariably the case that such records are rarely kept for small animals

<table>
<tr><td colspan="2">CHARING CROSS & WESTMINSTER
MEDICAL SCHOOL
ANIMAL UNIT</td><td colspan="2">ANIMAL
IDENTIFICATION</td></tr>
<tr><td colspan="4"></td></tr>
<tr><td colspan="2">Animal (and type)</td><td colspan="2">Identification code and methods:</td></tr>
<tr><td>Sex:</td><td colspan="2">Age (approx)</td><td>Body wt:</td></tr>
<tr><td colspan="2">Coat colour:</td><td colspan="2">Temperament:</td></tr>
<tr><td colspan="4">Date of Supply:</td></tr>
<tr><td colspan="4">Supplier:</td></tr>
<tr><td colspan="2">Veterinarian:</td><td colspan="2">Tel. Nos.
Dept:
Bleep:
Home:</td></tr>
<tr><td colspan="2">Experimenter:</td><td colspan="2">Tel. Nos.
Dept:
Bleep:
Home:</td></tr>
<tr><td colspan="2">Date of death:</td><td colspan="2">Autopsy performed:</td></tr>
<tr><td colspan="4">Autopsy results:</td></tr>
</table>

Fig. 15.1 Recording the 'vital identifying statistics' of an animal destined for surgery

CHARING CROSS & WESTMINSTER MEDICAL SCHOOL ANIMAL UNIT	Animal: Sex: Identification: Licensee: Dept: Examined by:

DATE	CLINICAL NOTES
	First contact with animal. **Recommendations:** **Before Surgery.** **Suitability for Surgery:**

Fig. 15.2 Recording clinical details on delivery of the animal and just prior to surgery

and consequently much information which could have been of value has been lost in the past. However, the keeping of individual surgical records for the small animals can be very much recommended.

Food and water

Withholding food from at least the night before the day on which the operation is scheduled is recommended for large laboratory animals. It is generally unnecessary to withhold water, though it may be an advantage to withhold it for two or three hours before actual surgery, though never longer otherwise dehydration may result and the animals may thus make a potentially poor patient.

There is no need to withhold food and water from rodents and rabbits and other small animals except in special circumstances such as surgery of the lower alimentary tract. If fasting is required it can be done overnight in large rodents or for 24 hours in the rabbit, which retains its food for long periods. Because of its high metabolic rate, the mouse should not be fasted for more than three or four hours, if it is thought necessary.

Premedication

Animals which are unusually nervous, agitated, fractious or just difficult to handle, particularly the larger species, can be given tranquillizers to calm them before the induction of anaesthesia. Less anaesthetic is usually required when such premedication is carried out and with some species, such as the dog, the induction of and recovery from anaesthesia can be much smoother. The routine use of tranquillizer premedication in the large species is invariably recommended. The use of such agents is covered in detail in Chapter 14. The use of antibiotics in surgery may have to be considered. Their indiscriminate employment as a blanket cover against possible surgical infection is generally ill advised for all animals and is particularly inappropriate for the smaller species. Such indiscriminate antibiotic prophylaxis may favour the colonization of wounds by resistant bacteria, with the result that a serious and possibly fatal infection develops — just the opposite of what is desired.

Administration of antibiotics, which ideally should be done immediately before surgery to build up peak tissue levels during surgery, may be indicated only when surgery will be prolonged or when high risk operations such as those on the alimentary tract are to be performed. Recommendations for prophylactic antibiotic therapy can be found in textbooks, but the type of antibiotic to use may depend on a knowledge of the likely contaminating micro-organisms that will be encountered. Also, some antibiotics are toxic to certain species and can be fatal (e.g. penicillin in conventional guinea pigs).

With some species, or when anaesthetics irritant to the mucous membranes are used, such as ether, excessive salivation may be a problem. Atropine sulphate can be administered up to 30 minutes before induction of anaesthesia, to reduce or abolish this complication. Some rats and rabbits have an atropinase in their circulation and greater than normal doses of atropine sulphate will be required. Again the use of such agents is discussed in Chapter 14.

The operating area

The physical environment in which surgery is to be performed can vary from a specially designed and sophisticated surgical suite to a small area on a laboratory bench. What is required will depend on the species being used, the type and extent of the surgical procedures and on whether the animal is to be allowed to recover from the anaesthetic. What is important is that the area is kept clean and uncluttered and is away from general human traffic. The liberal use of a disinfectant on all surfaces can be useful to reduce the room population of unwanted micro-organisms and dust.

Equipment for surgery

Instruments

An enormous variety of instruments are available and these have been developed mainly for human surgery. Certain specialist instruments have been developed uniquely for veterinary work. Instruments for rodents, rabbits and other small animals are usually chosen from those used in human paediatric, ophthalmic and neurosurgery. However, in experimental surgery, particularly with the small animals, ingenuity is often required on the part of the investigator to fashion the right instrument and apparatus for the experimental study to be undertaken. Novel instruments and apparatus for small animals have also recently become commercially available.

What instruments to use will depend on the type of operation being carried out. For large animals, the veterinary surgical textbooks should be consulted (see References). For small animals, a basic set of instruments is usually required for all operations. This can consist of a pair of blunt-ended scissors, a pair of pointed dressing scissors, three pairs of forceps, two with serrated tips and one with rat-toothed tips, a skin clip applicator and forceps with skin clips attached. Some operations will require specialized instruments in addition. Instruments have to be carefully prepared for surgery and methods for doing this are available in surgical textbooks (see References). Those used for large animal work are always sterile and this is commonly achieved by autoclaving. Sterile instruments can also be used for small animal surgery but in many establishments sterilizing facilities are not readily available and instruments are often used after only disinfecting them. For this, clean instruments are immersed in disinfectant for the required time and then used directly or after rinsing in sterile saline if the disinfectant is likely to be irritating for the tissues. Common disinfectants are 70 per cent ethyl or isopropyl alcohol, 0.5 per cent aqueous chlorhexidine and 0.1 per cent aqueous benzalkonium chloride. Boiling instruments for about ten minutes is equally effective. Both methods destroy bacteria but not bacterial spores which have the potential to induce infection, though rarely seem to do so in small animal surgery.

Operating table

Special veterinary surgical tables are available for use with large animals. These can also be used merely for supporting small animals where they are required to be free-lying and their limbs not restrained. However, specially designed operating tables, or boards, made from metal, plastic or wood are available and considerably aid surgery of animals such as rats, mice and rabbits. Such tables may incorporate a heating facility which is very useful and is able to maintain the animals' temperature. Unfortunately, the most helpful accessories for surgery of small animals, notably small skin and tissue retractors, tooth and leg holders, which would have to be specially designed to fit the small operating boards, are mostly unavailable from commercial sources and at present one often needs to improvise suitable devices.

For simple operations on small animals a cork board or a wooden board covered by a sheet of rubber may suffice, providing they are kept clean or, in the case of a corkboard, renewed frequently since they quickly become ingrained with dirt and blood and tissue residues.

Lighting

In small confined operating areas, a 100 W anglepoise lamp may be adequate for surgery of small animals. However, a portable cold-illumination lighting system which can be regulated to give diffuse or pinpoint lighting is more effective. Proper lighting is of enormous help for surgery of all animals and the use of special surgical theatre lights is highly recommended. Such units can be fitted even in confined spaces and their extra expense will pay dividends in better surgery.

Other apparatus and accessories

The apparatus required for surgery will depend on the type of surgery being carried out, the species being used and whether the surgery is an active part of an experiment. Similarly, surgical accessories such as gauze, drapes, cotton-wool, etc., will be variously required. All apparatus should be clean and dustfree and preferably wiped down with a detergent or disinfectant. Some apparatus and accessories will be required sterile and this can be achieved by autoclaving or alternatively with suitable glass and metal apparatus, by dry heat in an oven at 160°C for at least one hour. Plastic items can only satisfactorily be sterilized by γ irradiation but in many cases it is acceptable to use them after they have been disinfected in, for example, 0.5 per cent aqueous chlorhexidine. Plastic cannulas disinfected in this way and rinsed with sterile saline have been kept in various parts of the body of both large and small animals for long periods of time without producing infection.

CONSIDERATIONS REQUIRED DURING SURGERY

Aseptic technique

This can be defined as the removal of micro-organism from the site of surgery by the use of antiseptics and subsequently taking measures to prevent recontamination. Aseptic technique is always used for the large animals such as cats, dogs and pigs when they are to recover from the anaesthetic. For surgery on animals which are not to be allowed to recover from the anaesthetic, clean surgical conditions without full aseptic precautions are generally acceptable. With rodents and other small species, and in simple operations in rabbits, asepsis has traditionally rarely been used. The reason for this is that in those myriad of past cases where only clean but not aseptic technique has been employed, little or no infection has apparently developed. This implies that small animals, rodents in particular, have a unique ability to resist surgically induced infection. This has never been scientifically substantiated and the little evidence there is does not support this implication. Nevertheless, on subjective assessment, there do seem to be few reports of surgically induced infection in rodents. One reason for this may be the attitude of investigators towards rodents and the other small species. It cannot be doubted that such animals do not command the same humane respect as the dog, for example. They are also frequently regarded as expendable because they are available in large numbers. Consequently, little meticulous record-keeping is maintained on their surgical and postsurgical progress and thus data on surgical infection are not generated. However, equally valid may be a second reason why small animals do not appear to get surgically infected. For many pathogenic bacteria, their ability to induce infection depends on their numbers at the surgical site (sometimes estimated as 10^5 bacteria per gramme of tissue or per millilitre of fluid). It is known that a number of features of surgery will enhance the ability of bacteria to grow to the numbers required to produce a surgical infection. These features may be very much less in evidence during surgery in small animals. Thus, the majority of operations in small animals are carried out through relatively small incisions, they are completed in a relatively fast time, they produce little haemorrhage because of the small size of peripheral blood vessels — which also simplifies the process of stopping bleeding when it does occur — and in many cases there is less trauma to the tissues by mechanical damage and dehydration.

Advice from various authors (see References) on the precautions to be taken with small animals ranges from the use of the 'no-touch' technique where, in addition to a sterile preparation, the body is only ever in contact with sterile instruments and never with the surgeon's hands, to the use of clean but not aseptic conditions. This suggests that there is insufficient evidence to justify the revision of the present dichotomy often practised in surgical standards, i.e. full asepsis for large animals and only clean antiseptic surgery for small animals — though there is nothing to prevent the use of aseptic technique if so desired — and the following description will elaborate on their dual nature.

Preparation of the incision site

It is necessary to remove hair from the site of incision and from a sufficient area around it so that hair does not get into the wound. Apart from the possibility of causing infection by virtue of associated bacteria, hair can irritate the wound and delay healing. However, since short hair can be adequately disinfected, it is not always necessary to remove it for simple operations in rats and mice. Removal of hair should be carried out with clippers and a razor can then be used if required, though clipping suffices in most cases. The incision site must then be wiped liberally with an antiseptic such as 70 per cent ethyl or isopropyl alcohol, or 0.5 per cent chlorhexidine or 0.1 per cent benzalkonium chloride in 70 per cent alcohol. Also very commonly used for large animals in particular are tincture of iodine containing at least 1 per cent free iodine and 10 per cent povidone–iodine solution.

In large animals the incision area is first washed with a detergent, dried and then wiped from the actual site of the incision outwards, with antiseptic. In small animals, it is sufficient to douse the incision area with antiseptic and to remove the excess with gauze. If iodine is used, it can be followed by 70 per cent alcohol to dilute it, to reduce any possibility of it irritating the wound. However, povidone–iodine is non-irritating and has much to recommend it. When the wound has been prepared with antiseptic, the precise site of incision in large animals is surrounded by sterile drapes which are attached to the skin with special clips. This prevents contamination by bacteria living within the skin of the surrounding area. Drapes can be used in a similar manner in small animals if asepsis is deemed to be important (e.g. as in hysterectomy or hysterotomy — derivation of germfree or specified pathogen free rodents and rabbits). Special adhesive plastic drapes which can be put over the animal and the incision made through the drape are particularly useful for mice and rats. In all subsequent procedures which involve the operation itself, bacterial contamination of the wound is prevented by the use of sterile instruments and by the surgeon wearing sterile gowns and gloves and by donning clean face masks and caps. Where surgery is required only to be clean and not aseptic, the use of special gowns, gloves, masks and caps may not be necessary, though washing of the hands with attention to the nails should be done both before and after surgery.

The incision

Making the incision is carried out with a scalpel and a sharp blade, though scissors are more frequently employed for the small animals. Some knowledge of anatomy is useful in determining where incisions into skin and muscle should be made. The linea alba, for example, which is a white line running along the midline of the abdominal muscle, is a relatively bloodless area and is used therefore as much as possible when gaining entrance to the abdominal cavity. Another example is seen when obtaining access to the trachea or associated organs as the thyroid gland. These lie beneath the sternohyoid muscle, which is in two halves kept together by overlying connective tissue. Merely tearing the connective tissue will allow separation of the two muscle halves without damaging them and thus permit an approach to the underlying tissues. The sequences to be used when making incisions are well explained in veterinary texts (see References) and will depend where the wound is and how deep it is to be. These sequences are condensed when using the smaller animals, for anatomical reasons.

Surgical methods

Once the incision has been made, subsequent steps will depend on what the investigator is trying to achieve. It may be a routine surgical procedure, such as adrenalectomy, or it may be a novel experimental technique. A treatise on surgical methods in experimental animals cannot be written here but textbooks on veterinary surgery are available, though there is much less information on the small animals in comparison to the cat and dog for example (see References). Moreover, the information for the small animals is often only to be found scattered in individual scientific research papers.

Temperature

Animals undergoing surgery lose heat. The smaller the animal the faster the heat loss and for the mouse this can be alarming, with a 7°C drop in body temperature within the first ten minutes being quite possible. Methods of monitoring body temperature vary from sophisticated automatic recording apparatus to the simple oral thermometer placed in the rectum. Any of these can be used for any of the common laboratory species.

Since a substantial drop in body temperature puts the animal at risk the temperature should be maintained near normal during the operation. This can be done using special heating pads on which the animal lies or by placing a low-wattage lamp near to the animal.

Fluid balance

Animals may lose substantial amounts of water and minerals during surgery, by evaporation of tissue fluid, loss of blood and during respiration. Dehydration puts the animal at considerable risk, and if much fluid is lost it must be promptly replaced. Mainteance of fluid balance is a complex matter, but an interim measure is to give warm physiological saline or a 4 per cent dextrose + 0.18 per cent saline solution at a rate of 4 per cent of body weight per 24 hours. Any route of administration can be used and although the intravenous route is to be preferred, the fluid can be given intraperitoneally or subcutaneously as these routes are more accessible in the small animals. Giving the fluid as a continuous i.v. infusion is common practice in the large animals.

Drying of tissues from prolonged exposure during surgery must be avoided and this can be prevented by applying warm saline or a saline-soaked gauze pad to the tissue.

Haemorrhage

Some bleeding during surgery is inevitable but blood loss at all times must be minimized and even mild loss can result in the animal showing signs of shock. Persistent bleeding, which occurs even from peripheral vessels in large animals though rarely in the small species, must be stopped promptly. There are several methods to effect haemostasis, including clamping, ligation, twisting and electrocautery. If bleeding has occurred, excess blood must be removed before the wound is closed, otherwise it acts as an ideal medium for bacterial growth.

Trauma

Gentle handling of tissues is a major requirement of good surgery. Excessive retraction or inexpert handling with instruments may damage tissues which then could become necrotic. This not only provides an excellent medium in which bacteria can grow but wound healing is delayed. Blunt dissection is to be preferred to the cutting of tissues whereever possible and the investigator should not be afraid to use his fingers to manipulate tissues, unless he is employing a 'no touch' technique. There are only a relatively few instances in which the 'no touch' technique might be used as a matter of preference.

Antibiotics

Where it is thought necessary to given an antibiotic, it is preferable to do so with it in solution rather than it being dispensed as a powder into the wound. The latter could result in it caking, delaying its absorption and promoting adhesions.

Suturing

The closing of the incision at the conclusion of surgery is an art which requires practice to achieve properly. The methods used to do this are found in veterinary texts (see References). It is necessary to consider several aspects: the proper apposition of wound edges, use of the right needle and thread and use of the proper knot and stitch. Wound edges can be apposed in several ways depending on the type of tissue. In intestinal anastomoses, for example, the outside layer of the cut edges of the intestine should generally be in apposition. When suturing skin wounds the inner aspects of the skin should be together. To do these requires a knowledge of what stitch to use. The most common stitches are the continuous stitch, which runs through and over the wound edges from one end of the wound to the other, and the interrupted stitch, of which there are several types which allow inversion, eversion or natural apposition of wound edges. The interrupted stitch is a single stitch, a row of which is placed along the wound at precise distances from each other. Although a disadvantage of a continuous stitch is that if it breaks it can unravel completely, this rarely occurs if the thread of the right calibre is used. The continuous stitch is regularly used for stitching up all layers of the wound in small animals but is used in conjunction with interrupted stitches in large animals. When suturing, the needle should pierce the tissue several millimetres below the wound edge if the stitch is not to tear away. As a rule of thumb, the needle should pierce as far down the side of the wound edge as the distance between each suture (e.g. if the interval between interrupted stitches is to be 5 mm, then the needle should pierce the tissue about 5 mm below the wound edge).

Threads can be made from silk, cotton, catgut, plastic or metal and can be absorbed by the body or are non-absorbable. They come in various diameters. Sizes 2/0 and 3/0 are commonly used for general surgery of small animals and sizes 0, 1 or 2 for large animals. Non-absorbable threads, for example, silk, can be used and can remain within the body without causing complications. For the purposes of cleanliness or sterility, threads can be treated as for instruments.

Needles are obtained in a variety of curved shapes, or are straight. Their tips are usually either round or triangular. The latter have cutting edges and are used mainly for piercing tough tissue such as skin. Round-tipped needles should be used for soft tissue such as muscle. Needles may have eyes, in which case they have to be threaded manually, or they may be bonded to the thread. The latter type are now very common. Curved needles need to be stabilized during suturing by employing a needle holder and practice is required in using these. Straight needles are easier to use and seem to find favour particularly with relatively inexperienced investigators using small animals. The knot used in surgery is the reef knot though there are variations on this, such as the surgeon's knot, in which the first half hitch contains an extra twist to resist slipping. The art of making knots using forceps, and particularly with respect to the beginning and the completion of a continuous stitching sequence, requires practice. Small animals have a propensity for removing thread which has been used for suturing, where the thread is accessible to the animal's teeth. In these cases, it is usual to close skin wounds with metal clips (e.g. Michel clip). They are applied at close intervals (about 8–10 mm apart) using special application forceps. Metal clips come in various sizes and the 12 mm or smaller sizes are adequate for rats, mice and guinea pigs. Recently, special steel staples applied by an automatic device have become available and may prove useful. One apparent advantage of automatic staples is that whereas Michel clips can be applied too tightly, causing local tissue damage, this cannot occur with the staples. Although metal clips usually resist the efforts of rats, mice and guinea pigs to remove them, rabbits may be more successful. With these animals, it may be found necessary to apply 'hidden' interrupted stitches, where the stitches are placed on the inside of the skin edge without piercing the skin externally.

CONSIDERATIONS REQUIRED AFTER SURGERY

The period of recovery

Animals awake from the effects of the anaesthetic at varying times depending on the type of anaesthetic used and the length of the operation. After inhalent anaesthetics recovery is generally fast, while with most parenteral anaesthetics recovery is slow and may not be complete for several hours. The recovery period can be a particularly hazardous one and animals must be observed frequently and in some cases continuously until they regain consciousness. Particular attention must be given to neonatal rodents since maternal cannabalism is a frequent occurrence and steps must be taken to prevent this. If recovery has not occurred by night-time, the investigator should arrange for someone to be present during the night in case of complications. Animals should not be allowed to recover during the night period unattended.

Fig. 3   Drug administration record card

DRUG ADMINISTRATION RECORD

REGULAR DRUGS

DATE COMMENCED	DRUG	DOSE	ROUTE	ADMINISTRATION TIMES	SIGNATURE	DATE CANCELLED	PHARMACY NOTES
A							
B							
C							
D							
E							
F							
G							
H							
I							
J							
K							
L							
M							
N							
O							
P							
Q							

ONCE ONLY DRUGS

DATE COMMENCED	DRUG	DOSE	ROUTE	TIME OF ADMINISTRATION	SIGNATURE	GIVEN BY

DIET

ANIMAL:

IDENTIFICATION:

EXPERIMENTER:

During recovery, animals should be placed on dry bedding and in the best position to aid normal breathing. A number of nursing procedures may need to be carried out, such as removal of the tube, if the animal's trachea has been intubated, attention to i.v. drip lines if these are present, turning the animal at intervals to avoid bruising and vascular and respiratory problems, recording of pulse rate, rate and depth of breathing and temperature, etc. Ideally, recovery should be allowed to take place in a specially designated area which should be quiet and have subdued lighting. Hypothermia must be prevented and heating aids such as lamps, heating pads, hot water bottles, fan heaters, electric fires and even aluminium foil placed over the body can be found useful. Animals recovering within minutes of the completion of surgery may not require extra heat but this must always be a consideration for animals undergoing prolonged recovery. After the animal has recovered initially, it should be under close observation for at least the next one or two days, irrespective of the species. It is during this time that the majority of complications will occur. In the postrecovery period, routine nursing is necessary, especially for the large animals, and a meticulous record should be kept of vital signs and also whether the animal is eating, urinating and defaecating normally and whether its behaviour is normal. Records should also be kept of the drugs and other treatments given and at what time of the day – see Fig. 15.3.

Fluid therapy

The possible need to maintain the animal in fluid balance has been discussed above and a similar need may be required postoperatively.

Dressings

If these have been used they will need to be changed if they become soiled, or replaced if they are removed by the animal. Small animals persistently remove dressings with their teeth and their use in rats and mice, for example, is very limited. A plastic dressing applied as an aerosol may be useful in all species to keep out dust and dirt.

Medication

Three types of medication may commonly be required postoperatively.

Antibiotics

These may be given to prevent an infection whether real or potential. If treatment is to be prophylactic and the organisms which might be encountered are not positively known, broad-spectrum antibiotics should be used. On the other hand, if an infection develops postoperatively, the organism(s) causing it should be isolated and its antibiotic sensitivity determined so that the most effective antibiotic can be given. Since the process of bacterial isolation may take 24 hours or more, a broad spectrum antibiotic can be given to cover this intervening period if the infection is severe. It should be noted again that some antibiotics are toxic to some animals and care must be taken in deciding which antibiotic to use. Antibiotics given to prevent infection during surgery should be administered as a single large dose either just before or during surgery. Some workers recommend three doses, immediately before, during and after surgery, but it is not considered helpful to prolong treatment further than this since resistant bacteria are likely to emerge. If an infection develops postoperatively, then antibiotic therapy should be given for a full course of about five days.

Drugs required as supplements

After some operations, certain chemicals may be required to preserve the life or quality of the animal. For example, adrenalectomized animals often need injections of mineralocorticosteroids or a drinking regime of physiological saline to keep them alive. Pancreatectomized animals will require insulin and a special diet for their survival. Such treatments are an integral part of the surgical protocol necessitating a precise knowledge of the relevant physiology of the animal.

Analgesics and tranquillizers

Pain or distress is a very likely result of all surgery. After simple operations, the pain or distress may be so slight or momentary that the animal can be expected to tolerate it well and no action is needed on the part of the investigator. In other cases, the postoperative pain might be expected to be more severe and possibly even extremely severe, and here it is imperative, and possibly legally mandatory, to relieve the pain. The common human analgesics, such as pethidine (meperidine hydrochloride), morphine sulphate, codeine, paracetamol, etc., are effective in laboratory animals and doses and treatment schedules are given in Chapter 14 (also see References). A major disadvantage with these analgesics is their short action (2–4 hr), which necessitates repeated doses. Thus to be effective they may need to be administered throughout the night as well as during the day. A recent drug, buprenorphine hydrochloride, has been shown to be highly effective in the common laboratory animals and may have a useful duration of action of up to twelve hours. This is particularly useful for overnight cover.

Animals which are distressed, though not necessarily in pain, can be treated with tranquillizers and it is a matter of good humane practice to do so when required. Although the treatment of pain in laboratory animals now seems very feasible, a major problem has always been in the assessment of animal pain. In small animals which rarely vocalize their pain, as do human beings for example, it is easy for the relatively inexperienced investigator to recognize only severe pain. Here the animal usually assumes a hunched posture, its fur 'stands on end' (staring coat) and they 'look miserable'. The larger animals can give more recognizable signs but even here it may be difficult to recognize moderate pain. Animals in pain often variously exhibit restlessness and various forms of abnormal behaviour, abnormal posture, abnormal vocalization and unusual aggression, and careful observations by the investigator may help in suggesting whether treatment of the animal is recommended. In the last analysis, judgement should be made anthropomorphically on the basis of 'would I be feeling much pain and want an analgesic after such surgery'? If the answer is yes, then the animal should be treated accordingly. If the answer is equivocal, then the animal should be given the benefit of the doubt.

Sutures

These should be removed after about seven days. Sutures and skin clips are a potential site of infection and should not be left in the skin for long periods.

Emergencies

If there has been careful nursing and observation, emergencies will not take the investigator too much by surprise. Animals can haemorrhage, go into shock, become infected, the wound can open and other complications may occur. The investigator should know how to deal with common emergencies and should be able to do so promptly.

Restraint

This is most often required if the animal is carrying apparatus in or on its body (e.g. radiotelemetry sensors, exteriorized cannulas, infusion pumps), or if it is necessary to prevent the animal injuring itself. A range of apparatus and methods are available to provide effective restraint and therefore protection. Apparatus which allows the maximum amount of freedom of movement while yet providing protection should be used. Restraint inevitably stresses and distresses animals with consequential change in certain normal physiological values and every step taken to minimize these changes contributes to a more successful outcome of experimentation on the animal. It is helpful if the animal is trained in the restraint procedure for a few days before surgery. This may involve, for example, wearing an Elizabethan collar or a restraint jacket tethered to a fixed point via a flexible spring, or being confined to a special chair (e.g. for primates). Such training often ameliorates the stressful effects of restraint and encourages the animal to be more physiologically normal. During prolonged restraint, and particularly where the animal is not allowed very much movement (e.g. primates in restraint chairs), the animal must be closely watched so that ill-effects such as weight loss, abnormal bowel and bladder function, loss of muscle tone and muscle atrophy can be detected and prevented or treated.

REFERENCES

No references have been cited in the text but the following list contain a selection of the more important sources of information relevant to the topics discussed.

Archibald, J. (ed.) (1964). *Canine Surgery*, 2nd edn., American Veterinary Publications, California.

Bickhardt, K., Buttner, D., Muschen, U. and Plonait, M. (1983). Influence of bleeding procedure and some environmental conditions on stress dependent blood constituents of laboratory rats, *Lab. Animals,* **17**, 161.

Bojrab, M. J., Crane, S. W. and Arnoczky, S. P. (1983). *Current Techniques in Small Animal Surgery*, 2nd edn, Lea & Febriger, Philadelphia.

Bollman, J. L. (1948). A cage which limits the activity of rats, *J. Lab Clin. Med,* **33**, 1349.

Canadian Council on Animal Care (1980). *Guide to the Care and Use of Experimental Animals*, Vol. 1, CCAC, Ottawa.

Catacott, E. J. (ed.) (1964) *Feline Medicine and Surgery*, American Veterinary Publications, California.

Condon, P. E. (ed.) (1974). Symposium on surgical infections and antibiotics. In *The Surgical Clinics of North America*, 55, No. 6., W. B. Saunders, Philadelphia.

Dalton, R. G., Touraine, J. L. and Wilson, T. P. (1969). A simple technique for continuous intravenous infusion in rats, *J. Lab. Clin. Med.,* **74**, 813.

DeBoer, J., Archibald, J. and Downie, H. G. (eds) (1975). *An Introduction to Experimental Surgery*, Excerpta Medica, Amsterdam.

Flecknell, P. A. (1984). The relief of pain in laboratory animals, *Lab. Animals,* **18**, 147.

Gartner, K., Buttner, D., Dohler, P., Friedal R., Lindena, J. and Trautschold, I. (1980). Stress response of rats to handling and experimental procedures, *Lab. Animals,* **14**, 267.

Hard, G. C. (1975). Thymectomy of the neonatal rat, *Lab. Animals,* **9**, 105.

Hayek, A. and Kuehn, C. (1982). A method of preventing maternal cannabalism after neonatal surgery. *Lab. Animal Sci.,* **32**, 171.

Hickman, J. and Walker, R. (1980). *An Atlas of Veterinary Surgery*, 2nd edn, John Wright & Sons, Bristol.

Horton, M. L., Harris, A. M., Van Stee, E. W. and Back, K. C. (1975). Versatile protective jacket for chronically instrumented dogs, *Lab. Animal Sci.,* **25**, 500.

Hurov, L. (1978). *Handbook of Veterinary Surgical Instruments and Glossary of Surgical Terms*, W. B. Saunders, Philadelphia.

Karl, A. and Kissen, A. T. (1978). Durable jackets for non-human primates to protect chronically implanted instrumentation, *Lab. Animal Sci.,* **28**, 103.

Kirk, R. M. (1978). *Basic Surgical Techniques*, Churchill Livingstone, Edinburgh.

Lambert, R. (1965). *Surgery of the Digestive System of the Rat*, C. C. Thomas, Springfield.
Lennox, M. S. and Taylor, R. G. (1983). A resistant chair for primates, *Lab. Animals*, **17**, 225.
Leonard, E. P. (1968). *Fundamentals of Small Animal Surgery*, W. B. Saunders, Philadelphia.
Markowitz, J., Archibald, J. and Downie, H. G. (1959). *Experimental Surgery*, 4th edn, Bailliere, Tindall & Cox, London.
Steffens, A. B. (1969). A method of frequent sampling of blood and continuous infusion of fluids in the rat without disturbing the animal, *Physiol. Behaviour*, **4**, 833.
Strachan, C. J. L. and Wise, R. (1979). *Surgical Sepsis*, Academic Press, London.
Swain, S. F. (1980). *Surgery of Traumatized Skin: Management and Reconstruction in the Dog and Cat*, W. B. Saunders, Philadelphia.
Taylor, P. (1985). Analgesia in the dog and cat, *In Practice*, **7**, 5.
Waynforth, H. B. (1980). *Experimental and Surgical Technique in the Rat*, Academic Press, London.
Williams, D. J. and Harris, M. M. (1975). *Fundamental Techniques in Veterinary Surgery*, W. B. Saunders, Philadelphia.
Wingfield, W. E. and Rawlings, C. (1979). *Small Animal Surgery. An Atlas of Operative Techniques*, Holt Saunders, London.

CHAPTER 16

Laws Relevant to Animal Research in the United States

BERNARD E. ROLLIN
Colorado State University

INTRODUCTION

The development of legislative constraints on the care and use of animals in research in the United States has been significantly slower and more tentative than in the United Kingdom. Indeed, the first significant piece of federal legislation affecting the use of animals in science was not promulgated until 1966, almost a century after the pioneering British Act of 1876. This difference reflects in part the long history of general British concern with animal welfare, a concern not equally manifest until very recently in US society. In addition, US science has, again until very recently, remained relatively insulated from public accountability, and unconstrained use of animal subjects in the research process has been widely viewed by scientists as a corollary of academic freedom. The fact that invasive research on animals raises moral questions of precisely the sort that legislation should address has been by and large ignored by the research community, which has been buffered from such concerns both by public apathy on these issues and by widespread ideological dicta prevalent in the scientific community. These latter notions include the belief that science as value-free has nothing to do with ethics, the strongly held conviction that the human benefits flowing from research and the intrinsic value of new knowledge override any animal-welfare based thrusts for regulating science, and the belief that science can only proceed if it is left untrammeled and unfettered. It is only in the last decade that public concern about animal use and public sophistication about the moral issues involved have forced legislators and regulatory agencies to address the issue in a meaningful way.

ANTI-CRUELTY LAWS

Despite the aforementioned fact that animal welfare has not historically been the social issue in the US which it was in Britain, the first law discussing animal cruelty ever drafted was promulgated in the US in 1641 in the Massachusetts Bay Colony. The first modern anti-cruelty laws, specifying offenses and punishment, were passed in Britain in 1822, and in the US in 1828, in New York State. By the late nineteenth century, all states had anti-cruelty laws, modeled on modifications made in the New York law by Henry Bergh in 1867 (Leavitt, 1978).

Anti-cruelty laws have not historically proven to be an effective or viable mechanism for regulating the care and use of animals in research. In the first place, in 21 states and the District of Columbia, scientific research on animals is specifically exempted from the purview of the anti-cruelty laws. Secondly, in 25 states the laws are written in language requiring that, for successful prosecution of an offender, it must be shown that the cruel act was performed 'willfully,' 'recklessly,' 'maliciously,' or 'unnecessarily' (Leavitt, 1978). These qualifying phrases have the effect of removing research animals from the scope of these laws. Even in states which do not employ such statutory phrases, judicial precedent has been to exempt from these laws actions performed in the service of human 'necessity,' where almost any standard practice involving animals has been accepted as necessary.

Further, it is well known that the anti-cruelty laws are not widely enforced. Police, prosecutors, and judges are reluctant to press such cases. In Denver, Colorado, for example, a large jurisdiction, the district attorney had not prosecuted a single case of animal cruelty during 1978 and such a state of affairs is not unusual. Furthermore, even when such laws are enforced, the penalties are minimal, since the crime is a misdemeanor. For example, fines of $100–$500 for offenders are not uncommon (Leavitt, 1978). Most important to our discussion, the anti-cruelty laws are in fact virtually never applied to scientific research or teaching (US Congress, 1986). In those rare cases where attempts have been made to prosecute scientists under these laws, courts have underscored the exclusion of scientific research from the intent of such laws. (One notable and well-publicized exception is the Taub case, where a researcher was prosecuted successfully in 1982 under the Maryland anti-cruelty statutes but was eventually acquitted by Maryland's high court, which affirmed exemption from the law for scientific research.) Thus, for both *de facto* and *de jure* reasons, these laws have historically been essentially irrelevant to regulating the care and use of animals in research, though some people have argued both that they could be amended to serve as viable vehicles for such regulation, and that even the extant laws could be employed more effectively than they have been. Towards the latter end, various animal welfare organizations have been attempting to educate their constitutencies in using the law, and to find research cases where the human need and welfare defenses would be implausible. It has also been suggested as a way of strengthening these laws that local animal welfare organizations be given the right to seek injunctions against violators of anti-cruelty laws (hitherto, such groups have not been allowed legal standing in such cases). In Massachusetts, officers of some private animal welfare organizations such as the Massachusetts SPCA are granted limited authority to enforce the anti-cruelty law. Such changes could increase the role of state anti-cruelty laws in regulating laboratory animal care and use in the future (US Congress, 1986).

OTHER STATE AND LOCAL LAWS RELEVANT TO RESEARCH ANIMALS

States have been reluctant to pass specific laws regulating animal research. The one exception to this generalization has to do with the use of pound animals — dogs and cats — in research. Following World War II, when federal funding for biomedical research increased dramatically, so too did the demand for research animals. Increasing numbers of municipalities found it feasible to sell their unwanted dogs and cats — an ever-growing problem — to research laboratories (Rowan, 1984). States and municipalities passed laws and ordinances permitting or, in some cases, requiring the release of such animals to research laboratories (US Congress, 1986). By the mid to late 1970s, however, growing social concern with the welfare of laboratory animals, led to campaigns against the use of pound animals in research. During the ensuing decade, a number of laws permitting or requiring the release of pound animals were repealed, the most famous being the New York Metcalf–Hatch Act of 1952, repealed in 1979. Currently, eleven states prohibit the release of pound animals for research. Six states mandate the release of pound animals for research — the rest neither prohibit nor mandate (National Association for Biomedical Research, 1985). The strongest of the laws prohibiting the use of pound animals is that of Massachusetts, which went into effect in 1984. The law repeals old statutes permitting release of pound animals, prohibits pounds from releasing animals to either research laboratories or animal dealers, and, beginning in 1986, prohibits the use for research of pound animals acquired from sources outside of Massachusetts. Movements to pass similar laws exist in more than 20 states.

Eliminating the use of pound animals in research remains a major priority for substantial numbers of people concerned about the welfare of laboratory animals. In 1985, 11 major animal welfare organizations chartered and funded a new organization — the National Coalition to Protect our Pets (Pro-Pets) — whose exclusive concern is eliminating the use of pound animals in research, teaching, and testing. Attempting to pass a federal law against use of pound animals is a major goal of this and other organizations.

In addition to the state laws just discussed, a number of municipalities, most prominently Los Angeles, have promulgated ordinances prohibiting the release of pound animals for research. Furthermore, as municipalities across the US have begun to engage issues and pass ordinances traditionally not within the purview of city governments, dealing with such matters as nuclear testing, genetic engineering, and smoking in public places, they have also begun to focus on animal research. As of early 1987, an ordinance has been proposed in Cambridge, Massachusetts putting strong constraints on biomedical research done by institutions within the city, namely Harvard and MIT (Knobelsdorff, 1987). It is likely that such efforts will be attempted elsewhere as well. State legislatures are likely to be a major vehicle for animal welfare proponents in the future. According to a spokesman for the Federation of American Societies for Experimental Biology, speaking in October 1986, 'the state legislature has been the venue of greatest legislative activity concerning the use of animals in research. In 1985 legislation . . . affecting the availability or use of animals in research was introduced in 21 state legislatures' (O'Connor, 1986).

FEDERAL LAW — THE ANIMAL WELFARE ACT

Following the passage of the British law of 1876, there were attempts made in the US to pass similar federal legislation, but they always failed, in part because animal research was at first far less advanced in the US than in Britain, and later because those opposing animal use were poorly organized and ill-informed (Turner, 1980). At the same time, the organized research community was solidifying its power, prestige, and influence. It was not until 1966 that legislation was passed. As is too often the case, the legislation was promulgated not in response to reasoned argument, but in the wake of two highly emotional stories appearing in the national press. The first piece concerned the kidnapping of a family dog, which ended up dead in a research laboratory. The second story was a photographic essay in *Life* magazine showing the appalling conditions under which dogs were kept by some animal dealers. Public outrage prompted the passage of the Laboratory Animal Welfare Act of 1966, which was later renamed the

Animal Welfare Act, and amended in 1970, 1976, and 1985.

While the Animal Welfare Act was unquestionably a major step forward for laboratory animal welfare, it contained many deficiencies and incoherences. These resulted from two sources: First, the power of the research community, which had no commitment whatever to laws regulating any area of research, and which pushed for as weak a law as possible. Second, the fact that the majority of the public pressure for change was primarily motivated by emotion and sentimental attachment to certain favored animals, and had little understanding of the moral or scientific elements involved in animal research. Even a cursory examination of the 1966 Act reveals those difficulties.

The 1966 Act first and foremost concerns itself with regulating dealers in dogs and cats to prevent kidnapping of pets and bad treatment of dogs and cats by dealers in response to the publicity mentioned above. Individuals or organizations buying or selling dogs and cats for laboratory use were required to be licensed and were held to certain standards of care and housing for the animals promulgated by the Secretary of the US Department of Agriculture (USDA), the federal agency charged with enforcing the Act. Laboratories using these animals were required to register with the USDA and to identify and keep records on each dog and cat used. In addition, laboratories using dogs and cats were held to USDA standards of care and housing, not only for dogs and cats, but hamsters, guinea pigs, rabbits, and monkeys as well. This law covered only 'care, treatment, and transport,' etc., before and after research, but specifically disavowed any concern with the actual conduct or design olf research. The law was to be enforced by federal inspectors of USDA's APHIS (Animal and Plant Health Inspection Service).

The amendments of 1970 were a response to continued concern about laboratory animals. These amendments broadened the definition of animal to 'any live or dead dog, cat, monkey, guinea pig, hamster, rabbit, or other such warm-blooded animal as the Secretary may determine is being used, or intended for use, for research, testing, experimentation, or exhibition purposes, or as a pet.' Specifically designated as non-animals by statute were horses not used for research or any farm animals. Regulations promulgated by the Secretary have thus far continued to exclude rats and mice. Thus for purposes of the Act, a dead dog is an animal, while a live mouse is not. The definition of research facility was broadened to include those using animals other than dogs and cats, but the Secretary was permitted to exempt such facilities if they didn't use 'substantial numbers' of animals. The definition of dealer was expanded, as was the definition of research facilities. Penalties both civil and criminal were increased for violation of the Act (research laboratories are not subject to criminal penalties). The revised law also defined more clearly the sorts of standards to be promulgated by the Secretary — they were to cover housing, food and water, handling, sanitation, ventilation, shelters from extremes of weather and temperature, separation of species, and adequate veterinary care 'including the appropriate use of anaesthetic, analgesic or tranquilizing drugs, when such use would be proper in the opinion of the attending veterinarian' of research facilities. An annual report must be made to USDA regarding the latter. Detailed regulations were promulgated by the USDA regarding most of the categories specified in the Act (Animal and Plant Health Inspection Service, 1985) (the regulations are currently being revised in the wake of the 1985 amendments).

The amendments of 1976 did not affect research, serving primarily to bring common carriers such as airlines under the Act, and to cover animal-fighting ventures.

THE 1985 AMENDMENTS

Beginning in the mid-1970s, the US public became increasingly conscious of issues pertaining to the welfare of animals, especially animals used in research. This was due to an fueled by a variety of factors (Rollin 1981, Rowan and Rollin, 1983):

(1) Generalized social disillusionment with science as a curer of all ills.
(2) The rise of sophisticated thinking about human moral obligations to animals, as represented by the work of philosophers like Singer (1975), Clark (1977), Rollin (1981), and Regan (1983).
(3) Increased activism on the part of those concerned with animals.
(4) Increased press coverage of these issues.

With this new consciousness came increasing awareness of basic inadequacies in the Animal Welfare Act: The exclusion of rats, mice, and farm animals from protection even though they constituted some 85 percent of the animals used in research; the disavowal of concern with the actual conduct of research; the failure of researchers or the Act to control pain and distress and suffering for animals (attending veterinarians typically simply reported that such use was not deemed necessary and little or no use was made of analgesics); the isolation of research laboratories from public scrutiny; the USDA emphasis on facilities not on animal pain and suffering; the frivolous uses of animals in such areas as cosmetics testing and psychology; the multiple use of the same animals for teaching surgery. Beginning in 1976 with a bill drafted for the state of Colorado by a group at Colorado State University (Rollin, 1981), attempts were made to codify these concerns in law. A key feature of these efforts was a nabdate to develop local review committees consisting of scientists and lay persons, including animal welfare advocates, to ensure compliance with law. In the ensuing decade, various versions of the law were introduced in Congress, and the law was passed as an Amendment to the Animal Welfare Act in 1985, entitled 'The Improved Standards for Laboratory Animals Act.' Although still officially disavowing concern with the actual design and conduct of research, the amendment significantly strengthened the emphasis on pain and suffering, in fact the major thrust of the new law. The law went into effect January 1, 1987. Unfortunately, as of this writing, the USDA has not issued its regulations interpreting the law. The main statutory provisions of the 1985 amendment are as follows:

(1) Establishment of an institutional animal care committee to monitor animal care and inspect facilities. Members must include a veterinarian and a person not affiliated with the research facility.
(2) Standards for exercise of dogs are to be promulgated by the Secretary.
(3) Standards for a physical environment which promotes 'the psychological well-being of primates' are to be promulgated.
(4) Standards for adequate veterinarian care including use of anaesthetics, analgesics, and tranquilizers to be promulgated.
(5) No paralytics are to be used without anaesthetics.
(6) Alternatives to painful procedures must be considered by the investigator.
(7) Multiple surgery is prohibited except for 'scientific necessity.'
(8) the Animal Care Committee must inspect all facilities semi-annually, review practices involving pain, review the conditions of animals, and file an inspection report detailing violations and deficiencies. Minority reports must also be filed.
(9) The Secretary is directed to establish an information service at the National Agricultural Library which provides information aimed at eliminating duplication of animal experiments, reducing or replacing animal use, minimizing animal pain and suffering, and at aiding in training of animal users.
(10) The facility must provide for training for all animal users and caretakers on human practice and experimentation, research methods that limit pain, use of the information service of the National Agricultural Library, and methods of reporting deficiencies in animal care and treatment.
(11) A significant penalty is established for any animal care committee member who reveals trade secrets.
(12) The Secretary is directed to consult with the Department of Health and Human Services (under which falls biomedical research funding

responsibility through the National Institute of Health) in establishing the standards described.

(13) New civil penalties are provided for violation of the Act.

OTHER FEDERAL LAWS — THE GOOD LABORATORY PRACTICES ACT OF 1978 AND THE HEALTH RESEARCH EXTENSION ACT OF 1985

Two other federal laws are relevant to laboratory animals. In 1978, concern on the part of the Food and Drug Administration (FDA) about slipshod work in toxicology in laboratories across the country led to passage of the Good Laboratory Practices Act (GLP) — essentially a list of proper scientific procedures which the toxicological researcher should be doing for the sake of proper science, also used by the Environmental Protection Agency (EPA). Included in this list are detailed specifications of Standard Operating Procedures for animal care, feeding, handling, separation of animals, disease control and treatment, sanitation, food and water inspection, bedding, and pest control.

In June of 1985, the Congress passed the Health Research Extension Act of 1985, popularly known as the NIH Authorization Bill, which established provisions for the care and use of animals used in research funded by the Public Health Service, under whose aegis the National Institutes of Health, the major source of biomedical funding in the US, operate. The bill essentially made NIH guidelines, (to be discussed in the next section) hitherto a matter of contractual agreement between the NIH and the researcher and institution, a matter of legal requirement. Violation of these rules now requires seizure of federal funding from the offending institution.

The law, like the 1985 Animal Welfare Act amendments, mandates institutional training of animal users. In addition, it directs NIH to establish a plan for research into biomedical experimentation that does not use animals, that reduces the number of animals used, and that minimizes pain and suffering. These methods should be validated, and valid methods developed and taught to scientists.

FEDERAL REGULATIONS

As mentioned above, NIH has long made contractual agreement to abide by NIH policies for animal care and use a pre-condition of NIH funding. Grantees are currently required to follow the revised Public Health Service (PHS) Policy on Humane Care and Use of Laboratory Animals, which took effect December 31, 1985 (US Department of Health and Human

Services, 1985a). This policy in turn relies on the *Guide for the Care and Use of Laboratory Animals* (US Department of Health and Human Services, 1985b), a document which first appeared in 1963, and has been revised periodically. The newest edition appeared in 1985, and the 83-page work covers institutional policies, husbandry, veterinary care, and physical plant. Finally PHS policy is based on nine principles of animal care and use prepared by the US Interagency Research Animal Committee (US Department of Health and Human Services. 1985a, b).

Although many of the NIH policies, even in earlier editions of the *Guide*, were quite good, both from a scientific and welfare point of view, there was no enforcement structure, and research institutions often did not comply with the rules. For example, although the 1978 edition of the *Guide* specifically forbade multiple survival surgery for teaching, almost all veterinary schools were engaged in this practice for economic reasons. (When I challenged NIH in the early 1980s as to why they did not enforce their own rules, a senior official told me, 'We are not in the enforcement business.') In the face of mounting public concern and pressure, however, NIH was forced to change its *laissez faire* attitude, notably in the wake of the severe violations made public in the Taub case and head injury laboratory tapes taken from the University of Pennsylvania. Before Taub, NIH, although empowered to do so, had never seized funds from any institution. After Taub and other cases revealing flagrant violations of its rules, NIH was moved to ensure that institutions were in compliance with its policies. The NIH Office of Protection from Research Risks (OPRR) has been funded to make both unannounced and announced site visits to ensure compliance, both in terms of the *Guide* and in terms of institutional animal care committee functioning.

The heart of the new (1985) NIH policy parallels the 1985 amendments to the Animal Welfare Act, and is the animal care and use committee, which must consist of at least five members, one a veterinarian with laboratory animal medicine background, one a scientist experienced in animal research, one member whose primary concerns are in a non-science area, and one individual not affiliated with the institution. The committee must monitor animal care and use, meet regularly, and review and approve or disapprove research protocols before they are submitted for funding. The committee may suspend any activity involving animals not in compliance with policy.

The new rules require that institutions designate clear lines of authority and responsibility in animal care and use. Record-keeping requirements have also been strengthened. In addition, the new *Guide* has strengthened the requirements for adequate veterinary care, makes reference to the social environment for laboratory animals, requires aseptic surgery for rodents, and strengthens euthanasia requirements.

In conjunction with the laws discussed above, the new NIH policy ensures that the vast majority of animals used in research in the US fall under the oversight of an Animal Care Committee. Unlike the Animal Welfare Act, the NIH policy does not exclude rats, mice, or farm animals used in biomedical research. Committees must review facilities and care for *all* animals used in biomedical research and review protocols using all animals. Although NIH policy states only that committees should review protocols for animal welfare related issues, not judgments of scientific merit, in practice it is often difficult to make that distinction — for example in the case of the question of whether the proper number of animals are being used. (Similar problems have also existed for committees reviewing human subjects use.)

The final strengthening of federal animal welfare policy is embodied in a 1983 memorandum of understanding between APHIS, NIH, and FDA, agreeing to share information on institutions visited by each agency. In effect, this memorandum empowers the USDA inspectors, who regularly inspect institutions, to report violations of NIH policy to NIH, even if such violation concerns areas or animals not covered by the Animal Welfare Act.

CONCLUSION

The new and growing social concern with the welfare and neglect of laboratory animals which has developed in the past decade has helped to erode the traditional US attitude of *laissez faire* towards animals in research. The most powerful vehicles for change in the new laws and guidelines are the attention paid therein to controlling pain, suffering, and distress, and the placing of responsibility in the hands of local animal care committees made up of diverse members. Hopefully, the protocol review and monitoring functions of the committees will serve an educational as well as a regulatory role, making researchers more sensitive than they have been to the moral status of animals. The campaign for legislation is not over. The near future will very likely see thrusts for legislation which limits, bans, or regulates toxicity testing for cosmetics and other inessential products; military research on animals; and psychological research, all of which are viewed as both hurtful and useless by increasing numbers of people.

REFERENCES

Animal and Plant Health Inspection Service (1985). *Regulations*. US Government Printing Office, Washington, DC.

Clark, S.R.L. (1977). *The Moral Status of Animals*. Oxford University Press, Oxford.

Knobelsdorff, K.E. (1987). Stricter regulation sought over labs using animals in research. *The Christian Science Monitor*. January 8, p. 7.

Leavitt, E.S. (1978). *Animals and Their Legal Rights*. Animal Welfare Institute, Washington, DC.

National Association for Biomedical Research (1984). *State Laws Concerning the Use of Animals in Research*. Foundation for Biomedical Research, Washington, DC.

O'Connor, K. (1986). Laws, regulation, and legislation affecting the use of animals in research. *Public Affairs: Federation of American Societies for Experimental Biology Proceedings*, **45**, No. 11, 9a.

Regan, T. (1983). *The Case for Animal Rights*. University of California Press, Berkeley, California.

Rowan, A.N. (1984). *Of Mice, Models, and Men*. SUNY Press, Albany, New York.

Rowan, A.N. and Rollin, B.E. (1983). Animal research — for and against: a philosophical, social, and historical perspective. *Perspectives in Biology and Medicine*, **27**, No. 1.

Rollin, B.E. (1981). *Animal Rights and Human Morality*. Prometheus Books, Buffalo, New York.

Singer, P. (1975). *Animal Liberation*. New York Review Press, New York.

Turner, J. (1980). *Reckoning with the Beast*. Johns Hopkins University Press, Baltimore.

US Congress Office of Technology Assessment (1986). *Alternatives to Animal Use in Research, Testing, and Education*. US Government Printing Office, Washington, DC.

US Department of Health and Human Services (1985a). *NIH Guide for Grants and Contracts*. US Government Printing Office, Washington, DC.

US Department of Health and Human Services (1985b). *Guide for the Care and Use of Laboratory Animals*. National Institutes of Health, Bethesda, Maryland.

Bibliography

The following is a selected list of sources which are of a more general nature than the references quoted in the preceding chapters. The pre-1980 books listed are still valuable sources.

JOURNALS

Laboratory Animals (1967). Published quarterly by Laboratory Animals Ltd, 20 Queensbury Place, London SW7 2DZ, UK.
Laboratory Animal Science (1950). Published bimonthly by the American Association for Laboratory Animal Science, 210 N. Hammes Ave, Ste 205, Joliet, Ill., USA.
Animal Technology (1950). Journal of the Institute of Animal Technicians, 5 South Parade, Summertown, Oxford OX2 7JL, UK.
Zeitschrift für Versuchstierekunde (1961). Published bimonthly by Gustav Fischer Verlag, Jena, E. Germany.
Sciences et Techniques de l'Animal de Laboratoire (Stal) (1976). Published by SFEA, Centre d'Experimentation Animale et de Recherches Chirurgicales, 6 rue de General-Sarrail, 94000 Creteil, France.
Experimental Animals (1952). Japanese Association for Laboratory Animal Science, c/o Laboratory Animals Centre, Keio University School of Medicine, Shinanomachi, Shinjuku, Tokyo 160, Japan (Japanese, English summaries).

BOOKS

Baker, H. J., Lindsey, J. R. and Weisbroth S. H. (Eds) (1979–80). *The Laboratory Rat*, Vol. I, Biology and Diseases; Vol. II, Research Applications, Academic Press, London and New York.
Buckland, M. D., Hall, L., Mowlem A. and Whatley, B. F. (1981). *A Guide to Laboratory Animal Technology*, William Heinemann Medical Books, London.
Canadian Council on Animal Care (1980–4). *Guide to the Care and Use of Experimental Animals*, 2 vols, CCAC, Ottawa.
Clarke, H. E., Coates, M. E., Eva, J. J., Ford, D. J., Milner, C. K., O'Donaghue, P. N., Scott, P. P. and Ward, R. J. (1977). Dietary standards for laboratory animals; report of the Laboratory Animals Centre Diets Advisory Committee, *Lab. Anim.*, 11, 1–28.
Farris, E. J. and Griffith, J. Q. *The Rat in Laboratory Investigation*, 2nd edn, Lippincott Co., Philadelphia.

Foster, H. L., Small, J. D. and Fox, J. G. (1981–83). *The Mouse in Biomedical Research*, Vol. 1, History, Genetics, and Wild Mice (1981); 2 Diseases (1982); 3 Normative Biology, Immunology and Husbandry (1983); 4 Experimental Biology and Oncology (1982), Academic Press, London and New York.

Gay, W. I. (ed.) (1965–1981). *Methods of Animal Experimentation*, Vols I–VI, Academic Press, London and New York.

Green, E. L. (ed.) (1966). *Biology of the Laboratory Mouse*, 2nd edn, The Blakiston Divn, McGraw-Hill Book Co., New York.

Hafez, E. S. E. (1970). *Reproduction and Breeding Techniques for Laboratory Animals*, Lea & Febiger, Philadelphia.

Harkness, J. E. and Wagner, J. E. (1983). *The Biology and Medicine of Rabbits and Rodents*, Lea & Febiger, Philadelphia.

Hoffman, R. A., Robinson, P. F. and Magelhaes, H., (eds) (1968). *The Golden Hamster, its Biology and Use in Medical Research*, Iowa State University Press, Aemes, Iowa.

Home Office (1986). *Experiments on Living Animals Statistics* (published annually), HMSO, London.

Inglis, J. K. (1980). *Introduction to Laboratory Animal Science and Technology*, Pergamon, Oxford.

Institute of Laboratory Animal Resources (1985). *Guide for the Care and Use of Laboratory Animals*, US Dept of Health, Education & Welfare, NIH, Bethesda.

Kaplan, H. M. and Timmons, E. H. (1979). *The Rabbit. A Model for the Principles of Mammalian Physiology and Surgery*, Academic Press, London and New York.

Lane-Petter, W. and Pearson, A. E. G. (1971). *The Laboratory Animal — Principles and Practice*, Academic Press, London and New York.

Melby, E. C. and Altman, M. H. (eds) (1974–76). *Handbook of Laboratory Animal Science*, Vols 1, 2 (1974); 3 (1976), CRC Press, Ohio.

Mitruka, B. M. and Rawnsley, H. M. (1977). *Clinical, Biochemical and Haematological Reference Values in Normal Experimental Animals*, Masson Publishing USA Inc., New York.

Petty, C. (1982). *Research Techniques in the Rat*, C. C. Thomas, Springfield.

Royal Society/UFAW (in preparation). *Draft Guidelines on the Care of Laboratory Animals and their use for Scientific Purposes*.

Sanderson, J. H. and Phillips, C. E. (1982). *An Atlas of Laboratory Animal Haematology*, Oxford University Press, Oxford.

Short, D. J. and Woodnott, D. P. (1969). *IAT Manual of Laboratory Animal Practice and Technique*, Crosby, Lockwood & Son, London.

University Federation for Animal Welfare (1978) (ed. Scott, W. M.). *Handbook on the Care and Management of Farm Animals*, Bailliere Tindall, London.

University Federation for Animal Welfare (1984). *Standards in Laboratory Animal Management*, UFAW, 8 Hamilton Close, South Mimms, Potters Bar, Herts EN6 3QD.

University Federation for Animal Welfare (1986). *UFAW Handbook on the Care and Management of Laboratory Animals*, Longmans, London and New York (in press).

Wagner, J. E. and Manning, P. J. (eds) (1976). *The Biology of the Guinea Pig*, Academic Press, London and New York.

Weisbroth, S. H., Flatt, R. E. and Kraus, A. L. (eds) (1974). *The Biology of the Laboratory Rabbit*, Academic Press, London and New York.

Wyatt, H. V. (ed.) (1980). *Handbook for the Animal Licence Holder*, Inst. Biology, London.

Index

acclimatization, 305
acepromazine maleate, 269
acid–base imbalance, 294
air movement, 88, 94
air quality, 88, 94
alphaxalone–alphadolone, 272
alternatives, 1, 17, 28
anaesthesia, 261, 273
    complications, 287
    recommendations, 265, 282
    signs of, 274
anaesthetic chambers, 274
anaesthetic machines, 275
anaesthetics, dissociative, 268
    general, 262, 268
    response to, 263, 294
analgesic (*see also under species*), 262, 268, 319
animal behaviour, 129
animal care committees, 34, 37, 40
animal consciousness, 130
animal disease, 84, 198
animal health, 119
animal liberation, 11
animal models, 54, 193
Animal Procedures Committee, 28
Animal restraint, 84, 103, 198, 268, 320
Animal Scientific Procedures Act (1986), 22, 26, 56
animals, choice of, 79, 192
    defined, 81
    farm, 119
    genetics of, 82
    quality of, 55, 65, 82, 85
    reuse of, 30
    use of, 53
antibiotics, 308, 315, 318
anticholinergics, 268
anticoagulants, 246
antiseptics, 312
antivivisection, 7, 10
aseptic technique, 311
atropine, 268
autopsy, 176
axenic animals, 85, 105

barrier maintained animals, 85
barrier units, 105
bedding, 101
behaviour, animal, 129
    avoidance, 146
    environmental influences, 135, 137, 141
    experience, 135
    genetic influences, 134
    measures of, 139
    methods of study, 138
    phylogenetic influences, 132
    physical form influences, 134
blood, normal values, 247
blood collection, 246
blood samples, 247
blood volumes, 247
breeding, 65, 70
    cats, 74

dogs, 74
  guinea-pigs, 74
  hamsters, 73
  mice, 73
  rabbits, 74
  rats, 73
breeding cycle, 65
breeding season, 65, 66
Bruce effect, 67

cage, 56, 100
  exercise, 103
  metabolism, 104
  restraining, 103
  size, 102
carcase disposal, 176
cardiac arrest, 292
cardiac arrhythmias, 291
cardiovascular failure, 290
cat, anaesthesia, 285
  analgesia, 285
  bleeding, cardiac puncture, 254
  bleeding, cephalic vein, 254
  bleeding, jugular vein, 254
  cystocentesis, 256
  euthanasia, 174
  handling, 164
  injection, intradermal, 241
  injection, intramuscular, 242
  injection, intraperitoneal, 242
  injection, intravenous, 243
  injection, subcutaneous, 241
  oral dosing, 240
  restrain, 164
  sedation, 285
  urethral catheterization, 256
cerebrospinal fluid samples, 259
certificates, 1876 Act, 23, 24
chemical restraint, 268
chicken, bleeding, cardiac puncture, 254
  bleeding, jugular vein, 254
  bleeding, wing vein, 254
  handling, 169
  injection, intradermal, 245
  injection, intramuscular, 245
  injection, intraperitoneal, 245
  injection, intravenous, 246
  injection, subcutaneous, 245
  oral dosing, 245
chloroform, 173
closed colony, 72

Code of Practice, 1986 Act, 29
coefficient of inbreeding, 71
co-isogenic strains, 72
colostrum, 70
conditioning, classical, 136
  operant, 136,
  143, 144, 145
  Pavlovian, 136
conditions, 1876 Act, 25
controls, voluntary, 41
Convention, Council of Europe, 42, 56, 102
conventional units, 105
coprophagy, 111
corpus luteum, 67, 70
cost/benefit analysis, 27, 42, 40
costs, 74
Council of Europe Convention, 42, 52, 102
Ccuelty to Animals Act (1876), 22
cycle, breeding, 65
cycle, oestrus, 66

dehydration, in anaesthesia, 293
diazepam, 270
diet, additives, 217
  contaminants, 217
  experimental, 219
  fixed formula, 203
  formulation, 204
  microbiological contaminants, 218
  preparation, 213
  standardization of, 120
  sterilization, 215
  storage, 218
  tumours and, 219
  variable formula, 203
disease, 84, 198
disease-free animals, 85
dog, anaesthesia, 285
  analgesia, 284
  bleeding, cardiac puncture, 253
  bleeding, cephalic vein, 253
  bleeding, jugular vein, 253
  cystocentesis, 257
  euthanasia, 174
  handling, 165
  injection, intradermal, 244
  injection, intramuscular, 244
  injection, intraperitoneal, 244
  injection, intravenous, 245

injection, subcutaneous, 244
oral dosing, 244
restraint, 168
sedation, 285
urethral catheterization, 257
droperidol, 270
drugs, absorption of, 181, 184
  administration, 179
  administration, intragastric, 183
  administration, intramuscular, 182
  administration, intraperitoneal, 182
  administration, intravenous, 182
  administration, oral, 183
  administration, percutaneous, 183
  behaviour and, 197
  administration, subcutaneous, 182
  distribution of, 184
  environment and, 196, 197
  excretion of, 187
  lipid solubility, 180
  metabolism of, 186
  pH, 181, 187, 190

EEC Directive, 181, 187, 190
EDTA (ethylene diametetra-acetic acid), 246
electrolyte imbalance, 294
emotional reactivity, 142
endangered species, 76
endotracheal tube, 278
energy, dietary, 205
enflurane, 174
environment, 93, 86, 112, 137
ether, 172
ethical committees, 15, 38, 39, 40, 149
euthanasia, 171
  carbon dioxide, 173
  carbon monoxide, 173
  chloroform, 173
  enflurane, 174
  ether, 172
  halothane, 174
  pentobarbitone, 174
  physical methods, 172

faecal samples, 257
farm animals, 119
feeding, 109, 203, 308
feeding supplements, 111
fentanyl, 271
fibre, dietary, 206

fire precautions, 123
first aid, 122
fluid balance, 314

gerbil, anaesthesia, 283
  analgesia, 283
  sedation, 283
germfree animals, 85, 105, 210
gestation, 70
gnotobiotic animals, 85, 105
gonadotrophin-releasing hormone, 66
Graafian follicle, 66
guidelines, 17, 29, 30, 33, 34, 36, 38, 39 40, 44, 102
guinea-pig, anaesthesia, 284
  analgesia, 284
  bleeding, cardiac puncture, 252
  bleeding, ear veins, 252
  bleeding, exanguination, 252
  bleeding, jugular vein, 252
  euthanasia, 175
  handling, 163
  injection, footpad, 235
  injection, intradermal, 235
  injection, intramuscular, 235
  injection, intraperitoneal, 235
  injection, intravenous, 235
  injection, subcutaneous, 234
  oral dosing, 234
  sedation, 284

haemorrhage, 314
halothane, 174
Halsbury Bill, 26
hamster, anaesthesia, 283
  analgesia, 283
  bleeding, cardiac puncture, 251
  bleeding, exsanguination, 251
  bleeding, retro-orbital, 251
  euthanasia, 175
  injection, intradermal, 240
  injection, intramuscular, 240
  injection, intraperitoneal, 240
  injection, intravenous, 240
  injection, subcutaneous, 240
  oral dosing, 240
  sedation, 283
hand mating, 73
harems, 73, 74
hazards, 117

Health and Safety at Work Act (1974), 123
health status, 84, 92, 107, 264, 305
heat, 66, 113
heparin, 246
hybrid strains, 60, 82
hygiene, 104
hypnotic, 267
hypothermia, 292

inbreeding, coefficient of, 71
 depression, 71
 rate of, 71
 strains, 60, 71, 72, 82
infective animals, 120
injection, needle size, 191
 rate of, 191
 volumes, 190
isogenicity, 71
isolators, 105

ketamine, 272

Laboratory Animal Accreditation Scheme, 86
Laboratory Animal Breeders Association, 86
Laboratory Animal Welfare Act (1966) (USA), 30
law, UK, 21, 22, 26, 30, 31, 47, 76, 124
law, USA, 30, 323
learning, 142
Lee–Boot effect, 67
leutenizing hormone, 66
licence, 1876 Act, 23
 personal, 27
 project, 27
light, 89, 94, 114, 310
Littlewood Committee, 26

metabolic body size, 205
metabolizable energy, 205
midazolam, 270
milk samples, 258
minerals, 211
minimal disease animals, 85
minimal inbreeding, 72
models, animal, 54
monitoring, environment, 93
 genetic, 71, 91
 health, 92
 nutrition, 95

monoestrus, 66
monogamous pairs, 73
motor activity, study of, 141
mouse, anaesthesia, 282
 analgesia, 282
 bleeding, cardiac puncture, 251
 bleeding, exanguination, 251
 bleeding, tail vein, 251
 euthanasia, 176
 handling, 156
 injection, intradermal, 232
 injection, intramuscular, 233
 injection, intraperitoneal, 232
 injection, intravenous, 233
 injection, subcutaneous, 232
 oral dozing, 232
 restraint, 157
 sedation, 282
muscle relaxants, 263
mutant strains, 'nude', 83

narcotic, 267
needle, size of, 191, 225
nesting material, 102
neuroleptanalgesic, 268
noise, 115
nude mouse, 83
nutrition, 90, 95

oestrus, postpartum, 67
oestrus cycle, 66, 70
operating table, 310
oxalate, 247

pain, 10, 11, 16, 27, 29
palpebral reflex, 274
paturition, 70
pedal reflex, 274
pentobarbitone, 174
peritoneal fluid samples, 258
permanent mating, 73
pheromone, 67, 114
philosophy and ethics, 5
pituitary, 66
polygamous groups, 73
polyoestrus, 66
postanaesthetic management, 265, 295, 316
postmortem procedure, 121
postoperative management, 264, 316
postpartum oestrus, 67
preanaesthetic medication, 268

premedication, 308
primate units, 120
progesterone, 70
protective clothing, 119
protein, dietary, 204
protein/energy ratio, 206
protocols, 59, 194
pseudopregnancy, 67
psychology research, 148

quality control of animals, 82, 91
quarantine, 106

rabbit, anaesthesia, 284
  analgesia, 284
  bleeding, cardiac puncture, 252
  bleeding, ear vein, 252
  cystocentesis, 256
  euthanasia, 175
  handling, 161
  injection, intradermal, 238
  injection, intramuscular, 239
  injection, intraperitoneal, 238
  injection, intravenous, 239
  injection, subcutaneous, 238
  oral dosing, 237
  restraint, 163
  sedation, 284
  urethral catheterization, 255
random bred, 71, 72, 73
rat, anaesthesia, 283
  analgesia, 283
  bleeding, cardiac puncture, 249
  bleeding, exsanguination, 249
  bleeding, jugular vein, 250
  bleeding, tail vein, 250
  euthanasia, 175
  handling, 153
  injection, intradermal, 229
  injection, intramuscular, 230
  injection, intraperitoneal, 229
  injection, intravenous, 231
  injection, neonatal, 232
  restraint, 155
  sedation, 283
rederivation programmes, 107
reduction, 16
refinement, 16
relative himidity, 87, 94
replacement, 16
reproduction, 65
respiratory failure, 289

respiratory reflex, 274
restraint, cat, 164
  dog, 166
  guinea-pig, 163
  mouse, 157
  rabbit, 163
  rat, 155
righting reflex, 274
rights, animal, 8, 9, 12, 14
ringtail, 114

safety, 117, 261
safety policy, 123
saliva samples, 257
sedative, 267
semen samples, 259
sentinel animals, 58
smell, 114
social interactions, 135, 142
sodium fluoride, 247
sound, 88, 94
sources, of animals, 64, 105
species, choice of, 55
specified pathogen free animals, 85, 105, 210
standards (*see also* guidelines), 65
strain, choice of, 55, 106
stress, 56, 146, 247
suckling, 70
supply animals, 30, 64, 105
surgery, 303
  aseptic technique, 311
  body temperature, 313
  equipment, 309
  incision, 312, 313
sutures, 315
swallowing reflex, 274

temperature, 87, 93
temporary mating, 73
termination condition, 29
three Rs, 1, 16
tissue culture, 54
tranquillizer, 268, 269, 308, 319
trauma, 314

urine samples, 255

vaginal samples, 258
vehicle, injection, 189
ventilation (*see also air movement*), 113, 114

veterinary support, 57
vitamins, 207
vomiting, 294

watering, 108, 203, 212, 308

weaning, 70
welfare, 15, 150, 261
Whitton effect, 67

xylazine, 270